MATHEMATICS FOR THE PHYSICAL SCIENCES

Mathematics for the Physical Sciences

LAURENT SCHWARTZ

DOVER PUBLICATIONS, INC.
Mineola, New York

Bibliographical Note

This Dover edition, first published in 2008, is an unabridged republication of the English translation of *Méthodes mathématiques pour les sciences physiques,* originally published in 1966 by Editions Hermann, Paris, and Addison-Wesley Publishing Company, Reading, Massachusetts. This Dover edition is published by a special arrangement with Editions Hermann, 6 rue de la Sorbonne, 75005 Paris, France.

Library of Congress Cataloging-in-Publication Data

Schwartz, Laurent.
[Méthodes mathématiques pour les sciences physiques. English]
Mathematics for the physical sciences / Laurent Schwartz. — Dover ed.
p. cm.
Originally published: Mathematics for the physical sciences. Rev. ed. Paris : Hermann ; Reading, Mass. : Addison-Wesley, 1966.
Includes index.
ISBN-13: 978-0-486-46662-0
ISBN-10: 0-486-46662-0
1. Mathematical physics. I. Title.

QA401.S383 2008
530.15—dc22

2007046748

Manufactured in the United States of America
Dover Publications, Inc., 31 East 2nd Street, Mineola, N.Y. 11501

CONTENTS

Preface 11

CHAPTER I. Preliminary results in the integral calculus: series and integrals

1. Preliminary Results on Series 13
   1) Summable series 13
   2) Semi-convergent series 23
2. Preliminary Results on Integration 26
   1) The Lebesgue integral 26
   2) Improper semi-convergent Lebesgue integrals 43
3. Functions Represented by Series and Integrals 46
   1) Functions represented by series 46
   2) Functions represented by integrals 57

Exercises for Chapter I 65

CHAPTER II. Elementary theory of distributions

1. Definition of Distributions 71
   1) The vector space $\mathfrak{D}$ 71
   2) Distributions 73
   3) The support of a distribution 79
2. Differentiation of Distributions 80
   1) Definition 80
   2) Examples of derivatives in the one-dimensional case 82
   3) Examples of derivatives in the case of several variables 85
3. Multiplication of Distributions 91
4. Topology in Distribution Space. Convergence of Distributions. Series of Distributions 94
5. Distributions with Bounded Supports 98

Exercises for Chapter II 100

CONTENTS

CHAPTER III. CONVOLUTION

1. TENSOR PRODUCT OF DISTRIBUTIONS 110
   1) Tensor product of two distributions 110
   2) Tensor product of several distributions 112

2. CONVOLUTION 112
   1) Convolution of two distributions 112
   2) Définition of the convolution product of several distributions. Associativity of convolution 121
   3) Convolution équations 123

3. CONVOLUTION IN PHYSICS 134

EXERCISES FOR CHAPTER III 140

CHAPTER IV. FOURIER SERIES

1. FOURIER SERIES OF PERIODIC FUNCTIONS AND DISTRIBUTIONS 145
   1) Fourier series expansion of a periodic function 145
   2) Fourier series expansion of a periodic distribution 149

2. CONVERGENCE OF FOURIER SERIES IN THE DISTRIBUTION SENSE AND IN THE FUNCTION SENSE 152
   1) Convergence of the Fourier series of a distribution 152
   2) Convergence of the Fourier series of a function 154

3. HILBERT BASES OF A HILBERT SPACE. MEAN-SQUARE CONVERGENCE OF A FOURIER SERIES 158
   1) Definition of a Hilbert space 158
   2) Hilbert basis 159
   3) The space $L^2(T)$ 160

4. THE CONVOLUTION ALGEBRA $\mathcal{D}'(\Gamma)$ 164

EXERCISES FOR CHAPTER IV 170

CHAPTER V. THE FOURIER TRANSFORM

1. FOURIER TRANSFORMS OF FUNCTIONS OF ONE VARIABLE 178
   1) Introduction 178
   2) Fourier transform 179
   3) Fundamental relations and inequalities 180
   4) Spaces $\mathcal{S}$ of infinitely differentiable functions with all derivatives decreasing rapidly 182

2. FOURIER TRANSFORMS OF DISTRIBUTIONS IN ONE VARIABLE 187
   1) Definition 187
   2) Tempered distributions : the space $\mathcal{S}'$ 188
   3) Fourier transforms of tempered distributions 189
   4) The Parseval-Plancherel equation. Fourier transforms in L^2 194
   5) The Poisson summation formula 196
   6) The Fourier transform : multiplication and convolution 197
   7) Other expressions for the Fourier transform 199

3. Fourier Transforms in Several Variables 200
4. A Physical Application of the Fourier Transform : Solution of the Heat Conduction Equation 205

Exercises for Chapter V 208

CHAPTER VI. The laplace transform

1. Laplace Transforms of Functions 215

2. Laplace Transforms of Distributions 217
1) Definition 217
2) Examples of Laplace transforms 218
3) The Laplace transforms and convolution 222
4) Fourier and Laplace transform. Inversion of the Laplace transform 224

3. Applications of the Laplace Transform. Operational Calculus 230

Exercises for Chapter VI 235

CHAPTER VII. The wave and heat conduction equations

1. Equation of Vibrating Strings 242
1) Physical problems associated with the equation of vibrating strings. 242
2) Solution of the equation of vibrating strings by the method of travelling waves. Cauchy's problem 251
3) Solution of Cauchy's problem by Fourier analysis 270

2. Vibrating Membranes and Waves in Three Dimensions 280
1) The solution of the vibrating membrane equation and the wave equation in three dimensions by the method of travelling waves. Cauchy problems 281
2) Solution of the Cauchy problem for vibrating membranes by the method of harmonics 291
3) Particular cases of rectangular and circular membranes 293
4) The wave equation in R^n 298

3. The Heat Conduction Equation 298
1) Solution by the method of propagation. Cauchy's problem 298
2) The solution of Cauchy's problem by the method of harmonics 301

Exercises for Chapter VII 303

CHAPTER VIII. The gamma function

1. The Function $\Gamma(z)$ 311

2. The Function B (p, q) 313

3. The Complementary Formula 315

4. Generalization of the Beta Function 317

5. Graphical Representation of the Function $y = \Gamma(x)$ for Real x 318

6. Stirling's Formula 320
7. Application to the Expansion of $1/\Gamma$ as an Infinite Product ... 322
8. The Function $\Psi(z) = \Gamma'(z)/\Gamma(z)$ 326
9. Applications 327
Exercises for Chapter VIII 330

CHAPTER IX. Bessel functions

1. Definitions and Elementary Properties 334
   1) Definitions of the Bessel, Neumann and Hankel functions 334
   2) Integral representations of Bessel functions 341
   3) Recurrence relations 343
   4) Other properties of Bessel functions 345
2. Formulae 350
Exercises for Chapter IX 353
Index 357

PREFACE

This work is concerned with the mathematical methods of physics. In this sense it is not a collection of " recipes " to be applied with an imperfect knowledge of the background. The concepts to be discussed are mathematical entities defined with care. Their elementary properties are demonstrated, and examples, taken from the physical sciences, show how they should be used.

This elementary and concise work makes no pretence of being a comprehensive treatise, and in subjects where an extensive amount of preparation is required, the essential results only are stated without proof.

The previous knowledge necessary for the book to be read profitably corresponds to the current first-year university course (propédeutique) augmented with some ideas from linear algebra and functions of a complex variable.

At the end of each chapter, exercises are suggested to the reader. These are, as far as possible, arranged in increasing order of difficulty. Some of them are direct applications of the text; others touch on new subjects. However, all exercises are at the level of the Mathematical Methods of Physics certificate of the Faculté des Sciences de Paris.

I wish to thank Mlle Denise Huet who has kindly supplied enough additional matter to the previous draft of my notes on the Mathematical Methods of Physics to completely rewrite Sections 2 and 3 of Chapter VII, Chapter IX and all the exercises. Some of the statements of these exercises are due to Messrs Gourdin, Hennequin and Trèves.

LAURENT SCHWARTZ

Preliminary results in the integral calculus: series and integrals

1. Preliminary results on series

1. SUMMABLE SERIES

The idea of summability is an extension of the idea of absolute convergence to the case where the terms of the series depend on any set of indices I whatsoever and where consequently the terms of the series are not ordered.

The terms of the series can be either real or complex numbers; for simplicity they will be supposed real.

The set theory notations used here are :

$x \in \mathrm{A}$ — the element x belongs to the set A.

$\mathrm{A} \subset \mathrm{B}$ or $\mathrm{B} \supset \mathrm{A}$ — the set A is included in the set B; i.e. if $x \in \mathrm{A}$ then $x \in \mathrm{B}$.

$\mathrm{A} \cup \mathrm{B}$ — the union of the sets A and B; i.e. the set of elements belonging either to A or to B.

$\mathrm{A} \cap \mathrm{B}$ — the intersection of the sets A and B; i.e. the set of elements belonging both to A and to B.

$\emptyset$ — the empty or null set; i.e. the set containing no elements. Thus if A and B have no common element, we may put $\mathrm{A} \cap \mathrm{B} = \emptyset$.

$]a, b[$ — the open interval from a to b; i.e. the set of points x such that $a < x < b$.

$\mathrm{A} \times \mathrm{B}$ — the direct product of sets A and B; i.e. the set of pairs (x, y) where $x \in \mathrm{A}$ and $y \in \mathrm{B}$.

DEFINITION. *Let* I *be any set of indices whatsoever and* $(u_i)_{i \in \mathrm{I}}$ *a family of real numbers parametrised by the set of indices* I. *Then the series*

$$\sum_{i \in \mathrm{I}} u_i$$

is said to be summable with sum S *and is written*

$$\sum_{i \in \mathrm{I}} u_i = \mathrm{S} \qquad \text{(I, 1; 1)}$$

if, for any $\varepsilon > 0$, *there is a* finite *sub-set of indices* $\mathrm{J} \subset \mathrm{I}$ *such that, for any finite subset of indices* $K \supset J$

(I, 1; 2) $$|S - S_K| \leqslant \varepsilon,$$

where

(I, 1; 3) $$S_K = \sum_{i \in I} u_i.$$

Note. Whatever the set of indices I may be, if all the u_i are zero, then $\sum_{i \in I} u_i$ is summable and the sum S is zero.

Properties of summable series

THEOREM 1. *Uniqueness of the sum. If*

$$\sum_{i \in I} u_i = S \quad \text{and} \quad \sum_{i \in I} u_i = S',$$

then

$$S = S'.$$

For any $\varepsilon > 0$ there exists $J_1 \subset I$ such that $K \supset J_1$ implies $|S - S_K| \leqslant \varepsilon$ and also $J_2 \subset I$ such that $K \supset J_2$ implies $|S' - S_K| \leqslant \varepsilon$. Then if $K = J_1 \cup J_2$ we have both $|S - S_K| \leqslant \varepsilon$ and $|S' - S_K| \leqslant \varepsilon$. Thus $|S - S'| \leqslant 2\varepsilon$ and, since ε is arbitrary, it follows that $S = S'$. Q.E.D.

THEOREM 2. *If* $\sum_{i \in I} u_i$ *and* $\sum_{i \in I} v_i$ *are two summable series whose sums are respectively* S *and* T, *then the series* $\sum_{i \in I} (\lambda u_i + \mu v_i)$, λ *and* μ *being constants, is summable and has the sum* $\lambda S + \mu T$.

For any given $\varepsilon > 0$, there exists a finite set of indices $J_1 \subset I$ such that, for any finite set of indices $K_1 \supset J_1$,

$$|S - S_{K_1}| \leqslant \frac{\varepsilon}{(|\lambda| + |\mu|)}.$$

There also exists a finite set of indices $J_2 \subset I$ such that, for any finite set of indices $K_2 \supset J_2$,

$$|T - T_{K_2}| \leqslant \frac{\varepsilon}{(|\lambda| + |\mu|)}.$$

It follows that, for any finite set of indices $K \supset J_1 \cup J_2$,

$$|\lambda S - \lambda S_K| \leqslant \frac{|\lambda|\varepsilon}{(|\lambda| + |\mu|)} \quad \text{and} \quad |\mu T - \mu T_K| \leqslant \frac{|\mu|\varepsilon}{(|\lambda| + |\mu|)},$$

and hence

$$|(\lambda S + \mu T) - (\lambda S_K + \mu T_K)| \leqslant \varepsilon. \qquad \text{Q.E.D.}$$

THEOREM 3. *Series of positive terms. If all the u_i are $\geqslant 0$, then*

$$\sum_{i \in I} u_i$$

is summable if and only if all the partial sums S_K, corresponding to finite subsets K of I, are bounded by a fixed number $M > 0$; in this case *

$$\sum_{i \in I} u_i = \operatorname*{Sup}_{\text{finite } K} (S_K). \tag{I, 1; 4}$$

We start by assuming that the series is summable. Then, to $\varepsilon = 1$, for example, there corresponds a finite subset $J \subset I$ such that, for all finite $K \supset J$,

$$S_K \leqslant S + 1. \tag{I, 1; 5}$$

If K is any finite set of indices, then putting $K_1 = J \cup K$, we have

$$S_K \leqslant S_{K_1} \leqslant S + 1 \tag{I, 1; 6}$$

so that all sums S_K are bounded. Conversely, suppose all the S_K are bounded with B as their least upper bound. Then for any $\varepsilon > 0$, there exists a finite subset $J \subset I$ such that

$$B - \varepsilon \leqslant S_J \leqslant B. \tag{I, 1; 7}$$

For all finite $K \supset J$,

$$B - \varepsilon \leqslant S_K \leqslant B, \tag{I, 1; 8}$$

so that $\sum_{i \in I} u_i$ is summable with the sum B.

THEOREM 4. *Cauchy's criterion* (see proof below). $\sum_{i \in I} u_i$ *is summable if and only if the u_i satisfy Cauchy's criterion:*
For any $\varepsilon > 0$ there exists a finite $J \subset I$ such that, for every finite subset of indices K disjoint from J † *we have*

$$|S_K| \leqslant \varepsilon. \tag{I, 1; 9}$$

---

* Sup means supremum or least upper bound.
Inf means infimum or greatest lower bound (translator).

† That is, such that $K \cap J = \emptyset$.

THEOREM 5. *Consequence of Cauchy's criterion. If the series* $\sum_{i \in I} u_i$ *is summable, then every partial series* $\sum_{i \in J} u_i$, *where* J *is any sub-set of* I, *is summable.*
If J *is a finite subset of* I *such that, for each finite* $K \supset J$,

$$|S - S_K| \leqslant \varepsilon, \tag{I, 1; 10}$$

then the same inequality is true for any infinite $K \supset J$.
If $J_1, J_2, \ldots, J_n \subset I$ *are mutually disjoint and if* $J = J_1 \cup J_2 \cup \cdots \cup J_n$, *then*

$$S_J = S_{J_1} + S_{J_2} + \cdots + S_{J_n} \tag{I, 1; 11}$$

provided that one of the two sides has a meaning.

THEOREM 6. *Corollary to Cauchy's criterion. For* $\sum_{i \in I} u_i$ *to be summable it is necessary and sufficient that* $\sum_{i \in I} |u_i|$ *be summable; furthermore,*

$$\left|\sum_{i \in I} u_i\right| \leqslant \sum_{i \in I} |u_i|. \tag{I, 1; 12}$$

If $\sum_{i \in I} v_i$ *is a summable series of positive terms and if*

$$|u_i| \leqslant v_i, \tag{I, 1; 13}$$

then the series $\sum_{i \in I} u_i$ *is summable and*

$$\left|\sum_{i \in I} u_i\right| \leqslant \sum_{i \in I} v_i. \tag{I, 1; 14}$$

Proofs of Theorems 4, 5 *and* 6

1. Cauchy's criterion is satisfied if $\sum_{i \in I} u_i$ is summable. For any $\varepsilon > 0$ there exists a finite $J \subset I$ such that if K is finite and has no element in common with J, then

$$|S_J - S| \leqslant \frac{\varepsilon}{2} \quad \text{and} \quad |S_{J \cup K} - S| \leqslant \frac{\varepsilon}{2}, \tag{I, 1; 15}$$

and hence

$$|S_{J \cup K} - S_J| \leqslant \varepsilon. \tag{I, 1; 16}$$

However,

$$S_{J \cup K} - S_J = S_K \tag{I, 1; 17}$$

so that we find

(I, 1; 18) $$|S_K| \leqslant \varepsilon.$$

2. If Cauchy's criterion is satisfied by a series then it is evident *a fortiori* that it is satisfied by any partial series. In particular this includes the partial series formed by the $u_i \geqslant 0$ or the partial series formed by the $u_i \leqslant 0$.

3. A series of terms $\geqslant 0$ (or of terms $\leqslant 0$) is summable if it satisfies Cauchy's criterion. In fact, it can be seen at once that its partial sums are bounded and summability then follows from Theorem 3.

Combining 2 and 3, it follows immediately that if a series satisfies Cauchy's criterion, then the series of its positive terms and the series of its terms $\leqslant 0$ are both summable. Define, for any real number x,

(I, 1; 19) $$\begin{aligned} x^+ &= x \quad \text{if} \quad x \geqslant 0, \quad x^+ = 0 \quad \text{if} \quad x \leqslant 0, \\ x^- &= -x \quad \text{if} \quad x \leqslant 0, \quad x^- = 0 \quad \text{if} \quad x \geqslant 0, \end{aligned}$$

so that

(I, 1; 20) $$\begin{aligned} x^+ \geqslant 0, &\qquad x^- \geqslant 0, \\ x = x^+ - x^-, &\qquad |x| = x^+ + x^-. \end{aligned}$$

Then the results just derived can be expressed by saying that, if $\sum_{i \in I} u_i$ satisfies Cauchy's criterion, $\sum_{i \in I} u_i^+$ and $\sum_{i \in I} u_i^-$ are both summable. It follows by Theorem 2 that their difference $\sum_{i \in I} u_i$ is summable. Together with 1, this proves Theorem 4. It also follows that the sum $\sum_{i \in I} |u_i|$ is summable and that the inequality (I, 1; 12) is satisfied.

Conversely, if $\sum_{i \in I} |u_i|$ is summable, then the partial sums are bounded and hence $\sum_{i \in I} u_i^+$, $\sum_{i \in I} u_i^-$ and their difference $\sum_{i \in I} u_i$ are summable, which proves the first part of Theorem 6; the second part is then obvious, for if $\sum_{i \in I} v_i$ is summable, its partial sums are bounded and hence, *a fortiori*, so are those of $\sum_{i \in I} u_i$. The inequality (I, 1; 14) then follows from (I, 1; 12).

4. It still remains to prove Theorem 5. If the series $\sum_{i \in I} u_i$ is summable, it satisfies Cauchy's criterion and hence, *a fortiori*, the partial series must do so (see 2) and is thus summable. Let J be a set of indices such that, for each finite $K_1 \supset J$,

(I, 1; 21) $$|S - S_{K_1}| \leqslant \varepsilon.$$

Then if K is an infinite set of indices containing J, there exists for any $\eta > 0$ a finite set of indices K_1 satisfying $J \subset K_1 \subset K$, such that

$$|S_K - S_{K_1}| \leqslant \eta \tag{I, 1; 22}$$

$\left(\text{definition of summability of } \sum_{i \in K} u_i\right)$. Then

$$|S_K - S_{K_1}| \leqslant \eta, \qquad |S - S_{K_1}| \leqslant \varepsilon, \tag{I, 1; 23}$$

and thus

$$|S - S_K| \leqslant \varepsilon + \eta. \tag{I, 1; 24}$$

Hence, since η is arbitrary,

$$|S - S_K| \leqslant \varepsilon \tag{I, 1; 25}$$

(with infinite K).

Finally, the part of Theorem 5 dealing with a decomposition

$$J = J_1 \cup J_2 \cup \ldots \cup J_n$$

is now obvious.

Note. Let k be an element of I such that $k \notin J$. By applying Cauchy's criterion to the set $K = \{k\}$, we have $|S_K| = u_k \leqslant \varepsilon$. Hence :

If $\sum_{i \in I} u_i$ *is summable then for any* $\varepsilon > 0$ *there exists a finite subset* J *of* I *such that, for each* $k \notin J$, $|u_k| \leqslant \varepsilon$.

This can be expressed by saying that :

If $\sum_{i \in I} u_i$ *is summable, the general term* u_i *tends to zero.* This idea is thus independent of the order of the terms.

Theorem 7. *If* J_n $(n = 0, 1, 2, \ldots)$ *is a sequence of finite or infinite subsets of* I *such that, for any finite* $J \subset I$, *the subset* J_n *contains* J *for all large* n, *and if* $\sum_{i \in I} u_i$ *is summable with sum* S, *then* S_{J_n} *converges to* S *as* $n \to \infty$.

For any $\varepsilon > 0$, there exists a finite $J \subset I$ such that, for all finite or infinite $K \supset J$,

$$|S - S_K| \leqslant \varepsilon. \tag{I, 1; 26}$$

By hypothesis, there is a number n_0 such that for $n \geqslant n_0$, J_n contains J. Thus for $n \geqslant n_0$,

$$|S - S_{J_n}| \leqslant \varepsilon. \tag{I, 1; 27}$$

Q.E.D.

THEOREM 8. *Summability and absolute convergence. If the set of indices* I *is the set* N *of the positive integers and zero, then the series* $\sum_{i\in I} u_i$ *is summable if and only if it is absolutely convergent, and the sum as defined in the theory of summable series is then identical with the sum as defined in the theory of convergent series.*

Assume that $\sum_{i\in I} u_i$ is summable. Then $\sum_{i\in I} |u_i|$ is also summable. Denoting by J_n the subset of integers 0, 1, 2, ..., n, Theorem 7 applied to $\sum_{i\in I} |u_i|$ shows that the numbers $\sum_{0\leq\nu\leq n} |u_\nu|$ tend to a limit as $n \to \infty$ so that the series is absolutely convergent.

The same Theorem 7 applied to $\sum_{i\in I} u_i$ gives

$$\lim_{n\to\infty} S_{J_n} = S, \tag{I, 1; 28}$$

showing that the sums as defined in the two theories are the same. (S_{J_n} is denoted by S_n in the theory of convergent series.)

Conversely, assume that the series is absolutely convergent. Then the partial sums $\sum_{0\leq\nu\leq n} |u_\nu|$ are bounded and it follows immediately that all the partial sums S_J (finite $J \subset I$) relative to the series $\sum_{i\in I} |u_i|$ are bounded. Thus the series $\sum_{i\in I} |u_i|$ is summable and it follows that $\sum_{i\in I} u_i$ is summable.

COROLLARY. *If a series is absolutely convergent it remains absolutely convergent and has the same sum if the order of the terms is changed.*
For it is then summable and the property of summability as well as the value of the sum is independent of the order of the terms.

Repeated summation or associativity

THEOREM 9. *Assume that the set of indices* I *is the union of a family of mutually disjoint * subsets* I_α $(\alpha \in A)$:

$$I = \bigcup_{\alpha \in I} I_\alpha. \tag{I, 1; 29}$$

Then, if the series $\sum_{i\in I} u_i$ *is summable, each of the series* $\sum_{i\in I_\alpha} u_i$ *is summable with*

* This may also be expressed as follows : $I_\alpha \cap I_\beta = \emptyset$ whenever $\alpha \neq \beta$.

sum σ_α, *the series* $\sum_{\alpha\in A}\sigma_\alpha$ *is summable and*

(I, 1; 30) $$\sum_{i\in I} u_i = \sum_{\alpha\in A}\sigma_\alpha = \sum_{\alpha\in A}\Big(\sum_{i\in I_\alpha} u_i\Big).$$

Proof. By Theorem 5 each series $\sum_{i\in I_\alpha} u_i$ is summable. Since $\sum_{i\in I} u_i$ is summable, there exists, for any $\varepsilon > 0$, a finite set of indices $J \subset I$ such that for any finite or infinite set of indices K containing J,

$$|S - S_K| \leqslant \varepsilon.$$

Let B be the finite sub-set of A consisting of all the indices $\alpha \in A$ such that $I_\alpha \cap J \neq \varnothing$. Then, for any finite subset C of A containing B, $\bigcup_{\alpha\in C} I_\alpha$ is a subset K of I containing J. Hence

(I, 1; 31) $$\Big|S - \sum_{i\in \bigcup_{\alpha\in C} I_\alpha} u_i\Big| \leqslant \varepsilon.$$

However, by the last part of Theorem 5, since C is finite,

(I, 1; 32) $$\sum_{i\in \bigcup_{\alpha\in C} I_\alpha} u_i = \sum_{\alpha\in C}\sigma_\alpha.$$

Hence

(I, 1; 33) $$\Big|S - \sum_{\alpha\in C}\sigma_\alpha\Big| \leqslant \varepsilon$$

so that $\sum_{\alpha\in A}\sigma_\alpha$ is summable with the sum S. Q.E.D.

Note. The converse is not true. It can happen that each series $\sum_{i\in I_\alpha} u_i$ is summable with sum σ_α, and $\sum_{\alpha\in A}\sigma_\alpha$ is summable but $\sum_{i\in I} u_i$ is not summable. In this case if B is another set of indices and $(I_\beta)_{\beta\in B}$ another partition of I into subsets, no two of which have a common element, it is possible that each of the two expressions

(I, 1; 34) $$\sum_{\alpha\in A}\Big(\sum_{i\in I_\alpha} u_i\Big), \quad \sum_{\beta\in B}\Big(\sum_{i\in I_\beta} u_i\Big)$$

is defined but that they are unequal.

Example. A is the set of integers $n \geqslant 0$. For each $\alpha = n$, I_α consists of two elements, n and $-n$ (except for $n = 0$ where there is only one element).

$I = \bigcup_\alpha I_\alpha$ is thus the set of all positive and negative integers. For each integer $i \in I$, put $u_i = i$. Then $\sum_{i \in I} u_i$ is not summable since

(I, 1; 35) $$\sum_{i \geq 0} u_i = 0 + 1 + 2 + \cdots = +\infty.$$

But

(I, 1; 36) $$\sum_{i \in I_\alpha} u_i = \alpha - \alpha = 0 = \sigma_\alpha.$$

Hence

(I, 1; 37) $$\sum_{\alpha \in A} \sigma_\alpha = \sum 0 = 0.$$

Each series $\sum_{i \in I_\alpha} u_i$ is thus summable with sum 0; hence $\sum_{\alpha \in A} \sigma_\alpha = \sum 0 = 0$ is summable although $\sum_{i \in I} u_i$ is not.

Theorem 10. *Particular case where the converse to Theorem 9 is valid. If all the u_i are ≥ 0 and if we agree to designate $+\infty$ as the sum of a divergent series of terms ≥ 0, then it is always true that*

(I, 1; 38) $$\sum_{i \in I} u_i = \sum_{\alpha \in A} \left(\sum_{i \in I_\alpha} u_i \right),$$

the value of the two sides being finite or $+\infty$.

Hence the following partial converse to Theorem 9: if $u_i \geq 0$, if $\sum_{i \in I_\alpha} u_i$ is summable with sum σ_α, and if $\sum_{i \in I_\alpha} \sigma_\alpha = \sigma$ is summable, then $\sum_{i \in I} u_i$ is summable. For, putting $\sum_{\alpha \in A} \sigma_\alpha = \sigma$, every finite sum $\sum_{\alpha \in C} \sigma_\alpha$ is bounded above by σ.

Let J be any finite subset of I, and C the (finite) set of elements α of A such that $I_\alpha \cap J \neq \emptyset$. Since C is finite, the last part of Theorem 5 gives

(I, 1; 39) $$\sum_{i \in \bigcup_{\alpha \in C} I_\alpha} u_i = \sum_{\alpha \in C} \sigma_\alpha.$$

But $S_J \leq \sum_{i \in \bigcup_{\alpha \in} I_\alpha} u_i$ and $\sum_{\alpha \in C} \sigma_\alpha \leq \sigma$, so that $S_J \leq \sigma$.

Thus all the sums S_J are bounded above by σ and, by Theorem 3, $\sum_{i \in I} u_i$ is summable.

Theorem 11. *Consequence of Theorem 9. If the series*

$$\sum_{i \in I} u_i \quad \textit{and} \quad \sum_{j \in J} v_j$$

are summable, with the respective sums U *and* V, *then the product series*

(I, 1; 40) $$\sum_{(i,j)\in I\times J} u_i v_j$$

is summable with the sum

(I, 1; 41) $$W = UV.$$

For repeated summation gives

$$\sum_{(i,j)\in I\times J} u_i v_j = \sum_{i\in I}\Big(\sum_{j\in J} u_i v_j\Big) = \sum_{i\in I}\Big(u_i \sum_{j\in J} v_j\Big) = \sum_{i\in I}(u_i V) = V\sum_{i\in I} u_i = VU,$$

which shows that *if repeated summation can be performed* then the series $\sum_{(i,j)\in I\times J} u_i v_j$ does indeed have the sum UV. But for repeated summation to be legitimate it is first necessary to know that the series $\sum_{(i,j)\in I\times J} u_i v_j$ is summable.

However, the theorem is true when all the u_i and v_j are $\geqslant 0$, since repeated summation is then always legitimate, by Theorem 10.

If the u_i and v_j can take any sign, it is necessary to consider $|u_i|$ and $|v_j|$. By Theorem 6, $\sum_{i\in I}|u_i|$ and $\sum_{j\in J}|v_j|$ are summable so that $\sum_{(i,j)\in I\times J}|u_i|\,|v_j|$ is summable with the sum $\Big(\sum_{i\in I}|u_i|\Big)\Big(\sum_{j\in J}|v_j|\Big)$. Hence $\sum_{(i,j)\in I\times J} u_i v_j$ is also summable.

Note. If I and J are each the set of integers $\geqslant 0$ the product series is a double series whose set of indices is the set of pairs (m, n) of integers $\geqslant 0$. Repeated summation of the series is often performed by putting

(I, 1; 42) $$W_n = \sum_{i+j=n} u_i v_j, \qquad UV = \sum_{n=0}^{\infty} W_n.$$

Summable series : numerical examples. The series

(I, 1; 43) $$\sum_{\substack{(m,\,n)\\ m \text{ integer} \geqslant 1\\ n \text{ integer} \geqslant 1}} \left(\frac{1}{m+n}\right)^{\alpha}$$

is summable for $\alpha > 2$ but not summable for $\alpha \leqslant 2$.

More generally the series

(I, 1; 44) $$\sum_{\substack{p_1,\,p_2,\,\ldots,\,p_n\\ p_i \text{ integers} \geqslant 1}} \left(\frac{1}{p_1 + p_2 + \cdots + p_n}\right)^{\alpha}$$

is summable for $\alpha > n$, not summable for $\alpha \leqslant n$. The expansion of $(p_1 + p_2 + \cdots + p_n)^n$ shows in fact that this quantity is

$$\geqslant p_1 p_2 \ldots p_n \, n! \geqslant p_1 p_2 \ldots p_n$$

so that

(I, 1; 45)

$$\left(\frac{1}{p_1 + p_2 + \cdots + p_n}\right)^\alpha = \left[\left(\frac{1}{p_1 + p_2 + \cdots + p_n}\right)^n\right]^{\frac{\alpha}{n}} < \left(\frac{1}{p_1 p_2 \ldots p_n}\right)^{\frac{\alpha}{n}}.$$

Thus the given series is bounded above by

(I, 1; 46)
$$\sum\left(\frac{1}{p_1}\right)^{\frac{\alpha}{n}} \sum\left(\frac{1}{p_2}\right)^{\frac{\alpha}{n}} \cdots \sum\left(\frac{1}{p_n}\right)^{\frac{\alpha}{n}}.$$

If $\alpha > n$, then $\alpha/n > 1$ and this series is summable.

Conversely, assume $\alpha \leqslant n$. Since the terms of the series are $\geqslant 0$, repeated summation gives a sum

(I, 1; 47)
$$\sum_{p_1=1}^{\infty}\left[\sum_{p_2, \ldots, p_n}\left(\frac{1}{p_1 + p_2 + \cdots + p_n}\right)^\alpha\right]$$

with

(I, 1; 48)
$$\sum_{p_2, \ldots, p_n}\left(\frac{1}{p_1 + p_2 + \cdots + p_n}\right)^\alpha \geqslant \sum_{\substack{1 \leqslant p_2 \leqslant p_1 \\ \cdots\cdots\cdots \\ 1 \leqslant p_n \leqslant p_1}}\left(\frac{1}{p_1 + p_2 + \cdots + p_n}\right)^\alpha \geqslant \left(\frac{1}{np_1}\right)^\alpha p_1^{n-1}.$$

Thus the series is dominated by $\left(\frac{1}{n}\right)^\alpha \sum_{p_1=1}^{\infty}\left(\frac{1}{p_1}\right)^{\alpha+1-n}$. As $\alpha \leqslant n$, $\alpha + 1 - n \leqslant 1$ and the series diverges.

Note. It can be shown that

(I, 1; 49)

$$\frac{1}{\sqrt{n}}(p_1 + p_2 + \cdots + p_n) \leqslant \sqrt{p_1^2 + p_2^2 + \cdots + p_n^2} \leqslant p_1 + p_2 + \cdots + p_n.$$

Thus the series $\sum_{p_1, p_2, \ldots, p_n}\left(\frac{1}{\sqrt{p_1^2 + p_2^2 + \cdots + p_n^2}}\right)^\alpha$ has the same property, being summable for $\alpha > n$, and not summable for $\alpha \leqslant n$.

2. SEMI-CONVERGENT SERIES

Assume that the set of indices I is the set **N** of integers $\geqslant 0$, so that the terms of the series have a natural order. When a series $\sum_{n=0}^{\infty} u_n$ is not absolutely convergent the most important criterion is the *alternating series theorem. This can be generalized to Abel's theorem :*

Abel's Theorem. *Let $u_n = a_n b_n$, where the a_n are $\geqslant 0$ and form a decreasing sequence tending to zero as n tends to ∞, and the b_n are complex numbers such that*

$$|\sigma_{m,n}| = |b_m + b_{m+1} + \cdots + b_n|$$

is bounded above by a constant $\sigma \geqslant 0$. Then the series Σu_n is convergent and the modulus of the remainder is bounded above by σ multiplied by the first neglected term of the sequence a_n.

Let us write

(I, 1; 50)

$$\begin{aligned} S_{0,n} &= a_0 b_0 + a_1 b_1 + \cdots + a_n b_n \\ &= a_0 \sigma_{0,0} + a_1(\sigma_{0,1} - \sigma_{0,0}) + \cdots + a_n(\sigma_{0,n} - \sigma_{0,n-1}) \\ &= (a_0 - a_1)\sigma_{0,0} + (a_1 - a_2)\sigma_{0,1} + \cdots + (a_{n-1} - a_n)\sigma_{0,n-1} + a_n \sigma_{0,n}. \end{aligned}$$

The last term $a_n \sigma_{0,n}$, which has a different form from the others, converges to zero as $n \to \infty$, since $a_n \to 0$ and $|\sigma_{0,n}| \leqslant \sigma$. It remains to show that

(I, 1; 51) $$\sum_{k=0}^{n-1} (a_k - a_{k+1}) \sigma_{0,k}$$

has a limit as $n \to \infty$. This is equivalent to proving that the series with the general term

(I, 1; 52) $$v_k = (a_k - a_{k+1}) \sigma_{0,k}$$

is convergent. Now this series is absolutely convergent, since

(I, 1; 53) $$|v_k| \leqslant (a_k - a_{k+1}) \sigma,$$

whence

(I, 1; 54)

$$\sum_{k=0}^{n-1} |v_k| \leqslant \sum_{k=0}^{k=+\infty} |v_k| \leqslant (a_0 - a_1)\sigma + \cdots + (a_{n-1} - a_n)\sigma + \cdots = a_0 \sigma.$$

This proves the convergence of the original series.

From the above we have the inequality

(I, 1; 55) $$|S_{0,n}| \leqslant a_n \sigma + a_0 \sigma,$$

giving

(I, 1; 56) $$|S_0| \leqslant a_0 \sigma, \qquad \text{where} \qquad S_0 = \sum_{n=0}^{\infty} a_n b_n.$$

The same reasoning, starting with the term u_{m+1} and defining

(I, 1; 57) $$R_m = \lim_{n \to \infty} S_{m+1, n},$$

gives

(I, 1; 58) $$|R_m| \leqslant a_{m+1}\sigma.$$ Q.E.D.

Examples

A) *Alternating series :*

(I, 1; 59) $$b_n = (-1)^n, \qquad \sigma = 1,$$

whence

(I, 1; 60) $$|R_m| \leqslant a_{m+1}.$$

It can be seen that in this case it is also true that the remainder has the same sign as the first neglected term.

B) *Trigonometric series :*

(I, 1; 61) $$\sum_{n=-\infty}^{n=+\infty} a_n e^{ni\theta}.$$

It can be seen that this series converges if each of the two series

(I, 1; 62) $$\sum_{n=0}^{\infty} a_n e^{ni\theta} \quad \text{and} \quad \sum_{n=-\infty}^{0} a_n e^{ni\theta}$$

converges separately. Here

(I, 1; 63) $$b_n = e^{ni\theta}, \qquad \sigma_{m,n} = e^{mi\theta} + e^{(m+1)i\theta} + \cdots + e^{ni\theta}.$$

On the one hand, when $e^{i\theta} \neq 1$,

$$\sigma_{m,n} = \frac{e^{(n+1)i\theta} - e^{mi\theta}}{e^{i\theta} - 1},$$

(I, 1; 64) $$|\sigma_{m,n}| \leqslant \frac{2}{|e^{i\theta} - 1|}.$$

On the other hand, when $e^{i\theta} = 1$,

(I, 1; 65) $$\sigma_{m,n} = n - m + 1$$

is not bounded.

Conclusion. If the decreasing sequence $a_n \geqslant 0$ tends to zero as $n \to \infty$, the series

(I, 1; 66) $$\sum_{n=0}^{\infty} a_n e^{ni\theta}$$

is convergent for $\theta \neq 2k\pi$ (where k is an integer) and the modulus of the remainder is bounded above by $a_{n+1} \dfrac{2}{|e^{i\theta} - 1|}$.

It can be seen, by taking real and imaginary parts, that the series

$$\text{(I, 1; 67)} \qquad \sum_{n=0}^{\infty} a_n \cos n\theta, \qquad \sum_{n=0}^{\infty} a_n \sin n\theta$$

have a similar property. However, the latter is convergent even for $\theta = 2k\pi$ since all its terms are then zero.

2. Preliminary results on Integration

1. THE LEBESGUE INTEGRAL*

Integral with respect to one variable

The Lebesgue integral is a generalization of the Riemann integral. It is a functional on a certain class of real or complex functions of the variable x, called the class of *summable functions,* and assigns to each summable function $f(x)$ a real or complex number called its *integral* and denoted by

$$\text{(I, 2; 1)} \qquad \int_{-\infty}^{+\infty} f(x)\, dx \quad \text{or} \quad \int_{\mathrm{R}} f(x)dx \quad \text{or} \quad \int f(x)dx \quad \text{or} \quad \int f.$$

As the theory is very difficult we shall state a certain number of essential results without proof. We shall not give necessary and sufficient conditions for a function to be summable, nor methods of calculating its integral.

Set of measure zero on the line R

Given a subset E of R, the function φ_{E} which is equal to 1 at each point $x \in \mathrm{E}$ and to 0 at each point $x \notin \mathrm{E}$, is known as the *characteristic function* of E.

DEFINITION 1. *The measure of an open set* E *is defined as the least upper bound of the integrals of the continuous functions* $\geqslant 0$ *which are zero outside a finite interval and bounded above by the characteristic function.*

For example, if E is the interval $]a, b[$, its measure is $(b - a)$.

---

*The Lebesgue integral is essential in Hilbert space theory, which is itself essential in wave mechanics and in the theory or partial differential and integral equations.

DEFINITION 2. *A set E on the line is said to be of measure zero if, for any $\varepsilon > 0$, there exists an open set of measure $\leqslant \varepsilon$ which contains the set E.*

Example. A point is of measure zero.

Properties of sets of measure zero

THEOREM 12. *Any set contained in a set of measure zero is itself of measure zero.*

THEOREM 13. *Any set E which is the union of a finite or a denumerably infinite family of sets of measure zero is itself of measure zero.*

Inference. Since a point is of measure zero any denumerable set of points is of measure zero. For example, the set of rational numbers, which is denumerable, is of measure zero.

Notes

1) Theorem 13 is no longer valid for the union of a non-denumerably infinite number of sets E_i. For example, $E = R$ is the union of the sets E_i, each consisting of a single point, so that the measure of E_i is 0. However, the measure of E is not zero and is, in fact, infinite.

This shows incidentally that R is not denumerable!

2) Non-denumerable sets exist which are nevertheless of measure zero.
Let P be a property relating to the points x of the line R. P is said to apply almost *everywhere* or at almost every point x if the set of points x where P does not apply is of measure zero. Thus almost all numbers are irrational.

From Theorem 13 it is also possible to deduce

THEOREM 14. *Let (P_i) be a finite or denumerable family of properties relating to the points x of R, each valid almost everywhere. Then if we denote by P the property which consists, for each point x, in simultaneously satisfying all the properties (P_i), P is valid almost everywhere.*

Measurable functions

DEFINITION 3. *A function f is said to be measurable if it is the limit of a sequence of functions continuous almost everywhere.*

This means that there is a sequence, $f_m(x)$, of continuous functions such that the relation $f(x) = \lim_{m\to\infty} f_m(x)$ is valid for all x, except for a set of points x of measure zero.

In particular, any function is measurable if it is continuous except at the points x of a set of measure zero.

Properties of measurable functions

THEOREM 15. *Any continuous function of a finite number of measurable functions is itself measurable. In particular, if f and g are measurable, then $f + g$, fg,* sup (f, g), inf (f, g), $|f|$ *and, if $g \neq 0$ everywhere, f/g are measurable.*

Any function f is measurable if it is the limit of a sequence of measurable functions f_n. Non-measurable functions are of such an irregular character that not one of them is known explicitly; their existence is demonstrated theoretically only by using an axiom called the axiom of choice, which does not allow one to construct such a function explicitly. This means that all the functions encountered by a physicist must be measurable. *Hence, in future it will always be assumed, without any explicit reference to the fact, that all functions considered are measurable.*

Summable functions

THEOREM 16. *For a function to be summable it must be measurable. Also, it must not be " too large."*

Properties of summable functions

THEOREM 17. *A function f is summable if and only if $|f|$ is summable. If g is summable and $\geqslant 0$ and if $|f| \leqslant g$, then f is summable.*

THEOREM 18. *If f and g are summable and if λ is a constant, then $f+g$ and λf are summable.* Also,

$$\text{(I, 2; 2)} \qquad \int (f+g) = \int f + \int g, \qquad \int \lambda f = \lambda \int f.$$

The set of summable functions thus forms a vector space denoted by $\mathfrak{L}^1$. On this space $\mathfrak{L}^1$ the integral is a *linear* form or functional.

THEOREM 19

(I, 2; 3) *If* $f \geqslant 0$, *then* $\int f \geqslant 0$. *If* $|f| \leqslant g$, *then* $\left|\int f\right| \leqslant \int |f| \leqslant \int g$.

The integral is then known as a functional $\geqslant 0$.

Note. If $f \geqslant 0$ is measurable but not summable, then

$$\int f = +\infty.$$

THEOREM 20. *A bounded function f which is zero outside a finite interval (a, b) is summable. If, in addition, f is integrable in the Riemann sense in (a, b) then its*

Lebesgue and Riemann integrals are identical in the interval (a, b). *If, in* (a, b),

(I, 2; 4) $$m \leqslant f(x) \leqslant M \quad [\text{resp. } |f(x)| \leqslant M],$$

then

(I, 2; 5) $$m(b-a) \leqslant \int f(x)\,dx \leqslant M(b-a),$$

(I, 2; 6) $$\left[\text{resp. } \left|\int f(x)\,dx\right| \leqslant M(b-a)\right].$$

THEOREM 21. *If f is a real or complex function and* A *and* M *are any two numbers* > 0, *then we shall denote by* $f_{A,M}$ *the function which is equal to* $f(x)$ *if* $|x| \leqslant A$ *and* $|f(x)| \leqslant M$, *and is zero otherwise. The function f is summable if and only if* $\int |f_{A,M}|$ *is bounded independently of* A *and* M.

A function f which has only a finite number of points of discontinuity is summable if and only if its integral, in the sense of the theory of "improper" Riemann integrals, is absolutely convergent. Its Lebesgue integral is then identical with its "improper integral."

THEOREM 22. *If f is a function which is zero almost everywhere, it is summable and its integral is zero. If two functions f and g are equal almost everywhere and if f is summable, then g is summable and* $\int f = \int g$.

This makes it possible to say whether a function f is summable and to obtain its integral, even if it is not defined everywhere on R or if it takes the value $\mp\infty$ at certain points. This follows from defining a function f_1 which is equal to f at every point x where $f(x)$ is defined and $\neq \mp\infty$ and which takes arbitrary finite values at other points. Provided that f is defined almost everywhere and is $\neq \mp\infty$ almost everywhere, the summability of f_1 and the value of its integral are independent of the element of arbitrariness in its definition. We can say that f is summable if f_1 is summable and set

$$\int f = \int f_1.$$

THEOREM 23. *Converse of Theorem 22. If* $f \geqslant 0$ *and if* $\int f = 0$ *then f is zero almost everwhere. If* $f \geqslant g$ *and if* $\int f = \int g$, *then f and g are equal almost everwhere.*

Two functions f and g which are defined almost everywhere and $\neq \infty$ almost everywhere are said to be equivalent if they are equal almost everywhere.

A *class* of functions is the set of functions (defined almost everywhere and finite almost everywhere) which are equal to a given finite function almost everywhere.

The vector space of the classes of summable functions is denoted by $\mathcal{L}^1$. This is the quotient of the vector space $\mathcal{L}^1$ of summable functions by the sub-space of functions which are zero nearly everywhere. As the integral of a summable function depends only on its class, the integral is a linear form or functional on the space $\mathcal{L}^1$.

Measure of sets on R

If E is the interval $]a, b[$, then its measure $(b - a)$ is also the integral of its characteristic function. In general a set E is said to be *measurable* if its characteristic function φ_E is measurable. The integral $\int \varphi_E \geqslant 0$ is called the measure of E and denoted by mes E. If φ_E is summable, then E is said to be *summable*. If φ_E is not *summable* then we put mes E $= +\infty$.

THEOREM 24. *The measure (finite or infinite) of the union* E *of a finite or denumerably infinite number of sets* E_i *is at most equal to the sum of their measures :*

$$\text{mes } E \leqslant \sum_i \text{mes } E_i \quad (\textit{series of positive terms} \geqslant 0).$$

If no two of the E_i *have a point in common then* mes E $= \sum_i$ mes E_i.

Integral over a set

Let E be a subset of R and f a real or complex function defined on E. If the function f_0 which is equal to $f(x)$ for $x \in E$ and to 0 for $x \notin E$ is summable over R, then f is said to be *summable over* E. The integral of f over E is then defined by

$$\text{(I, 2; 7)} \qquad \int_E f(x)\, dx = \int f_0(x)\, dx.$$

This is why it has not been necessary to develop a theory of integration over the finite interval (a, b) before discussing integration over R, since, by the above definition, the former is an obvious special case.

If $a \leqslant b$ it is usual to denote the integral of f over the interval (a, b) by $\int_a^b f(x)\, dx$. If $b \leqslant a$, the negative, $-\int_b^a f(x)\, dx$, of the integral over (b, a) is denoted by $\int_a^b f(x)\, dx$.

If f is defined on E the function f_0 is identical with $f\,\varphi_E$ and

$$\text{(I, 2; 8)} \qquad \int_E f(x)\, dx = \int_R f(x)\varphi_E(x)\, dx.$$

We have the following inequality : if, for au $x \in E$,

$$m \leqslant f(x) \leqslant M \quad [\text{resp. } |f(x)| \leqslant M],$$

then

$$m(\text{mes } E) \leqslant \int_E f(x)\, dx \leqslant M(\text{mes } E)$$

$$\left[\text{resp. } \left|\int_E f(x)\, dx\right| \leqslant M(\text{mes } E)\right]$$

(this inequality is, of course, only of interest if mes E is finite). If E is of measure zero, $\int_E f(x)\, dx = 0$ for any f.

Change of variables

Let (a, b) be a finite interval and for $\alpha \leqslant t \leqslant \beta$, let $x = \xi(t)$ be a continuous function of the variable t with a continuous derivative $\xi'(t)$. Assume that $\xi(\alpha) = a$ and $\xi(\beta) = b$. Then if $f(x)$ is a continuous function on (a, b) it is known that $\int_a^b f(x)\, dx$ can be expressed in the form $\int_\alpha^\beta f(\xi(t))\xi'(t)\, dt$. Assume that $\xi(t)$ is *monotonic*. Let T denote the interval (α, β) or (β, α) according as $\alpha \leqslant \beta$ or $\beta \leqslant \alpha$, and, X denote the interval (a, b) or (b, a) according as $a \leqslant b$ or $b \leqslant a$. Then, whatever the respective positions of a and b and of α and β,

$$\int_X f(x)\, dx = \int_T f(\xi(t))|\xi'(t)|\, dt. \tag{I, 2; 9}$$

Suppose, for example, $\alpha \leqslant \beta$. If ξ is an increasing function, then $a \leqslant b$ and $\int_X = \int_a^b$, $\int_T = \int_\alpha^\beta$; in this case $\xi'(t) \geqslant 0$ so that $|\xi'(t)| = \xi'(t)$. If, on the other hand, ξ is a decreasing function, then $b \leqslant a$,

$$\int = -\int_a^b, \quad \int_I = \int_\alpha^\beta \quad \text{and} \quad |\xi'(t)| = -\xi'(t).$$

We can then state the generalization :

THEOREM 25. *Let* T *be a subset of* R, $x = \xi(t)$ *a continuous function with a continuous derivative defined on an open set containing* T, *and* X *the set traversed by x as t traverses* T. *Assume that the mapping ξ is one-to-one on* T. *Then $f(x)$ is summable on* X *if and only if $f(\xi(t))|\xi'(t)|$ is summable on* T, *and in this case*

$$\int_X f(x)\, dx = \int_T f(\xi(t))|\xi'(t)|dt. \tag{I, 2: 10}$$

The multiple integral

This is a functional which to each summable function f of the n variables $x_1, x_2, \ldots, x_n$ assigns a number called its integral, denoted by

$$\iint \cdots \int_{R^n} f(x_1, \ldots, x_n)\, dx_1 dx_2 \cdots dx_n \tag{I, 2; 11}$$

or

$$\iint \cdots \int_{R^n} f \quad \text{or} \quad \iint \cdots \int f.$$

Denoting the point with coordinates $x_1, \ldots, x_n$ in the n-dimensional space R^n by x and putting $dx_1 dx_2 \cdots dx_n = dx$ (often termed the volume element) it is then possible to regard f as a function defined on R^n and to denote its integral by

$$\iint \cdots \int f(x)\, dx. \tag{I, 2; 12}$$

The multiple integral has properties analogous to those of the single integral (i.e. the integral with respect to one variable only). The measure of a set in R^n is volumetric in character. Thus the measure of a parallelepiped (which we shall call a "block") defined by the inequalities $a_i \leqslant x_i \leqslant b_i$, $i = 1, 2, \ldots n$, is

$$\prod_{\nu=1}^{\nu=n} (b_\nu - a_\nu). \tag{I, 2; 13}$$

If the value of $f(x)$ lies between m and M everywhere in the block the value of its integral over the block lies between

$$m \prod_{\nu=1}^{\nu=n} (b_\nu - a_\nu) \quad \text{and} \quad M \prod_{\nu=1}^{\nu=n} (b_\nu - a_\nu). \tag{I, 2; 14}$$

A curve, surface or any sub-manifold of dimension $k \leqslant n - 1$ is of measure zero if it is sufficiently regular (for example if it has a continuously varying linear tangential sub-variety); these are simple examples of non-denumerable sets of measure zero.

Change of variables in multiple integration

The rule given for one variable is generalized as follows :

THEOREM 26. *Let* T *be a set in* R^n *and* $x = \xi(t)$ *a mapping of* T *into* R^n *defined by the n functions* $x_i = \xi_i(t_1, \ldots, t_n)$, $i = 1, 2, \ldots, n$, *the* ξ_i *being defined, continuous and with continuous first-order partial derivatives in an open set of* R^n *containing* T. *It is also assumed that* ξ *is one-to-one on* T. *The Jacobian determinant of the mapping* ξ *(the determinant of the* $\partial\xi_i/\partial x_j$*) will be denoted by* $J(t) = J(t_1, \ldots, t_n)$.

Then $f(x)$ is summable over the set X *traversed by x when t traverses* T *if and only if $f(\xi(t))|J(t)|$ is summable over* T, *and in that case*

$$\text{(I, 2; 15)} \qquad \iint \cdots \int_{\mathrm{X}} f(x_1, \ldots, x_n)\, dx_1 dx_2 \cdots dx_n =$$

$$\iint \cdots \int_{\mathrm{T}} f[\xi_1(t_1, \ldots, t_n), \ldots, \xi_n(t_1, \ldots, t_n)]\, |J(t_1, \ldots, t_n)|\, dt_1 \ldots dt_n,$$

where

$$\text{(I, 2; 16)} \qquad J(t_1, \ldots, t_n) = \begin{vmatrix} \frac{\partial \xi_1}{\partial t_1} & \frac{\partial \xi_1}{\partial t_2} & \ldots & \frac{\partial \xi_1}{\partial t_i} & \ldots & \frac{\partial \xi_1}{\partial t_n} \\ \frac{\partial \xi_2}{\partial t_1} & \frac{\partial \xi_2}{\partial t_2} & \ldots & \frac{\partial \xi_2}{\partial t_i} & \ldots & \frac{\partial \xi_2}{\partial t_n} \\ \cdots & \cdots & \cdots & \cdots & \cdots & \cdots \\ \cdots & \cdots & \cdots & \cdots & \cdots & \cdots \\ \frac{\partial \xi_k}{\partial t_1} & \frac{\partial \xi_k}{\partial t_2} & \ldots & \frac{\partial \xi_k}{\partial t_i} & \ldots & \frac{\partial \xi_k}{\partial t_n} \\ \cdots & \cdots & \cdots & \cdots & \cdots & \cdots \\ \frac{\partial \xi_n}{\partial t_1} & \frac{\partial \xi_n}{\partial t_2} & \ldots & \frac{\partial \xi_n}{\partial t_i} & \ldots & \frac{\partial \xi_n}{\partial t_n} \end{vmatrix}.$$

Example. In the plane R^2 put $x = r\cos\theta, y = r\sin\theta$. Take $X = R^2$. For T, take the set of values $0 < r$, $0 \leqslant \theta < 2\pi$, together with the element $r = 0$, $\theta = 0$, so that the transformation $(r, \theta) \to (x, y)$ is one-to-one on T. In this case,

$$\text{(I, 2; 17)} \qquad J(r, \theta) = \begin{vmatrix} \cos\theta & -r\sin\theta \\ \sin\theta & r\cos\theta \end{vmatrix} = r,$$

so that $f(x, y)$ is summable over R^2 if and only if $f(r\cos\theta, r\sin\theta)r$ is summable over T. Then

$$\text{(I, 2; 18)} \qquad \iint_{R^2} f(x,y)\, dx\, dy = \iint_{\mathrm{T}} f(r\cos\theta, r\sin\theta)r\, dr\, d\theta.$$

The integral $\iint_{\mathrm{T}}$ can be replaced by the integral over the set $0 \leqslant r$, $0 \leqslant \theta \leqslant 2\pi$ which in the Cartesian coordinate plane r, θ is a semi-infinite strip of width 2π parallel to the r-axis. This is equivalent to adding to T the set $r > 0$, $\theta = 2\pi$ and the set $r = 0$, $0 < \theta \leqslant 2\pi$ which are respectively a half-line and a line segment and thus of measure zero.

Calculation of a double integral by repeated integration. Fubini-Lebesgue theorem

Let $f(x, y)$ be a continuous function of the two variables x and y in the rectangle $a \leqslant x \leqslant a'$, $b \leqslant y \leqslant b'$. For any fixed value of x the function $y \to f(x, y)$ is a continuous function of one variable y and may thus be integrated over the interval (b, b'). The integral obtained depends on x

and is thus a function $I(x)$ of x defined in the interval (a, a'). Then we have

THEOREM 27. *If f is continuous for $a \leqslant x \leqslant a'$, $b \leqslant y \leqslant b'$, the integral $I(x) = \int_b^{b'} f(x, y)\, dy$ is a continuous function of x in the interval $a \leqslant x \leqslant a'$. Its integral $\int_a^{a'} I(x)\, dx$ is identical with $\iint_{\substack{a \leqslant x \leqslant a' \\ b \leqslant y \leqslant b'}} f(x, y)\, dx\, dy$ so that*

(I, 2; 19)

$$\iint_{\substack{a \leqslant x \leqslant a' \\ b \leqslant y \leqslant b'}} f(x, y)\, dx\, dy = \int_a^{a'} dx \int_b^{b'} f(x, y)\, dy = \int_b^{b'} dy \int_a^{a'} f(x, y)\, dx.$$

This result will now be extended to the case where f is no longer necessarily continuous and where the rectangle is replaced by R^2 itself.

THEOREM 28. *Fubini's Theorem. If $f(x, y)$ is summable over R^2, the function $y \to f(x, y)$ for fixed x is summable with respect to y, except for certain special values of x forming a set of measure zero.*

The quantity $I(x) = \int_{-\infty}^{+\infty} f(x, y)\, dy$ is thus a function of x which is defined almost everywhere. It is summable and its integral $\int_{-\infty}^{+\infty} I(x)\, dx$ is identical with $\iint_{R^2} f(x, y)\, dx\, dy$.

It follows that

(I, 2; 20)

$$\iint_{R^2} f(x, y)\, dy = \int_{-\infty}^{+\infty} dx \int_{-\infty}^{+\infty} f(x, y)\, dy = \int_{-\infty}^{+\infty} dy \int_{-\infty}^{+\infty} f(x, y)\, dx.$$

Notes

1) If $f(x, y)$ is not summable it can happen that one of the two expressions

(I, 2; 21) $$\int_{-\infty}^{+\infty} dx \int_{-\infty}^{+\infty} f(x,y)\, dy, \qquad \int_{-\infty}^{+\infty} dy \int_{-\infty}^{+\infty} f(x,y)\, dx$$

is defined while the other is not. It can even happen that both are defined but have different values.

Example

(I, 2; 22)

$$\int_0^1 dx \int_0^1 \frac{x^2 - y^2}{(x^2 + y^2)^2}\, dy = \frac{\pi}{4},$$

$$\int_0^1 dy \int_0^1 \frac{x^2 - y^2}{(x^2 + y^2)^2}\, dx = -\frac{\pi}{4}.$$

2) It is not to be expected that $f(x, y)$ should be summable for every value of x. In fact, the values of $f(x, y)$ on any vertical line $x = a$ can be changed

without affecting its summability, since the line $x = a$ is a set of measure zero in R^2.

This makes it possible to select the function $y \to f(a, y)$ quite arbitrarily with no necessity for it to be summable over y. However, the set of values of x where this can be done is of measure zero and the essential point is that $I(x)$ is still defined almost everywhere. See for example equation (I, 2; 61, page 41).

3) Fubini's theorem is comparable with the theorem of repeated summation in summable series. Now, the converse to the latter theorem is valid for series with terms $\geqslant 0$. A similar result can be proved here:

THEOREM 29. *Let $f(x, y)$ be $\geqslant 0$ (f being as usual assumed measurable) and let the function $y \to f(x, y)$ for fixed x be summable over y, except for certain values of x forming a set of measure zero. Then, if the function $I(x) = \int_{-\infty}^{+\infty} f(x, y)\, dy$ thus defined almost everywhere is summable in x, it follows that $f(x, y)$ is summable and*

$$\int_{-\infty}^{+\infty} I(x)\, dx = \iint_{R^2} f(x, y)\, dx\, dy. \tag{I, 2; 23}$$

Alternatively, it may be stated that for $f(x, y) \geqslant 0$, the three integrals

$$\iint_{R^2} f(x, y)\, dx\, dy, \qquad \int_{-\infty}^{+\infty} dx \int_{-\infty}^{+\infty} f(x, y)\, dy, \qquad \int_{-\infty}^{+\infty} dy \int_{-\infty}^{+\infty} f(x, y)\, dx, \tag{I, 2; 24}$$

which are always defined and either finite or equal to $+\infty$, always have equal values.

COROLLARY. *If f is such that one of the integrals* (I, 2; 24) *with f replaced either by $|f|$ or by an upper bound $g \geqslant 0$ of $|f|$ is finite, then all three integrals* (I, 2; 24) *with f as integrand are defined and equal.*

In fact, if $|f|$ is summable then so is f.

Naturally, it is possible to apply Fubini's theorem to an integral of type $\iint_E f(x, y)\, dx\, dy$, E being any set of R^2, provided that f is summable over E or $f \geqslant 0$. As this amounts to evaluating $\iint_{R^2} f_0(x, y)\, dx\, dy$, where f_0 is the function defined over E and extended by defining it to be zero for $x \notin E$, no new result is required. Let $E(x)$ be the "section of E" by the vertical line of abscissa x; i.e. the set of points y such that $(x, y) \in E$. Then

$$\iint_E f(x, y)\, dx\, dy = \int_{-\infty}^{+\infty} dx \int_{E(x)} f(x, y)\, dy. \tag{I, 2; 25}$$

Example. The primitive of order n and the remainder in Taylor's theorem. For f a continuous function of one variable let us evaluate

$$\text{(I, 2; 26)} \qquad \int_0^x d\xi \int_0^\xi f(t)\,dt.$$

This is a second primitive of f which becomes zero for $x = 0$, as does its first derivative. To start with, assume $x > 0$. The integral can be written $\iint_E f(t)\,dt\,d\xi$ where E is the set of points (ξ, t) satisfying $0 \leqslant \xi \leqslant x$, $0 \leqslant t \leqslant \xi$. Hence it may also be expressed as

$$\text{(I, 2; 27)} \qquad \int_0^x dt \int_t^x f(t)\,d\xi = \int_0^x (x - t) f(t)\,dt.$$

Thus it has been transformed into an ordinary integral. It can easily be seen that this is still valid for $x < 0$.

More generally, if f is a continuous function of one variable, the n-th primitive which, with its first $n - 1$ derivatives, becomes zero at $x = 0$ may be written

$$\text{(I, 2; 28)} \qquad \int_0^x d\xi_1 \int_0^{\xi_1} d\xi_2 \cdots \int_0^{\xi_{n-1}} f(\xi_n)\,d\xi_n.$$

It can be shown by induction that this integral can be transformed to an ordinary integral

$$\text{(I, 2; 29)} \qquad \int_0^x \frac{(x-t)^{n-1}}{(n-1)!} f(t)\,dt.$$

This is obviously true for $n = 1$. Assuming it is true for $n \leqslant m - 1$, it will be shown to be true for $n = m$. The integral (I, 2; 28) can then be written

$$\text{(I, 2; 30)} \qquad \int_0^x d\xi_1 \int_0^{\xi_1} \frac{(\xi_1 - t)^{m-2}}{(m-2)!} f(t)\,dt.$$

By inverting the order of integration as in (I, 2; 26-27) it follows that

$$\text{(I, 2; 31)} \quad \int_0^x f(t)\,dt \int_t^x \frac{(\xi_1 - t)^{m-2}}{(m-2)!}\,d\xi_1 = \int_0^x \frac{(x-t)^{m-1}}{(m-1)!} f(t)\,dt. \quad \text{Q.E.D.}$$

Let f be a continuous function of one variable whose first $n + 1$ derivatives are also continuous. The function of x denoted by the integral

$$\text{(I, 2; 32)} \qquad \int_0^x \frac{(x-t)^n}{n!} f^{(n+1)}(t)\,dt$$

is an $(n + 1)$-th primitive of $f^{(n+1)}$ which, with its first n derivatives, takes the value zero at $x = 0$. Hence f, which is also an $(n + 1)$-th primitive of $f^{(n+1)}$, is obtained by adding to this integral a polynomial of degree $\leqslant n$ whose derivatives of order $\leqslant n$ at $x = 0$ are equal to those of f. According to Taylor's theorem for polynomials the required polynomial is

$$\sum_{\nu=0}^{\nu=n} \frac{f^{(\nu)}(0)}{\nu\,!} x^{\nu}. \tag{I, 2; 33}$$

Thus

$$f(x) = \sum_{\nu=0}^{\nu=n} \frac{f^{(\nu)}(0)}{\nu\,!} x^{\nu} + \int_0^x \frac{(x-t)^n}{n\,!} f^{(n+1)}(t)\, dt. \tag{I, 2; 34}$$

This is Taylor's theorem for functions in general, with the remainder expressed in integral form. If the modulus of the $(n + 1)$-th derivative of f is bounded above by M, then the remainder is bounded above by

$$\mathrm{M}\left|\int_0^x \frac{(x-t)^n}{n\,!}\, dt\right| = \mathrm{M}\left|\left[\frac{(x-t)^{n+1}}{(n+1)\,!}\right]_0^x\right| = \mathrm{M}\,\frac{|x|^{n+1}}{(n+1)\,!}\,. \tag{I, 2; 35}$$

This is the classical inequality for the remainder in Taylor's theorem.

Extension to multiple integrals of any order

If f is summable or if $f \geqslant 0$, a triple integral $\iiint_{\mathrm{R}^3} f(x, y, z)\, dx\, dy\, dz$ may be evaluated by three successive ordinary integrations

$$\int_{-\infty}^{+\infty} dx \int_{-\infty}^{+\infty} dy \int_{-\infty}^{+\infty} f(x, y, z)\, dz. \tag{I, 2; 36}$$

For fixed x and y the integral with respect to z exists except at certain special values of x and y which form a set of measure zero in the space R^2. The integral $\mathrm{I}(x, y) = \int_{-\infty}^{+\infty} f(x, y, z)\, dz$ is thus defined almost everywhere. It is summable in (x, y) and the triple integral is obtained by evaluating the double integral $\iint_{\mathrm{R}^2} \mathrm{I}(x, y)\, dx\, dy$ by two successive ordinary integrations. Hence the triple integral could also be written as

$$\iint_{\mathrm{R}^2} dx\, dy \int_{-\infty}^{+\infty} f(x, y, z)\, dz. \tag{I, 2; 37}$$

Another possibility would be to carry out first a double integration, then an ordinary integration

$$\int_{-\infty}^{+\infty} dx \iint_{\mathrm{R}^2} f(x, y, z)\, dy\, dz, \quad \text{etc.} \tag{I, 2; 38}$$

THEOREM 30. *If $f(x_1, \ldots, x_n)$ is a product $f_1(x_1) \cdots f_n(x_n)$ of functions of one variable, none of which is zero almost everywhere, it is summable if and only* *if each of the functions $f_i(x_i)$ is summable, and then the integral of f is the product of the integrals*

$$\iint \cdots \int_{R^n} f_1(x_1) f_2(x_2) \ \ldots \ f_n(x_n)\, dx_1 \cdots dx_n = \prod_{\nu=1}^{\nu=n} \left(\int_{-\infty}^{+\infty} f_\nu(x_\nu)\, dx_\nu \right).$$

If f is summable, this follows from Fubini's theorem. However, it is unnecessary to assume at the outset that f is summable. If each of the functions $f_\nu(x_\nu)$ is summable, then so is each of the functions $|f_\nu(x_\nu)|$, but since the latter are $\geqslant 0$, it can be deduced that $|f|$ and hence f itself is also summable.

Applications of integration over a sphere

It is required to evaluate $\iint_{R^2} f(x, y)\, dx\, dy$, f being supposed summable or $\geqslant 0$. By a change of variables this can always be transformed to

$$\text{(I, 2; 39)} \qquad \iint_{\substack{0 \leqslant r \\ 0 \leqslant \theta \leqslant 2\pi}} f(r \cos \theta, \ r \sin \theta)\, r\, dr\, d\theta.$$

This double integral can be evaluated by two successive ordinary integrations:

$$\text{(I, 2; 40)} \qquad \int_0^\infty dr \int_0^{2\pi} f(r \cos \theta, r \sin \theta)\, r\, d\theta.$$

Note that $r\, d\theta = ds$, an element of length on the circumference of a circle of radius r.

This formula can be generalized to n-dimensional space, by assuming that integrals over spheres * of centre O with respect to the $(n - 1)$ dimensional area element dS exist and possess properties similar to ordinary or multiple integrals. Thus

THEOREM 31. *Let $f(x_1, \ldots, x_n)$ be a function of n variables, summable or $\geqslant 0$. Denote by $I(r)$ the integral $\iint \cdots \int f(x)\, dS$ of the function f over the sphere of centre O and radius r, with respect to the surface element of measure dS. $I(r)$ is defined for the values of r for which f is summable with respect to dS over the sphere of radius r. It is thus defined for almost all values of r if f is summable. It is also defined and either finite or equal to $+\infty$ for all values of r if $f \geqslant 0$. In both cases,*

(I, 2; 41)

$$\int \cdots \iint_{R^n} f(x)\, dx = \int_0^\infty I(r)\, dr = \int_0^\infty dr \iiint_{\sum x_i^2 = r^2} \cdots \int f\, dS,$$

* Where $n > 3$ the term " hypersphere " is sometimes used for the $(n - 1)$ dimensional surface which is here simply called a sphere (translator).

all these quantities being finite if f is summable over R^n, *and finite* $\geqslant 0$ *or equal to* $+\infty$ *if* $f \geqslant 0$.

Particular case. Suppose that f is a function of the variable $r = \sqrt{\Sigma x_i^2}$: $f(x_1, x_2, \ldots, x_n) = f(r)$. Then $\iint \cdots \int f\, dS$ is equal to the product $S(r) f(r)$, where the area of the sphere of radius r is denoted by $S(r)$. By symmetry,

(I, 2; 42) $$S(r) = S_n r^{n-1},$$

S_n being the area of the unit sphere in R^n. Hence

THEOREM 32. *A function f, defined over R^n and depending only on $\sqrt{\Sigma x_i^2} = r$, is summable if and only if $f(r)\, r^{n-1}$ is summable over $(0, \infty)$, and then*

(I, 2; 43)
$$\iint \cdots \int f(x_1, x_2, \ldots, x_n)\, dx_1\, dx_2 \cdots dx_n = S_n \int_0^{+\infty} f(r) r^{n-1}\, dr.$$

Notes

1) In contrast to Fubini's theorem it is not necessary to suppose initially that f is summable over $\mathbf{R}^n$. Indeed if $f(r)\, r^{n-1}$ is summable over $(0, +\infty)$ then $|f(r)| r^{n-1}$ is also summable over $(0, +\infty)$, and as the latter is a function $\geqslant 0$, this implies that $|f|$ and thus f are summable over R^n.
2) Provided S_n is known, this formula reduces multiple integrals to ordinary integrals when only functions of r occur. Now, $S_2 = 2\pi$, $S_3 = 4\pi$. It can easily be shown that for the formula to be valid generally it is necessary to put $S_1 = 2$. We shall eventually calculate S_n for any value of n [Chapter VIII, equation (VIII, 1; 24)].

For future reference note that the volume of the ball of centre O and radius R is

(I, 2; 44) $$B_n(R) = \iint_{\sqrt{\Sigma x_i^2} \leqslant R} \cdots \int 1 \cdot dx_1, \ldots, dx_n = S_n \int_0^R r^{n-1}\, dr.$$

(I, 2; 45) $$B_n(R) = S_n \frac{R^n}{n}.$$

Example

$$B_2(R) = 2\pi \cdot \frac{R^2}{2} = \pi R^2,$$

(I, 2; 46) $$B_3(R) = 4\pi \cdot \frac{R^3}{3} = \frac{4}{3}\pi R^3$$

and also $B_1(R) = 2 \cdot R = 2R$.

THEOREM 33. *The integral*

(I, 2; 47) $$\iint_{r \geqslant 1} \cdots \int \frac{dx_1 \cdots dx_n}{r^\alpha}$$

is finite for $\alpha > n$ *but infinite for* $\alpha \leqslant n$; *the integral*

(I, 2; 48) $$\iint_{r\leqslant 1}\cdots\int \frac{dx_1\, dx_2 \cdots dx_n}{r^\alpha},$$

is finite for $\alpha < n$, *infinite for* $\alpha \geqslant n$.

This follows by writing these integrals as

(I, 2; 49) $$S_n\int_1^\infty r^{n-1-\alpha}\, dr; \qquad S_n\int_0^1 r^{n-1-\alpha}\, dr.$$

Example. The potential of a homogeneous electric charge distribution of density μ *in a ball of radius* R.

At a distance r from the charge e the field is e/r^2 which is directed outwards along the radius vector if positive and in the opposite sense if negative (i.e. the field is repulsive if $e \geqslant 0$). The potential due to the charge e is thus $e/r + C$ at a distance r; the constant C is fixed by taking the potential at infinity as zero, giving $C = 0$.

The potential $U(a)$ at a point a due to a distribution of arbitrary density $\mu(x)$ is the sum of the potentials due to charge elements $\mu(x)\, dx$ so that, by definition,

(I, 2; 50) $$U(a) = \iiint \frac{\mu(x)}{|x - a|}\, dx.$$

Applying this formula to the case where the density is equal to the constant μ in a ball of radius R and to 0 outside and where a is the centre of the sphere and is also taken as origin, we have

(I, 2; 51) $$\begin{aligned} U(0) &= \iiint_{|x|\leqslant R} \frac{\mu}{|x|}\, dx = \iiint_{r\leqslant R} \frac{\mu}{r}\, dx \\ &= S_3\mu\int_0^R \frac{r^2}{r}\, dr = 4\pi\cdot\frac{R^2}{2}\,\mu = 2\pi R^2\mu. \end{aligned}$$

This formula can also be obtained by applying Newton's law concerning the attraction of spherical shells.

At a point outside the sphere, at a distance r from the centre, the field is the same as if all the charge were concentrated at the centre of the sphere so that

(I, 2; 52) $$H(r) = \frac{4}{3}\pi R^3\frac{\mu}{r^2}, \qquad r \geqslant R,$$

which, if positive, is directed outwards along the radius vector. It follows that the potential, at a distance $r \geqslant R$ from 0, is

(I, 2; 53) $$U(r) = \frac{4}{3}\pi R^3\frac{\mu}{r} + C_1 = \frac{4}{3}\pi R^3\frac{\mu}{r}$$

since $U(\infty)$ is taken as zero.

For a point at a distance $r \leqslant R$ from 0, only the charge at distances $\leqslant r$ from 0 need be considered and the field is the same as if this charge were concentrated at the centre of the sphere. Thus

$$\text{(I, 2; 54)} \qquad H(r) = \frac{4}{3}\pi r^3 \mu \cdot \frac{1}{r^2} = \frac{4}{3}\pi r \mu$$

which is positive if directed in the sense of the radius vector. The corresponding potential at a distance $r \leqslant R$ from 0 is

$$\text{(I, 2; 55)} \qquad U(r) = -\frac{2}{3}\pi r^2 \mu + C_2, \qquad r \leqslant R.$$

The constant C_2 is determined by putting $U(R)$ equal to the value derived from the previous formula (I, 2; 53) so that

$$\text{(I, 2; 56)} \qquad -\frac{2}{3}\pi R^2 \mu + C_2 = \frac{4}{3}\pi R^2 \mu$$

whence

$$\text{(I, 2; 57)} \qquad C_2 = 2\pi R^2 \mu,$$

giving, for (I, 2; 55),

$$\text{(I, 2; 58)} \qquad U(r) = -\frac{2}{3}\pi r^2 \mu + 2\pi R^2 \mu, \qquad r \leqslant R.$$

It then suffices to put $r = 0$ to recover

$$\text{(I, 2; 59)} \qquad U(0) = 2\pi R^2 \mu,$$

a result identical to (I, 2; 51).

General remarks

1) Recalling that $\vec{H} = -\overrightarrow{\text{grad}}\, U$, it follows that

$$\text{(I, 2; 60)} \qquad H(r) = -\frac{dU}{dr}.$$

As a check (to avoid mistakes in the sign!) remember that the field always points in the direction of decreasing potential.

2) As $\mu(x)$ is bounded and zero outside a bounded set, the integral (I, 2; 50) is summable since it is thus bounded above by a multiple of $\iiint_{r \leqslant r_0} \frac{dx}{r}$; this is the particular case $\alpha = 1 < 3$ of Theorem 33.

Suppose that we evaluate (I, 2; 51) by Fubini's theorem. (This can be recommended as an exercise.) Then denoting the coordinates of $x \in R^3$ by ξ, η and ζ,

$$\text{(I, 2; 61)} \qquad U(0) = \mu \iint_{\xi^2+\eta^2<R^2} d\xi\, d\eta \int_{-\sqrt{R^2-\xi^2-\eta^2}}^{+\sqrt{R^2-\xi^2-\eta^2}} \frac{d\zeta}{\sqrt{\xi^2+\eta^2+\zeta^2}}.$$

For fixed ξ and η, the integral in ζ is summable except for $(\xi, \eta) = (0, 0)$, in which case it reduces to

$$\text{(I, 2; 62)} \qquad \int_{-R}^{+R} \frac{d\zeta}{|\zeta|} = +\infty.$$

This is a situation of a type mentioned in relation to Fubini's theorem (Theorem 28, Note 2). Here the function

$$I(\xi, \eta) = \int_{-\sqrt{R^2-\xi^2-\eta^2}}^{+\sqrt{R^2-\xi^2-\eta^2}} \frac{d\zeta}{\sqrt{\xi^2+\eta^2+\zeta^2}}$$

is defined at any point where $\xi^2 + \eta^2 < R^2$, except for $(\xi, \eta) = (0, 0)$, and must be taken as equal to 0 for $\xi^2 + \eta^2 \geqslant R^2$.

3) Equation (I, 2; 60) is valid only if $H(r)$ is a defined and continuous function of $r > 0$. This is the case here, according to equations (I, 2; 52) and (I, 2; 54). Hence it is permissible to state in advance that $U(r)$ is a continuous function of r with a continuous first derivative. It is this fact which justifies equating at $r = R$ the two functions $U(r)$ found for the cases $r \leqslant R$ and $r \geqslant R$ respectively.

If $H(r)$ is defined only for almost all values of r and is summable in r over any finite interval then equation (I, 2; 60) becomes inexact and must be replaced by

$$\text{(I, 2; 63)} \qquad U(r_1) - U(r_2) = -\int_{r_2}^{r_1} H(r)\, dr.$$

Here $U(r)$ is still a continuous function of r (see Theorem 34, below).

In general it may be shown from equation (I, 2; 50) that if $\mu(x)$ is bounded and zero outside a bounded set, then $U(a)$ is a continuous function of a, with continuous first-order partial derivatives (see page 57).

The indefinite integral as a function of its upper limit

THEOREM 34. *Lebesgue. If $f(x)$ is a function of a real variable x, defined almost everywhere and summable in the finite or infinite interval (a, b), then the indefinite integral $F(x) = \int_a^x f(\xi)\, d\xi$ is continuous for all $x \in (a, b)$ and has a derivative equal to $f(x)$ for almost all values of x.*

Note. It is not to be expected that $F(x)$ should always possess a derivative equal to $f(x)$, for F remains unchanged if f is changed at each point of a set of measure zero. At the points x (forming a set of measure zero) where it is not true that $F'(x) = f(x)$, F can either have a derivative different from $f(x)$ or no derivative at all.

2. IMPROPER SEMI-CONVERGENT LEBESGUE INTEGRALS

Let (a, b) be a finite or infinite interval of the real line and f a function summable over $(a, b-\varepsilon)$, for any $\varepsilon > 0$. Whether or not f is summable over (a, b) it can happen that

$$\int_a^{b-\varepsilon} f(x)\,dx \qquad \text{has a limit for} \quad \varepsilon \to 0;$$

the integral of f over (a, b) is then said to be convergent and is denoted by $\int_a^{\to b} f(x)\,dx$ (or by $\int_a^b f(x)\,dx$ as in the case of the integral of a summable function). If $\int_a^{\to b} |f(x)|\,dx$ is convergent, the integral is said to be absolutely convergent over (a, b), f is then summable and the converse applies. Otherwise the integral is said to be semi-convergent.

A similar definition is given for $\int_{\to a}^{b} f(x)\,dx$ and $\int_{\to a}^{\to b} f(x)\,dx$ (which exists if, for a point c between a and b, each of the integrals $\int_{\to a}^{c}$, $\int_c^{\to b}$ exists).

Abel's Theorem. *If, for $x > a$, $f(x)$ is the product $\alpha(x)\,\beta(x)$, where $\alpha(x)$ is decreasing function $\geqslant 0$, tending to 0 as $x \to +\infty$, and $\beta(x)$ is continuous, and if a the modulus of*

$$\sigma_{c,d} = \int_c^d \beta(x)\,dx \tag{I, 2; 64}$$

is bounded above by a fixed constant σ, then $\int_a^{\to +\infty} f(x)\,dx$ is convergent and the modulus of the "remainder" $\int_A^{\to +\infty} f(x)\,dx$ is bounded above by

$$\sigma\alpha(A). \tag{I, 2; 65}$$

We shall confine ourselves to proving this when $\alpha(x)$ is continuous and differentiable, thus possessing a derivative $\alpha' \leqslant 0$.

The function $\sigma_{a,x}$ is a primitive of $\beta(x)$ and, by integration by parts,

$$\int_a^B f(x)\,dx = [\sigma_{a,x}\alpha(x)]_{x=a}^{x=B} - \int_a^B \sigma_{a,x}\alpha'(x)\,dx \tag{I, 2; 66}$$
$$= \sigma_{a,B}\alpha(B) - \int_a^B \sigma_{a,x}\alpha'(x)\,dx.$$

As $B \to \infty$, $\sigma_{a,B}$ is bounded above by σ and $\alpha(B) \to 0$ so that the first term $\to 0$. It remains to show that as $B \to \infty$ the integral in the second term approaches a limit; that is to say that $\int_a^{\to +\infty} \sigma_{a,x}\alpha'(x)\,dx$ is convergent.

In fact, it is *absolutely* convergent, since $|\sigma_{a,x}||\alpha'(x)| \leqslant \sigma|\alpha'(x)|$ and

$$\text{(I, 2; 67)} \quad \int_a^B |\alpha'(x)|dx = -\int_a^B \alpha'(x)\, dx = \alpha(a) - \alpha(B) \to \alpha(a)$$

as $B \to \infty$. Thus $\int_a^{\to +\infty} f(x)\, dx$ exists and also

$$\text{(I, 2; 68)} \quad \left|\int_a^{\to +\infty} f(x)\, dx\right| \leqslant \int_a^{+\infty} |\sigma_{a,x}||\alpha'(x)|dx \leqslant \sigma\alpha(a).$$

The same inequality applied to $\int_A^{\to +\infty} f(x)\, dx$ gives $\sigma\alpha(A)$. Q.E.D.

Examples. Trigonometric integrals.

$$\text{(I, 2; 69)} \quad \int_a^{\to +\infty} \alpha(x) e^{i\lambda x}\, dx$$

exists for real $\lambda \neq 0$ if α is $\geqslant 0$, decreasing and $\to 0$ as $x \to \infty$. For here $\beta(x) = e^{i\lambda x}$, $\sigma_{c,d} = \dfrac{e^{i\lambda d} - e^{i\lambda c}}{i\lambda}$, $\sigma \leqslant \dfrac{2}{|\lambda|}$. Similarly the integrals

$$\text{(I, 2; 70)} \quad \int_a^{\to +\infty} \alpha(x) \cos \lambda x\, dx, \qquad \int_a^{\to +\infty} \alpha(x) \sin \lambda x\, dx$$

are convergent, with the same conditions on $\alpha(x)$. Also the latter integral still exists for $\lambda = 0$ since it is then zero. If $\alpha(x)$ is not summable these integrals do not converge absolutely. Consider, for example, the integrals

(I, 2; 71)

$$I = \int_0^{\to +\infty} \frac{e^{i\lambda x}}{\sqrt{x}}\, dx, \qquad C = \int_0^{\to +\infty} \frac{\cos \lambda x}{\sqrt{x}}\, dx, \qquad S = \int_0^{\to +\infty} \frac{\sin \lambda x}{\sqrt{x}}\, dx,$$

which converge for $\lambda \neq 0$. Carry out the change of variable $x = t^2$ (which is justified for a finite interval $\int_0^B$ and thus, by passage to the limit, for $\int_0^{\to +\infty}$) Hence it follows that the integrals

(I, 2; 72)

$$\frac{I}{2} = \int_0^{\to +\infty} e^{i\lambda t^2} dt, \qquad \frac{C}{2} = \int_0^{\to +\infty} \cos \lambda t^2\, dt, \qquad \frac{S}{2} = \int_0^{\to +\infty} \sin \lambda t^2\, dt$$

are semi-convergent (even though $|e^{i\lambda t^2}| = 1$). These are *Fresnel's integrals.* It can be shown that

$$\int_0^{\to +\infty} \cos t^2\, dt = \int_0^{\to +\infty} \sin t^2\, dt = \frac{\sqrt{\pi}}{2\sqrt{2}}.$$

Cauchy principal value

Let f be a function defined for $a \leqslant x \leqslant b$, summable over $(a, c - \varepsilon)$ and over $(c + \varepsilon, b)$ for any $\varepsilon > 0$, but not necessarily over (a, b). It is usually considered that $\int_a^b f(x)\, dx$ exists if each of the integrals $\int_a^{\to c}$, $\int_{\to c}^b$ exists. This amounts to saying that $\int_a^{c-\eta} + \int_{c+\varepsilon}$ possesses a limit when η and $\varepsilon \to 0$ independently of one another.

It can happen that this limit does not exist except in the special case $\eta = \varepsilon$, $\varepsilon \to 0$. It is then said that the integral of f converges to the *Cauchy principal value* and the notation $\text{pv}\int_a^b f(x)\, dx = \lim_{\varepsilon \to 0}\left(\int_a^{c-\varepsilon} + \int_{c+\varepsilon}^b\right)$ is used.

THEOREM 35. $\text{pv}\int_a^b f(x)\, dx$ *exists if and only if, in the neighbourhood of* $x = c$, f *is the sum of an antisymmetric function* f_1 $(f_1(c + u) = -f_1(c - u))$ *and a symmetric function* f_2 $(f_2(c + u) = f_2(c - u))$ *which is such that* $\int_{\to c} f_2(x)\, dx$ *exists.*

If $c = 0$, "symmetric" and "antisymmetric" functions are called "even" and "odd" respectively. It will be supposed that $c = 0$, since the general case can be reduced to this by the change of variables $x = c + u$. In the neighbourhood of $x = 0$ every function f is uniquely the sum of an even function and an odd function :

(I, 2; 73) $$f(x) = \frac{f(x) - f(-x)}{2} + \frac{f(x) + f(-x)}{2} = f_1(x) + f_2(x).$$

As f_1 is odd, its integral over an interval symmetric about $x = 0$ is zero. and hence, if α is chosen so that $a \leqslant -\alpha < 0 < \alpha \leqslant b$, then

(I, 2; 74) $$\int_{-\alpha}^{-\varepsilon} f_1(x)\, dx + \int_{+\varepsilon}^{+\alpha} f_1(x)\, dx = 0.$$

Thus the convergence in the pv sense over (a, b) or, what amounts to the same thing, over $(-\alpha, +\alpha)$, is the same for f as for f_2. Again, f_2 being even,

(I, 2; 75) $$\int_{-\alpha}^{-\varepsilon} f_2(x)\, dx + \int_{+\varepsilon}^{+\alpha} f_2(x)\, dx = 2\int_\varepsilon^\alpha f_2(x)\, dx,$$

so that the integral of f in the pv sense exists if and only if $\int_{\to 0}^a f_2(x)\, dx$ exists.

Example. Suppose that, in the neighbourhood of $x = c$, f can be expanded in a finite series $f(x) = c_{-1}/(x - c) + c_0 + c_1(x - c) + \cdots$ Then the integral exists in the vp sense because $1/(x - c)$ is antisymmetric. More generally, if $f(x) = \varphi(x)/(x - c)$, φ being continuous in the neighbourhood of $x = c$ and differentiable at $x = c$, then the integral of f in the pv sense exists since

$$f(x) = \frac{\varphi(c)}{x - c} + \frac{\varphi(x) - \varphi(c)}{x - c},$$

the first function being antisymmetric and the second bounded in the neighbourhood of $x = c$.

There is a similar definition of pv for the interval $(-\infty, +\infty)$. If f is summable in any finite interval, then pv $\int_{-\infty}^{+\infty} f(x)dx = \lim_{B\to\infty} \int_{-B}^{+B}$. This pv exists if and only if $f = f_1 + f_2$, f_1 being odd and f_2 even and such that $\int_0^{\to+\infty} f_2(x)\,dx$ exists.

It is easy to prove the following :

(I, 2; 76) $\quad \text{pv}\int_a^b \frac{dx}{x} = \log\left|\frac{b}{a}\right|$ provided that $a \neq 0,\ b \neq 0$.

(I, 2; 77) $\quad \text{pv}\int_{-\infty}^{+\infty} \frac{dx}{x} = 0 \qquad$ (odd function).

3. Functions represented by series and integrals

1. FUNCTIONS REPRESENTED BY SERIES

Let $\sum_{i\in I} u_i(x)$ be a series which is summable for all values of a parameter x belonging to a set E [$u_i(x)$ is a complex number for all i and x]. The sum $f(x)$ is then a complex-valued function of x. How are the properties of $f(x)$ (continuity, differentiability, integrability) related to those of $u_i(x)$?

The same question can be asked if $\sum_{n=0}^{\infty} u_n(x)$ is a series where the set of indices is the set N of integers $n \geqslant 0$ and which is assumed to be convergent for all $x \in$ E.

Finally, if $f_n(x)$ is a *sequence* of complex-valued functions and if, for each $x \in$ E, $f_n(x)$ has a limit as $n \to \infty$, this limit $f(x) = \lim_{n\to\infty} f_n(x)$ can also be studied in relation to the properties of the f_n.

Convergence and uniform convergence

The sequence $f_n(x)$ is said to *converge* to $f(x)$ as $n \to \infty$ if, for *any fixed* x, the numbers $f_n(x)$ converge to the number $f(x)$ as $n \to \infty$. This is equivalent to saying that, x being fixed and $\varepsilon > 0$ given, there exists an N such that $n \geqslant N$ implies $|f_n(x) - f(x)| \leqslant \varepsilon$. This number N depends on ε *and on* x and hence should be written $N(\varepsilon, x)$.

The sequence $f_n(x)$ is said to converge *uniformly* to $f(x)$ for $x \in E$ if, given $\varepsilon > 0$, there exists an integer $N(\varepsilon)$ *independent of* x and such that, $|f(x) - f_n(x)| \leqslant \varepsilon$ for all $n \geqslant N(\varepsilon)$ and all $x \in E$. Now we introduce the idea of the *distance* between two functions f and g, defined on the set E:

$$d(f, g) = \sup_{x \in E} |f(x) - g(x)| \leqslant +\infty. \tag{I, 3; 1}$$

Note that if f, g and h are three functions, $d(f, g) \leqslant d(f, h) + d(g, h)$ (triangle inequality). The sequence f_n converges to f uniformly as $n \to \infty$ if $d(f, f_n)$ tends to 0 as $n \to \infty$; the uniform convergence of a sequence of functions is thus equivalent to the convergence to zero of a sequence of numbers $d(f, f_n)$.

Examples

A) $$f_n(x) = x^n, \qquad 0 \leqslant x \leqslant 1.$$

$$\lim_{n \to +\infty} f_n(x) = \begin{cases} 0 & \text{for } x \neq 1, \\ 1 & \text{for } x = 1; \end{cases} \tag{I, 3; 2}$$

$$|f(x) - f_n(x)| = \begin{cases} x^n & \text{for } x \neq 1, \\ 0 & \text{for } x = 1; \end{cases} \tag{I, 3; 3}$$

$$|f(x) - f_n(x)| \leqslant \varepsilon \qquad \text{for } n \log x \leqslant \log \varepsilon, \tag{I, 3; 4}$$

i.e., for $n \geqslant \dfrac{\log \frac{1}{\varepsilon}}{\log \frac{1}{x}}$; but $\dfrac{\log \frac{1}{\varepsilon}}{\log \frac{1}{x}} \to \infty$ as $x \to 1$, so that the convergence is not uniform for $0 \leqslant x \leqslant 1$ nor is it uniform for $0 \leqslant x < 1$. However, it is uniform for $0 \leqslant x \leqslant 1 - \delta$, $\delta > 0$, since then $|f(x) - f_n(x)| \leqslant \varepsilon$ for $n \geqslant \dfrac{\log \frac{1}{\varepsilon}}{\log \frac{1}{1-\delta}}$, a quantity independent of x. Alternatively, note that $d(f(x), f_n(x)) = 1$.

B) $$f_n(x) = \frac{1}{1 + (x - n)^2} \tag{I, 3; 5}$$

which is the function $1/(1 + x^2)$ translated by $x \to x + n$. As $n \to +\infty$, $f_n(x)$ converges to zero uniformly in the finite interval (a, b) since, for large enough n,

$$|f_n(x)| \leqslant \frac{1}{1 + (b - n)^2}$$

which tends to 0 as $n \to \infty$; $f_n(x)$ even tends uniformly to zero on the whole of the half-line $(-\infty, b)$. However, it does not tend uniformly to zero on the whole of the real axis, since in this case $d(f_n(x), 0) = 1$.

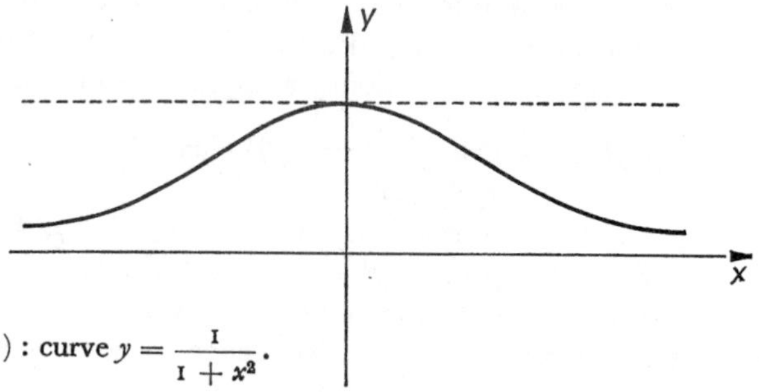

Figure (I, 1) : curve $y = \frac{1}{1+x^2}$.

C) A series $\sum_{n=0}^{\infty} u_n(x)$ converges uniformly for $x \in E$ if the partial sums $S_n(x)$ converge uniformly to the sum $S(x)$ as $n \to \infty$; or, if the remainder $R_n(x)$ converges uniformly to 0 as $n \to \infty$; or, again, if $d_n = d(R_n(x), 0) = \sup_{x \in E} |R_n(x)| \to 0$ as $n \to \infty$.

D) A series $\sum_{i \in I} u_i(x)$ is uniformly summable for $x \in E$ if, for any $\varepsilon > 0$, there exists a finite set J of indices, depending on ε but not on x, such that, for any finite or infinite K containing J, $|S_K(x) - S(x)| \leqslant \varepsilon$ for all $x \in E$.

E) If $f_\lambda(x)$ is a function of x, depending on a parameter λ (which will, for definiteness, be supposed real) then it is said that $f_\lambda(x)$ converges uniformly to $f_{\lambda_0}(x)$ as $\lambda \to \lambda_0$ for $x \in E$ if $d(f_\lambda(x), f_{\lambda_0}(x))$ tends to zero as $\lambda \to \lambda_0$.

Theorem 36. *Cauchy's criterion. A sequence $f_n(x)$ of complex-valued functions defined on a set* E *converges uniformly to a limit as $n \to \infty$ if and only if $d_{m,n} = d(f_m(x), f_n(x))$ converges to 0 as m and n tend to ∞. A series $\sum_{n=0}^{\infty} u_n(x)$ converges uniformly if and only if $d_{m,n} = d(S_m(x), S_n(x))$ converges to 0 as m and $n \to \infty$. It is thus necessary that the general term $u_n(x)$ tends to 0 as $n \to \infty$ uniformly for $x \in E$. A series $\sum_{i \in I}^{\infty} u_i(x)$ is uniformly summable if and only if, for any $\varepsilon > 0$, there exists a finite set of indices* J, *depending on ε but not on x, such that, for any set of indices* K *disjoint from* J, *we have $|S_K(x)| \leqslant \varepsilon$, for all x.*

A proof will be given only for the sequence $f_n(x)$. The case of the series $\sum_{n=0}^{\infty} u_n(x)$ can be deduced immediately by putting $f_n(x) = S_n(x)$ while the case $\sum_{i \in I} u_i(x)$ is recommended as an exercise for the reader.

1) Assume that the sequence f_n is uniformly convergent to f. Then, given $\varepsilon > 0$, there exists $N(\varepsilon)$ such that, for $n \geqslant N$, $d(f, f_n) \leqslant \frac{1}{2}\varepsilon$. Then if $m \geqslant N$ and $n \geqslant N$, it follows that $d(f, f_m) \leqslant \frac{1}{2}\varepsilon$ and $d(f, f_n) \leqslant \frac{1}{2}\varepsilon$, giving $d(f_m, f_n) \leqslant d(f, f_m) + d(f, f_n) \leqslant \varepsilon$. Thus $d(f_m, f_n) \to 0$ as m and $n \to \infty$.

2) Assume now that $\lim_{m, n \to \infty} d(f_m, f_n) = 0$. Given $\varepsilon > 0$, there exists $N(\varepsilon)$ such that, for $m \geqslant N$ and $n \geqslant N$, $| f_m(x) - f_n(x)| \leqslant \varepsilon$ for any $x \in E$. Now, for fixed x, the sequence of numbers $f_n(x)$ obeys Cauchy's criterion since $f_m(x) - f_n(x) \to 0$ as m and $n \to \infty$. Hence the sequence $f_n(x)$ tends to a limit. This limit depends on the value of x which was chosen and is thus a function $f(x)$. Since $\lim_{m \to \infty} f_m(x) = f(x)$, for fixed x, and $|f_m(x) - f_n(x)| \leqslant \varepsilon$ for $m \geqslant N(\varepsilon)$ and $n \geqslant N(\varepsilon)$, it follows that $|f(x) - f_n(x)| \leqslant \varepsilon$ for $n \geqslant N(\varepsilon)$ and fixed x. However, as this result is independent of x, it follows that $d(f, f_n) \leqslant \varepsilon$ for $n \geqslant N(\varepsilon)$ and thus, as $n \to \infty$, f_n converges uniformly to f for $x \in E$. Q.E.D.

Practical criteria for uniform convergence

DEFINITION 4. *If, for any x, $|u_i(x)| \leqslant v_i$, where $v_i \geqslant 0$ and $\sum_{i \in I} v_i$ is summable, then the series $\sum_{i \in I} u_i(x)$ (which is summable for any x by Theorem 6) is said to be normally summable.*

THEOREM 37. *Any normally summable series is uniformly summable.*

Let J be a finite set of indices such that $\sum_{i \in K} v_i \leqslant \varepsilon$ for all K disjoint from J. Then

$$\left| \sum_{i \in K} u_i(x) \right| \leqslant \sum_{i \in K} |u_i(x)| \leqslant \varepsilon,$$

so that, by Cauchy's criterion (Theorem 36), the series is uniformly summable.

Examples. A) Let $\sum_{n=0}^{\infty} a_n Z_n$ be a power series. Suppose that $\lim_{n \to \infty} \left| \frac{a_{n+1}}{a_n} \right| = k$, where k is finite. Then the series is normally summable if $|Z| \leqslant 1/k - \delta$, for any $\delta > 0$. In fact, in this case,

$$|a_n Z^n| \leqslant |a_n| \left(\frac{1}{k} - \delta \right)^n,$$

and this series of numbers $\geqslant 0$ is summable by d'Alembert's rule since

$$\text{(I, 3; 6)} \qquad \frac{u_{n+1}}{u_n} = \left| \frac{a_{n+1}}{a_n} \right| \left(\frac{1}{k} - \delta \right) \xrightarrow[n \to \infty]{} k \left(\frac{1}{k} - \delta \right) < 1.$$

B) Let $\sum a_n Z^n$ be a power series and suppose that $\lim_{n \to \infty} \sqrt[n]{|a_n|} = k$, where k is finite. Then the series is normally summable if $|Z| \leqslant 1/k - \delta$, for any $\delta > 0$. This follows because $|a_n Z^n| \leqslant |a_n| |1/k - \delta|^n$ and this series with terms $u_n \geqslant 0$ is summable by Cauchy's rule since

$$\text{(I, 3; 7)} \qquad \sqrt[n]{u_n} = \sqrt[n]{|a_n|} \left(\frac{1}{k} - \delta \right) \xrightarrow[n \to \infty]{} k \left(\frac{1}{k} - \delta \right) < 1.$$

C) The series $\sum_{n=1}^{\infty} \frac{1}{n^{\alpha}}$ is summable for* $\mathcal{R}\alpha > 1$. It is normally summable for the set of complex numbers α such that $\mathcal{R}\alpha \geqslant 1 + \delta$ for any $\delta > 0$.

In fact, for this case, $\left| \frac{1}{n^{\alpha}} \right| \leqslant \frac{1}{n^{1+\delta}}$, which is the general term of a summable series of numbers $\geqslant 0$. On the other hand, this series does not converge uniformly for $\alpha > 1$, since

$$\begin{aligned} \text{(I, 3; 8)} \quad & d(S_{2n}(\alpha), S_n(\alpha)) \\ & = \sup_{\alpha > 1} \left[\frac{1}{(n+1)^{\alpha}} + \frac{1}{(n+2)^{\alpha}} + \cdots + \left(\frac{1}{2n} \right)^{\alpha} \right] \\ & \geqslant \sup_{\alpha > 1} \frac{n}{(2n)^{\alpha}} = \sup_{\alpha > 1} \frac{1}{2^{\alpha}} \frac{1}{n^{\alpha - 1}} = \frac{1}{2} \end{aligned}$$

does not tend to zero as $n \to \infty$.

D) The trigonometric series $\sum_{n=-\infty}^{\infty} a_n e^{ni\theta}$. If the series $\sum |a_n|$ is summable then, since $|e^{ni\theta}| = 1$, the given series is normally summable for any real θ. The series $\sum_{n=0}^{\infty} a_n \cos n\theta$, $\sum_{n=1}^{\infty} a_n \sin n\theta$ are similar.

THEOREM 38. *Abel's criterion. Let $u_n(x) = a_n(x)\, b_n(x)$. Suppose that, for all x, the sequence $a_n(x)$ is decreasing, $\geqslant 0$ and, as $n \to \infty$, tends uniformly to 0*

---

*$\mathcal{R}\alpha$ = real part of α.

for $x \in E$. *Also, put*

$$\sigma_{m,n}(x) = b_m(x) + b_{m+1}(x) + \cdots + b_n(x) \tag{I, 3; 9}$$

and suppose that there exists a constant σ, *independent of* m, n *and* x *and such that* $|\sigma_{m,n}(x)| \leqslant \sigma$. *Then the series* $\sum_{n=0}^{\infty} a_n(x)\, b_n(x)$ *is uniformly convergent.*

The series converges for every value of x and it can be seen that

$$|R_n(x)| \leqslant a_{n+1}(x)\sigma \leqslant a_{n+1}\sigma,$$

where $a_{n+1} = \sup_{x \in E} a_{n+1}(x)$. Thus $d\,(R_n(x), 0) \leqslant a_{n+1}\sigma$ and, as the sequence a_n converges to 0, the theorem is proved.

Note. If $a_n(x) \leqslant a_n$ and $a_n \to 0$ as $n \to \infty$, it does not follow that, for any x, the sequence $a_n(x)$ is decreasing; this must be carefully verified.

Examples. A) Consider the series $\sum_{n=0}^{\infty} (-1)^n a_n(x)$, where $a_n(x)$ is $\geqslant 0$, decreasing for all x, and $a_n(x) \leqslant a_n$ with $\lim_{n \to \infty} a_n = 0$. This series is uniformly convergent.

For instance, the series $\sum_{n=1}^{\infty} \frac{(-1)^n}{n^\alpha}$ is uniformly convergent for real $\alpha \geqslant \delta > 0$, since $1/n^\alpha$ is a decreasing sequence for each fixed α and $1/n^\alpha \leqslant 1/n^\delta$ with $\lim_{n \to \infty} 1/n^\delta = 0$.

On the other hand, the series does not converge uniformly for $\alpha > 0$ since the modulus of the general term $1/n^\alpha$ does not tend uniformly to zero as $n \to \infty$: we have

$$d\left(\frac{1}{n^\alpha}, 0\right) = \sup_{\alpha > 0} \frac{1}{n^\alpha} = 1.$$

B) The trigonometric series $\sum_{n=-\infty}^{+\infty} a_n e^{ni\theta}$. Suppose that $a_n \geqslant 0$ for all n, and that a_n decreases steadily to 0 as $n \to \infty$ and as $n \to -\infty$. Then with $b_n = e^{ni\theta}$ we have $|\sigma_{m,n}(\theta)| \leqslant 2/|e^{i\theta} - 1| \leqslant 2/|e^{i\theta} - 1|$.

The series converges uniformly for $|\theta - 2K\pi| \geqslant \delta > 0$, since it is then possible to put $\sigma = 2/|e^{i\delta} - 1|$. The series $\sum_{n=0}^{\infty} a_n \cos n\theta$ behaves similarly.

The series $\sum_{n=1}^{\infty} a_n \sin n\theta$ is convergent for all values of θ, even for $\theta = 2K\pi$ since all terms are then zero. Nevertheless it converges *uniformly* only for $|\theta - 2K\pi| \geqslant \delta > 0$.

Continuity of the limit of a sequence or of the sum of a series

THEOREM 39. *If, as $n \to \infty$, the sequence $f_n(x)$ converges uniformly to $f(x)$ when x belongs to a set* E *in a Euclidean space* R^m *and if all the $f_n(x)$ are continuous at a point x_0 of* E, *then $f(x)$ is continuous at x_0, relative to* E.

We can write

$$(I, 3; 10) \qquad f(x) - f(x_0) = [f(x) - f_n(x)] + [f_n(x) - f_n(x_0)] + [f_n(x_0) - f(x_0)],$$

giving :

$$(I, 3; 11) \qquad |f(x) - f(x_0)| \leqslant |f(x) - f_n(x)| + |f_n(x) - f_n(x_0)| + |f_n(x_0) - f(x_0)|.$$

Given $\varepsilon > 0$, choose n large enough so that $|f(y) - f_n(y)| \leqslant \frac{1}{3}\varepsilon$ for all y belonging to E.

Then for any x the first and third terms of the right-hand side are each bounded above by $\frac{1}{3}\varepsilon$. The number n having thus been chosen, the function f_n is continuous, so that there exists $\eta > 0$ such that $|x - x_0| \leqslant \eta$ implies $|f_n(x) - f_n(x_0)| \leqslant \frac{1}{3}\varepsilon$. Hence, for $|x - x_0| \leqslant \eta$,

$$|f(x) - f(x_0)| \leqslant \varepsilon. \qquad \text{Q.E.D.}$$

If the sequence f_n is convergent but not uniformly convergent to f as $n \to \infty$, then f can be discontinuous even though all the functions f_n are continuous. Thus in example A, page 47, $f_n(x) = x^n$, $0 \leqslant x \leqslant 1$, the limit $f(x)$ is equal to 0 for $x \neq 1$ and to 1 for $x = 1$, so that it is discontinuous at the point $x_0 = 1$.

The theorem remains valid if the uniformly convergent sequence of continuous functions is replaced by a uniformly convergent or uniformly summable series of continuous functions. However, consider the series $\sum_{n=0}^{\infty} x^n (1 - x)$, which is summable for $0 \leqslant x \leqslant 1$ (for $x < 1$ as a geometric series and for $x = 1$, since all its terms are zero) but not uniformly. When $x < 1$ its sum is $\left(\frac{1}{1-x}\right)(1 - x) = 1$, and when $x = 0$ its sum is 1, so that the sum is discontinuous at the point $x = 1$.

Integrability of the limit of a sequence or of the sum of a series

THEOREM 40. *If, as $n \to \infty$, the sequence of bounded functions $f_n(x)$ converges to the function $f(x)$ uniformly on a subset* E *of finite measure in the Euclidean space* R^m, *then, as $n \to \infty$, $\iint \cdots \int_F f_n(x)\, dx$ converges to $\iint \cdots \int_F f(x)\, dx$ for any subset* F *of* E, *and the convergence is uniform on* F.

It can be seen that*

$$\left|\iint\cdots\int_F f(x)\,dx - \iint\cdots\int_F f_n(x)\,dx\right| \leqslant (\text{mes E})\, d(f, f_n).$$

Now, given $\varepsilon > 0$, there exists N such that, for $n \geqslant N$, $d(f, f_n) \leqslant \dfrac{\varepsilon}{\text{mes E}}$; then

$$\text{(I, 3; 12)} \qquad \left|\iint\cdots\int_F f(x)\,dx - \iint\cdots\int_F f_n(x)\,dx\right| \leqslant \varepsilon$$

for any $F \subset E$ and $n \geqslant N$.

Obviously the theorem remains valid for a uniformly convergent or a uniformly summable series :

$$\text{(I, 3; 13)} \qquad \iint\cdots\int_F \left(\sum_{n=0}^{\infty} u_n(x)\right) dx = \sum_{n=0}^{\infty}\left(\iint\cdots\int_F u_n(x)\,dx\right)$$

(integration under the Σ sign or inversion of the symbols $\int$ and Σ).

Theorem 40 can be generalized to

THEOREM 41. *Lebesgue's theorem, assumed without proof. If, as* $n \to \infty$, *the sequence of functions* $f_n(x)$ *defined on* R^m *converges to* $f(x)$, *and if the moduli of the* $f_n(x)$ *are bounded above by a certain non-negative summable function* g $\left(|f_n(x)| \leqslant g(x),\ g(x) \geqslant 0,\ \iint\cdots\int_{R^m} g(x)\,dx < +\infty\right)$, *then* $f(x)$ *is summable and*

$$\iint\cdots\int_{R^m} f_n(x)\,dx$$

converges to

$$\iint\cdots\int_{R^m} f(x)\,dx \qquad \textit{as} \qquad n \to \infty.$$

Let us now look again at example A on page 47. Although the convergence is not uniform, $\int_0^1 x^n\,dx = 1/(n+1)$ converges to $0 = \int_0^1 f(x)\,dx$ as $n \to \infty$. Here Theorem 41 applies because $|f_n(x)| \leqslant 1$. On the other hand, consider the following example :

$$\text{(I, 3; 14)} \quad f_n(x) = \begin{cases} 0 & \text{for } x \leqslant 0 \text{ and } x \geqslant \dfrac{1}{n}, \text{ see fig. (I, 2)} \\ n^2 & \text{for } 0 < x < \dfrac{1}{n}. \end{cases}$$

*It is obvious that f is also bounded.

As $n \to \infty$, $f_n(x)$ converges to zero for any x (but not uniformly). However $\int_0^1 f_n(x)\,dx = n^2/n = n$ tends to $+\infty$ with n. Theorem 41 does not apply because the modulus of $f_n(x)$ is not bounded above by a function $g(x) \geqslant 0$ which is summable and independent of n.

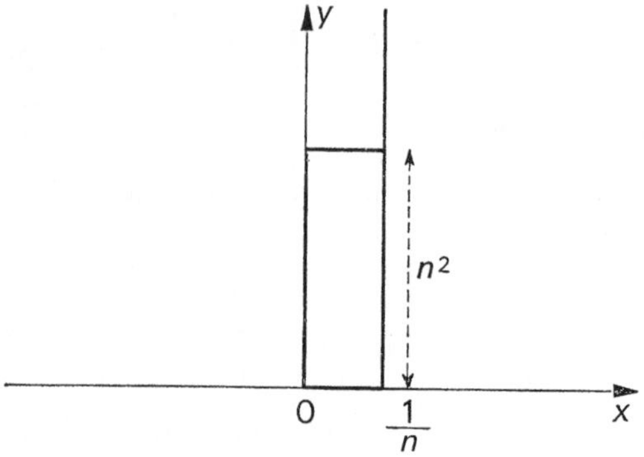

Figure (I, 2)

Note. It is enough to suppose that, as $n \to \infty$, $f_n(x)$ tends to $f(x)$ for almost all values of x (again with $|f_n(x)| \leqslant g(x)$, g summable and $\geqslant 0$).

COROLLARY. *If, as $n \to \infty$, $f_n(x)$ converges to $f(x)$ and if the moduli of the f_n are bounded above by a fixed constant $M \geqslant 0$, then $\iint \cdots \int_E f_n(x)\,dx$ converges to $\iint \cdots \int_E f(x)\,dx$ as $n \to \infty$, provided that the measure of the set E is finite.*

It suffices to apply Theorem 41, with $g(x) = M$, summable over E of finite measure. It can be seen that this corollary is more general than Theorem 40.

The condition that E must be of finite measure is essential in this corollary as in Theorem 40. Thus, taking

$$(I, 3; 15) \qquad f_n(x) = \begin{cases} \dfrac{1}{n} & \text{for } |x| \leqslant n^2 \\ 0 & \text{for } |x| > n^2 \end{cases}$$

f_n converges uniformly to 0 as $n \to \infty$, $(d(0, f_n) = 1/n)$. However, $\int_{-\infty}^{+\infty} f_n(x)\,dx = 2n^2(1/n) = 2n$ tends to $+\infty$ with n. For an infinite interval it must be carefully verified that the f_n are bounded above, not by a constant $M \geqslant 0$, but by a *summable* function $g \geqslant 0$.

Theorem 41 is not easy to use for series. There is, however, the following criterion.

THEOREM 42. 1) *If the set of indices* I *is denumerable and if* $\sum_{i\in I} u_i(x)$ *is a series of terms* $\geqslant 0$, *then*

$$(I,\ 3;\ 16)\qquad \iint\cdots\int_{R^m}\Big(\sum_{i\in I} u_i(x)\Big)dx = \sum_{i\in I}\iint\cdots\int_{R^m} u_i(x)\,dx,$$

the two sides being finite or equal to $+\infty$.

2) *If* I *is denumerable, if the* $u_i(x)$ *are any real or complex numbers and if*

$$\sum_{i\in I}\iint\cdots\int_{R^m}|u_i(x)|\,dx = \iint\cdots\int_{R^m}\Big(\sum_{i\in I}|u_i(x)|\Big)\,dx$$

is finite, then, on the one hand, each of the functions $u_i(x)$ *is summable with respect to* x *over* R^m *and the series* $\sum_{i\in I}\Big(\iint\cdots\int_{R^m} u_i(x)\,dx\Big)$ *is summable. On the other hand, for nearly all values of* x, *the series* $\sum_{i\in I} u_i(x)$ *is summable, and its sum is a function of* x, *defined almost everywhere, which is summable over* R^m; *finally,*

$$(I,\ 3;\ 17)\qquad \iint\cdots\int_{R^m}\sum_{i\in I} u_i(x)\,dx = \sum_{i\in I}\iint\cdots\int_{R^m} u_i(x)\,dx.$$

The first part of Theorem 42 is the most important. For terms $u_i(x) \geqslant 0$ it is always possible to invert $\int$ and Σ (just as it was possible above to invert the two symbols Σ in repeated summation of series with terms $\geqslant 0$ and to invert the two symbols $\int$ in Fubini's theorem concerning the double integral of a function $\geqslant 0$).

There is another generalization of Theorem 40, corresponding to semi-convergent improper Lebesgue integrals $\Big($referring, for example, to $\int_0^{\to+\infty}\Big)$; this will not be given here.

Differentiation of the limit of a sequence or of the sum of a series

It might be expected that if, as $n \to \infty$, f_n tends uniformly to f for $a \leqslant x \leqslant b$ (one-dimensional case, $m = 1$) and if the f_n are differentiable with continuous derivatives then the sequence f_n' would converge uniformly to a limit g as $n \to \infty$, f would be differentiable and f' would be equal to g. *This is not so.*

Examples. A) $\dfrac{e^{inx}}{\sqrt{n}}$ converges uniformly to 0 as $n \to \infty$ for real x. However, the derivative $i\sqrt{n}e^{inx}$ has the modulus $\sqrt{n}$ which tends to ∞ with n.

B) In the preceding example the limit f is, nevertheless, differentiable ($f \equiv 0$).

Consider, however, the series $\sum_{n=0}^{\infty} \frac{\sin(3^n\theta)}{2^n}$. The modulus of its general term is bounded above by $1/2^n$ so that it is uniformly summable for real θ. Its sum $f(\theta)$ is a continuous function. The differentiated series $\sum_{n=0}^{\infty} \frac{3^n}{2^n} \cos(3^n\theta)$ is divergent; Weierstrass has shown that $f(\theta)$ is a function everywhere continuous but nowhere differentiable.

The *correct* proposition is as follows :

THEOREM 43. *If $f_n(x)$ is a sequence of continuous functions, differentiable and with continuous derivatives, if $f'_n(x)$ converges as $n \to \infty$, to a limit $g(x)$ uniformly for $a \leqslant x \leqslant b$, a and b being finite, and if at a certain point x_0 of (a, b) the sequence $f_n(x_0)$ tends to a limit as $n \to \infty$, then the sequence $f_n(x)$ converges uniformly to a limit $f(x)$ as $n \to \infty$ for $a \leqslant x \leqslant b$, $f(x)$ is differentiable and $f'(x) = g(x)$.*

We have

(I, 3; 18) $$f_n(x) = f_n(x_0) + \int_{x_0}^{x} f'_n(\xi)\, d\xi.$$

By hypothesis $f_n(x_0)$ has a limit; according to Theorem 40 the integral has a limit which is, in fact, $\int_{x_0}^{x} g(\xi)\, d\xi$, and its convergence is uniform with respect to x. This shows that f_n converges uniformly to a limit f and, since

$$f(x) = f(x_0) + \int_{x_0}^{x} g(\xi)\, d\xi,$$

g being continuous, it follows that f is differentiable and $f' = g$. Q.E.D.

Thus if we know that a function is represented by a convergent or summable series $\sum_{n=0}^{\infty} u_n(x)$ and we wish to know whether it is differentiable we must obtain a formal derivative : $\sum_{n=0}^{\infty} u'_n(x)$. It is then necessary to check that this *derived series* $\sum_{n=0}^{\infty} u'_n(x)$ converges uniformly for $a \leqslant x \leqslant b$. If it does so then f is indeed differentiable and $f'(x) = \sum_{n=0}^{\infty} u'_n(x)$. (Differentiation of a series term by term or under the Σ sign.) On the other hand if the derived series $\sum_{n=0}^{\infty} u'_n(x)$ does not converge uniformly then it cannot be affirmed that f is differentiable. Even if it is, it does not follow that its derivative *at the point* x_0 *is equal to* $\sum_{n=0}^{\infty} u_n'(x_0)$ when the derived series $\sum_{n=0}^{\infty} u_n'(x_0)$ converges.

Example. The series $\sum_{n=-\infty}^{n=+\infty} a_n e^{in\theta}$ represents a continuous function, having continuous derivatives of order $\leqslant k$, if $|a_n| \leqslant \frac{1}{|n|^{k+2}}$ as $|n| \to \infty$.

2. FUNCTIONS REPRESENTED BY INTEGRALS

For simplicity we shall consider only ordinary integrals, the extension to multiple integrals being automatic. Let $F(x) = \int_a^b f(x, t)\, dt$. For any x of a set E (for simplicity, a subset of R), it will be supposed that the integral exists (summability, semi-convergence, etc.). It thus defines a function of x. The dependence of the continuity, integrability and differentiability of F on the analogous properties of f will be studied.

Continuity of the integral

We wish to see whether $\int_R f_x(t)\, dt \to \int_R f_{x_0}(t)\, dt$ as $x \to x_0$, where $f_x(t)$ denotes the function of t defined by f at a fixed value of x. All the results have already been met in connection with the integrability of the limit of a sequence (Theorems 40 and 41).

THEOREM 44 (*corresponding to Theorem* 40). *If, as* $x \to x_0$, $f_x(t)$ *converges uniformly to* $f_{x_0}(t)$ *where* t *belongs to* (a, b) *and if the interval of integration* (a, b) *is finite, then* $\int_{(a,b)} f_x(t)\, dt$ *converges to* $\int_{(a,b)} f_{x_0}(t)\, dt$.

For,

$$\left| \int_{(a,b)} (f_x(t) - f_{x_0}(t))\, dt \right| \leqslant (b-a)\, d(f_x(t), f_{x_0}(t)). \tag{I, 3; 19}$$

Since, given $\varepsilon > 0$, there exists $\eta > 0$ such that $|x - x_0| \leqslant \eta$ implies

$$d(f_x(t), f_{x_0}(t)) = \sup_{a \leqslant t \leqslant b} |f(x, t) - f(x_0, t)| \leqslant \frac{\varepsilon}{b-a}$$

it follows that

$$|F(x) - F(x_0)| \leqslant \varepsilon.$$

COROLLARY. *If the interval of integration* (a, b) *is finite and if for* $a \leqslant t \leqslant b$, $\alpha \leqslant x \leqslant \beta$, f *is continuous with respect to the set of variables* x, t *then* $F(x)$ *is a continuous function of* x *for* $\alpha \leqslant x \leqslant \beta$.

In fact, f is then uniformly continuous or alternatively it may be stated that, given $\varepsilon > 0$, there exists $\eta > 0$, such that $|x_2 - x_1| \leqslant \eta$, $|t_2 - t_1| \leqslant \eta$ implies $|f(x_2, t_2) - f(x_1, t_1)| \leqslant \varepsilon$. Then for $|x - x_0| \leqslant \eta$ and any t,

$$(I, 3; 20) \qquad |f_x(t) - f_{x_0}(t)| \leqslant \varepsilon \quad \text{or} \quad d(f_x(t), f_{x_0}(t)) \leqslant \varepsilon$$

and Theorem 44 can be applied.

Note. A function of x and t which is continuous with respect to each variable *separately* when the other is fixed is not necessarily continuous with respect to the set of two variables. Thus :

$$(I, 3; 21) \qquad f(x, t) = \begin{cases} \dfrac{xt}{x^2 + t^2} & \text{for} \quad (x, t) \neq (0, 0) \\ 0 & \text{for} \quad x = t = 0 \end{cases}$$

is continuous in x and in t separately at the origin. However, it is not continuous in (x, t) at $(0, 0)$ since on the line $x = mt$ it takes the value $m/(1 + m^2)$ for $t \neq 0$ and 0 for $t = 0$.

THEOREM 45. *Lebesgue, corresponding to Theorem 41. If f is separately continuous in x at the point $x = x_0$, for almost all values of t, the interval of integration being arbitrary, and if the modulus of $f(x, t)$ is bounded above by a summable function $g(t) \geqslant 0$, then F is continuous in x at the point x_0.*

Note. In certain cases there may be reason to combine the result given here with others. This happens if improper semi-convergent Lebesgue integrals are under discussion. Consider for example

$$(I, 3; 22) \qquad F(x) = \int_{\to -\infty}^{\to +\infty} a(t) e^{itx}\, dt.$$

If $a(t)$ is summable, Theorem 45 shows directly that F is continuous for all values of x, since $|a(t)e^{itx}| = |a(t)|$ is summable. Suppose, however, that $a(t)$ is not summable but continuous and that, for both $t \to +\infty$ and $t \to -\infty$, $a(t) \geqslant 0$ is monotonic and tends to zero. Then, according to Abel's theorem, $F(x)$ is defined for $x \neq 0$. Consider

$$(I, 3; 23) \qquad F_n(x) = \int_{-n}^{n} a(t) e^{itx}\, dt.$$

By Theorem 44 or 45, this is a continuous function of x, for fixed n.

When $n \to \infty$, $F_n(x)$ converges uniformly to $F(x)$ for $x \geqslant \delta > 0$, by Abel's theorem (page 43). In fact,

$$(I, 3; 24) \quad |F(x) - F_n(x)| \leqslant \left| \int_{n}^{\to +\infty} \right| + \left| \int_{\to -\infty}^{-n} \right| < \frac{2}{\delta}(a(n) + a(-n)).$$

As $a(t) \to 0$ for $t \to \pm\infty$, there exists N such that $n > N$ implies $a(n) \leqslant \varepsilon_1$,

$a(-n) \leqslant \varepsilon_1$. Hence $|F(x) - F_n(x)| \leqslant \frac{4\varepsilon_1}{\delta}$. Then, given $\varepsilon > 0$, choose $\varepsilon_1 = \frac{\delta}{4}\varepsilon$ giving, for $|x| \geqslant \delta$, $d(F(x), F_n(x)) \leqslant \varepsilon$. Theorem 39 then shows that $F(x)$ is continuous at any point $x_0 \neq 0$. The same is true for the integrals $\int_0^{\to +\infty} a(t) \cos tx \, dt$ and $\int_0^{\to +\infty} a(t) \sin tx \, dt$. The latter also exists for $x = 0$ but Abel's criterion can only be applied *uniformly for* $|x| \geqslant \delta > 0$, so that it may only be stated that the integral represents a function defined at any point x but continuous at any point $x_0 \neq 0$. There are many similar examples.

Integrability of the integral

To determine whether the integral is integrable, we must ascertain whether $F(x)$ is summable and whether

$$\int_\alpha^\beta F(x)\, dx = \int_\alpha^\beta dx \int_a^b f(x, t)\, dt$$

can be written

$$\int_a^b dt \int_\alpha f(x, t)\, dx.$$

However this has already been considered (Fubini's theorem).

Differentiability of the integral

Let E be the finite interval $\alpha \leqslant x \leqslant \beta$.

THEOREM 46. *Let f possess a partial derivative $f'_x(x, t)$ for nearly all values of t and let $f'_x(x, t)$ be separately continuous in x for nearly all values of t with a modulus bounded above by $g(t) \geqslant 0$ which is summable for any interval (a, b). Then, if $\int_a^b f(x, t)\, dt$ exists for a particular value $x = x_0$, it also exists for any $x \in E$, the integral $F(x)$ is continuous and differentiable and*

$$F'(x) = \int_a^b f'_x(x, t)\, dt, \tag{I, 3; 25}$$

(differentiation under the $\int$ sign.)

Take $(a, b) = (-\infty, +\infty)$. First, note that $G(x) = \int_{-\infty}^{+\infty} f'_x(x, t)\, dt$ is continuous. Next :

$$\int_{x_0}^{x} G(\xi)\, d\xi = \int_{x_0}^{x} d\xi \int_{-\infty}^{+\infty} f'_\xi(\xi, t)\, dt = \int_{-\infty}^{+\infty} dt \int_{x_0}^{x} f'_\xi(\xi, t)\, d\xi. \tag{I, 3; 26}$$

Inversion of the order of integration is permissible. For, $|f'_\xi(\xi, t)| \leqslant g(t)$

and $g(t)$ is summable in (ξ, t) over $\alpha \leqslant \xi \leqslant \beta$ (finite interval), $-\infty \leqslant t \leqslant +\infty$ since it is $\geqslant 0$, and $\int_\alpha^\beta d\xi \int_{-\infty}^{+\infty} g(t)\, dt$ is finite. Finally,

$$\int_{x_0}^{x} G(\xi)\, d\xi = \int_{-\infty}^{+\infty} [f(x, t) - f(x_0, t)]\, dt. \tag{I, 3; 27}$$

Since, by hypothesis, $\int_{-\infty}^{+\infty} f(x_0, t)\, dt$ exists, $\int_{-\infty}^{+\infty} f(x, t)\, dt$ also exists and defines a function $F(x)$, $x \in E$.

Again,

$$F(x) = F(x_0) + \int_{x_0}^{x} G(\xi)\, d\xi, \tag{I, 3; 28}$$

G being continuous, so that F is continuous and differentiable, and

$$F'(x) = G(x).$$

Examples. Here again it is often necessary, in discussing semi-convergent improper Lebesgue integrals, to combine this theorem with other results. Consider, for example,

$$F(x) = \int_{-\infty}^{+\infty} \frac{e^{itx}}{1 + t^2}\, dt. \tag{I, 3; 29}$$

This is continuous, since $\dfrac{e^{itx}}{1 + t^2}$ is continuous in (x, t) and has a modulus bounded above by the summable function $1/(1 + t^2)$. The formal derivative is

$$F'(x) = \int_{-\infty}^{+\infty} e^{itx} \frac{it}{1 + t^2}\, dt. \tag{I, 3; 30}$$

The function $t/(1 + t^2)$ is not summable but $\int_{\to -\infty}^{\to +\infty}$ exists for $x \neq 0$, by Abel's theorem. [$t/(1 + t^2)$ is in fact decreasing for $t \to +\infty$ and for $t \to -\infty$, since its derivative is $(1 - t^2)/(1 + t^2)^2$.]

Put $F_n(x) = \int_{-n}^{+n} \dfrac{e^{itx}}{1 + t^2}\, dt$; then F_n is continuous and differentiable for all x, and $F_n'(x) = \int_{-n}^{+n} \dfrac{it e^{itx}}{1 + t^2}\, dt$.

As $n \to \infty$, $F_n'(x)$ converges uniformly to the limit $G(x) = \int_{\to -\infty}^{\to +\infty} e^{itx} \dfrac{it}{1 + t^2}\, dt$ for $|x| \geqslant \delta > 0$. The modulus of the remainder $\int_{-\infty}^{-n} + \int_{+n}^{+\infty}$ is bounded above by $2 \cdot \dfrac{2}{\delta} \dfrac{n}{1 + n^2} \leqslant \varepsilon$ if n is chosen large enough for $\dfrac{n}{1 + n^2} \leqslant \dfrac{\varepsilon\delta}{4}$. Also, $F_n(x)$ converges uniformly to $F(x)$ for real x.

Thus, by the theorem on the differentiability of the limit of a sequence, F possesses a derivative at any point $x_0 \neq 0$ and this derivative is continuous :

$$\text{(I, 3; 31)} \qquad F'(x) = G(x) = \int_{\to -\infty}^{\to +\infty} e^{itx} \cdot \frac{it}{1 + t^2} \, dt.$$

No similar statement can be made about the second derivative since $\frac{e^{itx}(-t^2)}{1 + t^2}$ no longer gives a convergent integral. It will be seen later that

$$\text{(I, 3; 32)} \qquad F(x) = \pi e^{-|x|}.$$

Thus $F(x)$ is indeed continuous for all x and has a continuous derivative at any point $x \neq 0$. In fact it has continuous derivatives of all orders except at the origin $x = 0$, but this is not shown by the integral expression.

Applications. Continuity and differentiability of a volume potential

Let U be the Newtonian potential due to an electric charge distribution with a bounded density function $\mu(x)$, $\mu(x) \leqslant M$, zero outside a bounded et X. By (I, 2; 50)

$$\text{(I, 3; 33)} \qquad U(a) = \iiint_X \frac{\mu(x)\, dx}{|x - a|}.$$

We wish to show that $U(a)$ is a continuous function of a. This does not result directly from the theorems given above, for the integrand, $\frac{\mu(x)}{|x - a|}$, is infinite at $x = a$ and thus has a singularity at the variable point a. In particular there cannot be an inequality $\frac{|\mu(x)|}{|x - a|} \leqslant g(x)$, where g is summable and independent of a, since putting $a = x$ makes $g(x) = \infty$ for any x, [if $\mu(x) \neq 0$]. Here again a combination of given methods is necessary.

Note first that if a varies in an open set Ω of R^3 containing no charge [$\mu(x) = 0$ for $x \in \Omega$], then $U(a)$ is a continuous function and is indeed infinitely differentiable, all the partial derivatives being obtained by differentiation under the sign $\iiint$. For, if $a \in \Omega$ and $x \in \complement\Omega$, $\frac{1}{|x - a|}$ has partial derivatives of all orders which are continuous, and thus separately continuous in a, and are bounded provided a does not approach closer than an arbitrary positive distance d to the boundary of Ω. As μ is bounded and the volume of integration is bounded, Theorems 45 and 46 give the stated result. In particular, if the coordinates of a are denoted by a_i $(i = 1, 2, 3)$,

$$\text{(I, 3; 34)} \qquad \frac{\partial^2}{\partial a_i^2} U(a) = \iiint_X \mu(x) \frac{\partial^2}{\partial a_i^2}\left(\frac{1}{|x - a|}\right) dx$$

so that putting $\Delta U(a) = \sum_{i=1,2,3} \frac{\partial^2}{\partial a_i^2} U(a)$ (the differential operator Δ being termed the Laplacian), it follows that

$$\text{(I, 3; 35)} \qquad \Delta U(a) = \iiint_X \mu(x) \Delta_a \frac{1}{|x-a|} dx = 0,$$

since $\Delta_a \frac{1}{|x-a|} = 0$ for $x \neq a$. Thus the potential $U(a)$ is *harmonic* (which means $\Delta U = 0$) in an open set Ω containing no charge.

We now go over to the general case. Let $a_0 \in R^3$, $\rho > 0$. We find an upper bound for the integral

$$\text{(I, 3; 36)} \qquad \iiint_{|x-a_0| \leqslant \rho} \frac{\mu(x)\, dx}{|x-a|}.$$

For $|a - a_0| \leqslant 2\rho$, it is bounded above by

$$\text{(I, 3; 37)} \quad M \iiint_{|x-a| \leqslant 3\rho} \frac{dx}{|x-a|} = M \iiint_{r \leqslant 3\rho} \frac{dx}{r} = 18\pi M \rho^2$$

[equation (I, 2; 51)].

For $|a - a_0| \geqslant 2\rho$, $|x - a| \geqslant \rho$ in the volume of integration so that (I, 3; 36) is bounded above by

$$\text{(I, 3; 38)} \qquad \frac{M}{\rho} \iiint_{|x-a_0| \leqslant \rho} dx = \frac{M}{\rho} \frac{4}{3} \pi \rho^3 = \frac{4}{3} \pi M \rho^2.$$

It can thus be made arbitrarily small by taking ρ small enough, independently of the position of $a_0 \in R^3$. Alternatively it may be stated that the function of a depending on the parameter ρ,

$$\text{(I, 3; 39)} \qquad U_\rho(a) = \iiint_{\substack{|x-a_0| \geqslant \rho \\ x \in X}} \frac{\mu(x)\, dx}{|x-a|},$$

converges uniformly with respect to a to the function $U(a)$ as $\rho \to 0$.

By Theorem 39 a condition sufficient for U to be continuous at the point a_0 is that U_ρ should be continuous at the point a for $\rho > 0$. Now, for $\rho > 0$, U_ρ is the potential of a distribution of density zero in the ball $|x - a_0| \leqslant \rho$ and $\mu(x)$ outside this ball so that, by a result proved above, U_ρ is continuous at the point a_0.

Formal differentiation under the sign $\iiint$ yields the function

$$\text{(I, 3; 40)} \quad V_i(a) = \iiint_X \mu(x) \frac{\partial}{\partial a_i} \frac{1}{|x-a|} dx = \iiint_X \mu(x) \frac{x_i - a_i}{|x-a|^3} dx.$$

As $\left| \frac{x_i - a_i}{|(x-a)|^3} \right| \leqslant \frac{1}{|x-a|^2}$, this is an example of a singularity at a of type $1/r^\alpha$, $\alpha = 2 < 3$, and by Theorem 33 the integral of the second and third members of (I, 3; 40) thus exists. An argument similar in all respects to the one which has just been given shows that this integral represents a continuous function of a. It remains to show that this function $V_i(a)$ is indeed $\frac{\partial}{\partial a_i} U(a)$. For this it is necessary, as in Theorem 45, to use an integration method showing that U is in fact an indefinite integral of V_i with respect to a_i. More precisely, if b and c are two points (with coordinates $b_j, c_j, j = 1, 2, 3$) situated on the same line parallel to the x_i-axis ($b_j = c_j$, for $j \neq i$) then we have only to show that

$$\int_{b_i}^{c_i} V_i(a)\, da_i = U(c) - U(b), \tag{I, 3; 41}$$

it being understood that, on the left-hand side, a has all coordinates a_j ($j \neq i$) fixed and equal to $b_j = c_j$, the i-th, the only variable, being the variable of integration a_i. The left-hand side can be written

$$\int_{b_i}^{c_i} da_i \iiint_X \mu(x) \frac{x_i - a_i}{|x-a|^3}\, dx. \tag{I, 3; 42}$$

If inverting the order of integration is legitimate, this becomes

$$\iiint \mu(x)\, dx \int_{b_i}^{c_i} \frac{\partial}{\partial a_i} \frac{1}{|x-a|}\, da_i. \tag{I, 3; 43}$$

Now the second integral exists and has the value $|x-c|^{-1} - |x-b|^{-1}$ except when x lies on the segment (b, c), a set of measure zero in R^3. Hence (I, 3; 43) is equal to

$$\iiint_X \frac{\mu(x)}{|x-c|}\, dx - \iiint_X \frac{\mu(x)}{|x-b|}\, dx = U(c) - U(b), \tag{I, 3; 44}$$

giving (I, 3; 41).

All this depends on the possibility of inverting the order of integration in equation (I, 3; 42). By the corollary to Theorem 29 it suffices to show that

$$\iiint_X |\mu(x)|\, dx \int_{b_i}^{c_i} \left| \frac{\partial}{\partial a_i} \frac{1}{|x-a|} \right| da_i \tag{I, 3; 45}$$

is finite. Consider the inequality

$$\int_{b_i}^{c_i} \left| \frac{\partial}{\partial a_i} \frac{1}{|x-a|} \right| da_i \leqslant \int_{b_i}^{c_i} \frac{da_i}{|x-a|^2} \tag{I, 3; 46}$$

$$\leqslant \int_{-\infty}^{+\infty} \frac{da_i}{(x_i - a_i)^2 + \sum_{j \neq i} (x_j - b_j)^2} = \int_{-\infty}^{+\infty} \frac{dt}{t^2 + k^2} = \frac{\pi}{k},$$

where

$$k^2 = \sum_{j \neq i} (x_j - b_j)^2.$$

Thus (I, 3; 45) is bounded above by

$$\text{(I, 3; 47)} \qquad \pi M \iiint_X \frac{dx}{\sqrt{\sum_{j \neq i} (x_j - b_j)^2}}.$$

For definiteness, put $i = 1$. By Fubini's theorem, since the integrand is $\geqslant 0$, (I, 3; 47) may be expressed in the form

$$\text{(I, 3; 48)} \qquad \pi M \int_{x \in X} dx_1 \iint_{x \in X} \frac{dx_2 dx_3}{\sqrt{(x_2 - b_2)^2 + (x_3 - b_3)^2}}.$$

The double integral is taken over a bounded area (since the volume of integration X is bounded) and it has at the point (b_2, b_3), a singularity of type $1/r^\alpha$ with $\alpha = 1 < 2$, so that it is finite. The integral in x_1 is also finite since X is bounded, so that (I, 3; 48) is finite. Q.E.D.

If we now attempt to obtain a similar expression for the second-order partial derivatives of U, we encounter a difficulty. In fact, the second-order partial derivatives (with respect to a) of $1/|x - a|$ display, when x approaches a, a singularity of type $1/r^\alpha$ with $\alpha = 3$, so that the integrals obtained have no meaning. If no supplementary assumptions are made about the density μ, then it may be shown by examples that U *is not* twice differentiable.

Moreover, equation (I, 3; 35) is *always* false at points where $\mu \neq 0$ (although it would be true if U could be differentiated twice under the sign $\iiint$; this shows that the mathematical precautions taken in these operations are not excessive). It is in fact known that Poisson's equation

$$\text{(I, 3; 49)} \qquad \Delta U(a) = -4\pi\mu(a)$$

applies (at least if μ is continuously differentiable once). All these properties of the potential will be looked at from another point of view by making use of the convolution product.

Note. By a change of variables, (I, 3; 33) can be written

$$\text{(I, 3; 50)} \qquad U(a) = \iiint \frac{\mu(a - x)}{|x|} dx.$$

Suppose that μ is continuous and not merely bounded in R^3. Then $\mu(x + a)/|x|$ is separately continuous in a, except at $x = 0$; it is bounded by $M/|x|$ which is summable since $|x| = r^\alpha$ with $\alpha = 1 < 3$ and the domain of integration is bounded. (This domain is $X + a$, the set of points $x + a$ where x belongs to X; this set varies with a but when a remains bounded it is contained in a fixed bounded volume which can be taken as the domain of integration.) By Theorem 41 this shows that U is continuous. This proof, which assumes μ continuous, is simpler than the one given above for bounded μ. If μ is p-times continuously differentiable it can easily be seen that U is p times continuously differentiable, the derivatives being obtained by formal differentiation under the sign $\iiint$: thus

$$\frac{\partial U}{\partial a_i} = \iiint \frac{1}{|x|} \frac{\partial}{\partial a_i} \mu(a-x)\, dx. \tag{I, 3; 51}$$

EXERCISES FOR CHAPTER I

Exercise I-1

Show that if a series $\sum_{n=0}^{\infty} u_n$ is semi-convergent, then $\sum_{n=0}^{\infty} u_n^+$ and $\sum_{n=0}^{\infty} u_n^-$ are both divergent. Hence deduce that the series $\sum_{i \in \mathrm{N}} u_i$ does not satisfy Cauchy's criterion for summable series. (You may begin by considering explicitly the case of the alternating harmonic series.)

Exercise I-2

Show that if a series $\sum_{n=0}^{\infty} u_n$ is semi-convergent, it is possible by changing the order of its terms to make it converge to a sum arbitrarily fixed in advance and also to make it diverge to $+\infty$ or $-\infty$.

Exercise I-3

Show that if a series $\sum_{n=0}^{\infty} u_n$ is convergent and remains convergent whatever the order of its terms may be, then it is absolutely convergent.

Exercise I-4 (*Characteristic functions of sets*)

Denote by A and B two subsets of R^n and by A — B the subset of R^n consisting of the points belonging to A and not belonging to B. Let

$$A \Delta B = (A - B) \cup (B - A).$$

Denote by χ_A and χ_B the characteristic functions of A and B respectively. Prove that

$$(1) \quad \chi_{A \cap B} = \chi_A \chi_B$$
$$(2) \quad \chi_{A \cup B} = \chi_A + \chi_B - \chi_{A \cap B}$$
$$(3) \quad \chi_{A \Delta B} = |\chi_A - \chi_B|.$$

Equation (2) is related to Theorem 24.

Exercice I-5 (*Spherical integrals in* R^n)

Evaluate as a function of the area S_n of the unit sphere in R^n the volume of the ball of radius R and also the moment of inertia of a homogeneous ball of radius R with respect to its centre.

Exercise I-6

(*a*) Adopt the usual notations : $i = \sqrt{-1}$, $e^{it} = \cos t + i \sin t$,

$$|a + ib| = \sqrt{(a^2 + b^2)}.$$

Supposing x real, $0 < x < 2$, show that

$$|e^{in\pi x} + e^{i(n+1)\pi x} + \cdots + e^{im\pi x}| \leqslant \frac{1}{\sin\left|\frac{\pi x}{2}\right|}.$$

The obvious inequality $|b| \leqslant |a + ib|$ then gives

$$(1) \qquad |\sin n\pi x + \sin (n + 1)\pi x + \cdots + \sin m\pi x| \leqslant \frac{1}{\left|\sin \frac{\pi x}{2}\right|}.$$

By making use of the inequality (1) and of the equation

$$u_n v_n + u_{n+1} v_{n+1} + \cdots + u_m v_m = (u_n - u_{n+1}) v_n + (u_{n+1} - u_{n+2})(v_n + v_{n+1}) + \cdots + u_m(v_n + v_{n+1} + \cdots + v_m)$$

prove the inequality

$$(2) \quad \left|\frac{1}{n} \sin n\pi x + \frac{1}{n+1} \sin (n + 1)\pi x + \cdots + \frac{1}{m} \sin m\pi x\right| \leqslant \frac{1}{n} \frac{1}{\left|\sin \frac{\pi x}{2}\right|}.$$

(*b*) Show that the trigonometric series

$$\frac{1}{12} + \frac{4}{\pi^3}\left(\sin \pi x + \frac{1}{3^3} \sin 3\pi x + \cdots + \frac{1}{(2p + 1)^3} \sin (2p + 1)\pi x + \cdots\right) - \frac{1}{2\pi^2}\left(\cos 2\pi x + \frac{1}{2^2} \cos 4\pi x + \cdots + \frac{1}{p^2} \cos 2p\pi x + \cdots\right)$$

converges for all $-1 \leqslant x \leqslant 1$. In the interval $(-1, +1)$ it defines a real-valued function which will be denoted by $f(x)$.

(*c*) Show that the series obtained by term-by-term differentiation of $f(x)$ converges for all x such that $0 < |x| < 1$. Denote by $g(x)$ the series obtained by differentiating $f(x)$ term by term; up to now $g(x)$ is defined only for $-1 < x < 0$ and $0 < x < 1$. Assuming that the series with the general term $1/n^2$ has the sum $\pi^2/6$, evaluate $g(0)$. Prove that $f(x)$ is a continuous function of x at every point of the interval $(-1, +1)$.

(*d*) Consider the following function, defined on the interval $(-1, +1)$:

$$G(x) = \begin{cases} 0 & \text{if} \quad -1 \leqslant x < 0, \\ 1 - 2x & \text{if} \quad 0 < x \leqslant 1, \end{cases}$$

and set

$$G(0) = \frac{1}{2}(G(-0) + G(+0))$$

where

$$G(-0) = \lim_{\substack{x \to 0 \\ x < 0}} G(x) \qquad \text{and} \qquad G(+0) = \lim_{\substack{x \to 0 \\ x > 0}} G(x).$$

Write down the expansion of $G(x)$ in a Fourier series. Hence deduce an expression for $f(x)$ in the interval $(-1, 0)$ and an expression for $f(x)$ in the interval $(0, +1)$. From the value of $f(0)$ deduce that of the series with general term $1/n^2$. What can be said about the derivative of f at the origin?

Exercise I-7

(*a*) We define

$$e_0(x) = e^{-x^2}, \qquad e_n(x) = \int_{-\infty}^{x} e_{n-1}(t)\, dt.$$

(i) Show that these functions exist.

(ii) Let

$$f_1(x) = e_2(x)$$

and let $f_n(x)$ be defined for $n > 1$ by

$$f_n'''(x) + 2x f_n''(x) = \sum_{i=1}^{n-1} f_i(x) f_{n-i}''(x)$$

with

$$\lim_{x \to -\infty} f_n(x) = \lim_{x \to -\infty} f_n'(x) = \lim_{x \to -\infty} e^{x^2} f_n''(x) = 0.$$

Show that

$$f_n''(x) \leqslant e_0(x) e_3^{n-1}(x), \qquad f_n(x) \leqslant e_2(x) e_3^{n-1}(x)$$

and hence deduce that these functions exist.

(*b*)

(i) Let $A > 0$ be given. Show that when x is less than a certain value $X_1(A)$ the series

$$(1) \qquad -2x + \sum_{i=1}^{\infty} A^i f_i(x)$$

converges and its sum is an integral of

$$(2) \qquad y''' = yy''.$$

(ii) Let $Z_A(x)$ be the integral of (2) which coincides with the sum of (1) when $x < X_1(A)$. [The region where $Z_A(x)$ exists may not be identical with the region where (1) converges.]

Show by writing a first integral of (2) in the form $y'' = Ce^{\int y(x)\,dx}$ that either $Z_A(x)$ exists for all x or $Z_A(x)$ exists for $x < x_0(A)$ and becomes infinite as x tends to $x_0(A)$.

(iii) Show that, for any value of x where $Z_A(x)$ exists, the sum $S_n(x)$ of the first n terms of (1) is less than $Z_A(x)$. [It may be shown that $h(x) = Z_A(x) - S_n(x)$ satisfies a differential equation which implies $h''(x) \geqslant 0$.]

(iv) Hence deduce that the region where $Z_A(x)$ exists coincides with the region where (1) converges.

Exercise I-8

(*a*) It is proposed to obtain Feynman's method of integration for a product of factors. For this purpose two equations giving transformations of products of n factors into integrals containing an n-th power are derived.

Feynman's first equation:

$$\frac{1}{a_1a_2 \ldots a_n} = (n-1)! \int_0^1 dx_1 \int_0^{1-x_1} dx_2 \ldots \int_0^{1-x_1-x_2-\cdots x_{n-2}} dx_{n-1}$$
$$\times \frac{1}{[(a_{n-1}-a_n)x_{n-1} + (a_{n-2}-a_n)x_{n-2} + \cdots + (a_1-a_n)x_1 + a_n]^n} \quad (a_i > 0).$$

(i) Prove this equation directly by making use of the following result and the indicated changes of variable :

$$\frac{1}{a_p} = \int_0^\infty e^{-a_py_p}\,dy_p; \qquad X = \sum_{i=1}^n y_i; \qquad x_i = \frac{y_i}{X}.$$

An integral is obtained involving $n + 1$ variables linked by a relation. To derive the required result an integration with respect to X is performed.

(ii) Prove this equation by induction.

Feynman's second equation:

$$\frac{1}{a_1a_2 \ldots a_n} = (n-1)! \int_0^1 dx_1 \ldots \int_0^1 dx_{n-1}$$
$$\times \frac{x_1^{n-2}x_2^{n-3} \ldots x_{n-2}}{[(a_n-a_{n-1})x_{n-1}\ldots x_1 + (a_{n-1}-a_{n-2})x_{n-2}\ldots x_1 + \cdots + (a_2-a_1)x_1 + a_1]^n}.$$

(i) Prove this equation by induction.

(ii) Show that there is a change of variables which transforms the first expression into the second. Note that the limits of integration are variable in the first case and fixed in the second. This is of assistance in finding the new variables with the range 0 to 1, in terms of the first set of variables.

(*b*) Application to the evaluation of a four-dimensional space integral

$$\int d_4K \frac{(A\cdot K)\,(C\cdot(K+V))}{K^2(K^2+2(K\cdot V))\,(K^2+2(K\cdot W))},$$

where A, C, K, V and W are vectors with components A_i, C_i, K_i, V_i and W_i, $i = 1, 2, 3, 4$, such that

$$(A\cdot V) = (A\cdot W); \qquad (C\cdot V) = (C\cdot W); \qquad V^2 = W^2; \qquad (A\cdot C) = 0;$$

where
$$(A\cdot K) = A_1K_1 + A_2K_2 + A_3K_3 + A_4K_4,$$
$$K^2 = K_1^2 + K_2^2 + K_3^2 + K_4^2,$$
$$d_4K = dK_1\, dK_2\, dK_3\, dK_4.$$

The integration is carried over the whole of the space.

(i) Apply Feynman's first equation to the product

$$\frac{1}{K^2(K^2+2(K\cdot V))(K^2+2(K\cdot W))}.$$

(ii) Integrate with respect to K, making use of a linear transformation of variables and of the four-dimensional polar coordinates,

$$S_1 = R\cos a_1; \qquad S_2 = R\cos a_2 \sin a_1; \qquad S_3 = R\sin a_1 \sin a_2 \cos a_3;$$
$$S_4 = R\sin a_1 \sin a_2 \sin a_3.$$

(iii) A double integral in x_1 and x_2 is then obtained which may be evaluated by the change of variables $x_1 + x_2 = w$; $x_1 = uw$.

Exercise I-9 (*Convergence of sequences of functions*)

(*a*) Let E be a vector space over the complex numbers C.

A mapping $\vec{x} \to \|\vec{x}\|$ of E into R is said to be a *norm* if it possesses the following properties :

$$\|\vec{x}\| \geqslant 0 \text{ for all } \vec{x} \in E, \ \|\vec{x}\| = 0 \Longleftrightarrow \vec{x} = \vec{0},$$
$$\|\lambda\vec{x}\| = |\lambda|\,\|\vec{x}\| \text{ for all } \lambda \in C \text{ and all } \vec{x} \in E,$$
$$\|\vec{x} + \vec{y}\| \leqslant \|\vec{x}\| + \|\vec{y}\| \text{ for all } \vec{x}, \vec{y} \in E.$$

A sequence $\vec{x}_n$ in E is said to converge to $\vec{x}$ in E for the topology defined by the norm $\|\cdot\|$ if $\|\vec{x}_n - \vec{x}\| \to 0$ as $n \to \infty$, and this is written $\vec{x}_n \to \vec{x}$. A sequence $\vec{x}_n$ is said to be a *Cauchy sequence* for the topology defined by the norm $\|\cdot\|$ if, for any $\varepsilon > 0$, there exists an integer n_0 such that, for any pair of integers m and n, both greater than n_0,

$$\|\vec{x}_m - \vec{x}_n\| \leqslant \varepsilon.$$

(i) Show that if $\vec{x}_n \to \vec{x}$ and $\vec{y}_n \to \vec{y}$ in E, then $\vec{x}_n + \vec{y}_n \to \vec{x} + \vec{y}$.

(ii) Show that if $\vec{x}_n \to \vec{x}$ in E and if λ_n is a sequence of complex numbers which converge to λ, then $\lambda_n\vec{x}_n \to \lambda\vec{x}$ in E.

(iii) Show that any sequence $\vec{x}_n$ which converges in E is a Cauchy sequence. In general the converse of (iii) is false : a Cauchy sequence in E does not necessarily converge to a point of E.

When a vector space over C possesses a norm for which every Cauchy sequence converges to a point in E, it is said that E is *complete* for the topology defined by this norm, and E is then called a *Banach space.*

(*b*) Denote by $\mathcal{E}^0_{[0,1]}$ the vector space over C of the complex functions defined and continuous on [0, 1]. Put

$$\|f\|_1 = \int_0^1 |f(t)|\,dt, \qquad \|f\|_2 = \sup_{t\in[0,1]} |f(t)|.$$

(i) Show that $\|f\|_1$ and $\|f\|_2$ define two norms on $\mathcal{E}^0_{[0,1]}$.

(ii) Show that if a sequence of functions $f_n \in \mathcal{E}^0_{[0,1]}$ converges to a function $f \in \mathcal{E}^0_{[0,1]}$ for $\|\cdot\|_1$, then $f_n(t) \to f(t)$ in the elementary sense on [0, 1].

(iii) By constructing an example show that a sequence of functions $f_n \in \mathcal{E}^0_{[0,1]}$ which converges in the elementary sense on [0, 1] to a function $f \in \mathcal{E}^0_{[0,1]}$ does not necessarily converge to f for $\|\cdot\|_1$. (One example which could be taken is $f_n(t) = 0$ for $t = 0$ and $t \geqslant 1/n$, $= 2n^3 t$ for $0 < t \leqslant 1/2n$ and $= -n^3 2(t - 1/n)$ for $1/2n \leqslant t \leqslant 1/n$.)

(iv) Show that if $f_n \to f$ in $\mathcal{E}^0_{[0,1]}$ for $\|\cdot\|_2$, then $f_n \to f$ uniformly on [0, 1] and that, conversely, if $f_n \in \mathcal{E}^0_{[0,1]}$ converges uniformly to f on [0, 1], then $f_n \to f$ in $\mathcal{E}^0_{[0,1]}$ for $\|\cdot\|_2$.

(v) Given two norms $\|\cdot\|_1$ and $\|\cdot\|_2$ on a vector space E, it is said that $\|\cdot\|_2$ is finer than $\|\cdot\|_1$ if there exists a constant K such that, for all $\vec{x} \in$ E :

$$\|\vec{x}\|_1 \leqslant K\|\vec{x}\|_2.$$

Two norms are said to be equivalent if each is finer than the other.

Show that, on $\mathcal{E}^0_{[0,1]}$, $\|\cdot\|_2$ is finer than $\|\cdot\|_1$. By using the sequence of functions $f_n(t)$ defined on [0, 1] by

$$f_n(t) = \begin{cases} 0 & \text{for} \quad t \geqslant 1/n \\ -nt + 1 & \text{for} \quad 0 \leqslant t \leqslant 1/n \end{cases}$$

show that $\|\cdot\|_1$ and $\|\cdot\|_2$ are not equivalent.

(vi) Show that, for $\|\cdot\|_2$, $\mathcal{E}^0_{[0,1]}$ is a Banach space.

(vii) Consider the sequence of functions $f_n(t)$ defined on [0, 1] by

$$f_n(t) = \begin{cases} 0 & \text{for} \quad t \leqslant 1/2 \\ n(t - 1/2) & \text{for} \quad 1/2 \leqslant t \leqslant (1/2 + 1/n) \\ 1 & \text{for} \quad t \geqslant (1/2 + 1/n). \end{cases}$$

Show that for $\|\cdot\|_1$ this is a Cauchy sequence which does not converge to a function f of $\mathcal{E}^0_{[0,1]}$. For $\|\cdot\|_1$, $\mathcal{E}^0_{[0,1]}$ is thus not complete.

(viii) One of the above questions is false. Which one?

CHAPTER II

Elementary theory of distributions

1. Definition of distributions

1. THE VECTOR SPACE $\mathfrak{D}$

$\mathfrak{D}$ is a vector subspace of the vector space of complex-valued functions defined on R^n; such a function of $x \in R^n$ is also a function of the n real variables $x_1, x_2, \ldots, x_n$. $\mathfrak{D}$ is defined as follows :

A function φ on R^n belongs to $\mathfrak{D}$ if and only if it is infinitely differentiable and there exists a bounded set K of R^n outside which it is identically zero.

Obviously K is not the same for all functions φ of $\mathfrak{D}$. For each function φ, if K is the smallest *closed* set outside which φ is zero, then K is called the supporting set or *support* of φ. In other words, K is the closure of the set of points x for which $\varphi(x) \neq 0$. We can now say :

DEFINITION 1. *The space $\mathfrak{D}$ is the space of complex functions on R^n which are infinitely differentiable and have bounded supports.*

Example. If $n = 1$, the function φ defined by

$$\text{(II, 1; 1)} \qquad \begin{cases} \varphi(x) = 0 & \text{if} \quad |x| \geqslant 1, \\ \varphi(x) = \exp\left(-\dfrac{1}{1 - x^2}\right) & \text{for} \quad |x| < 1, \end{cases}$$

belongs to $\mathfrak{D}$. Its support, which is the interval $|x| \leqslant 1$, is obviously bounded. It is infinitely differentiable for $|x| > 1$, since it is then identically zero, and for $|x| < 1$, since it is then the exponential of an infinitely differentiable function. In fact it is easily shown to be infinitely differentiable everywhere since its derivatives of all orders are zero at $x = \pm 1$. It is as well to remark that any finite Taylor series for φ in the neighbourhood of $x = \pm 1$ has all terms zero except the remainder, while the infinite Taylor series in powers of $x - 1$ or $x + 1$ converges to 0, since all its terms are zero, and does not represent the positive function $\varphi(x)$ for $|x| < 1$. A similar situation occurs for any function $\varphi \in \mathfrak{D}$ which is not identically zero. For n dimensions an analogous example is

$$\text{(II, 1; 2)} \quad \begin{cases} \varphi(x) = 0 & \text{if} \quad r \geqslant 1 \\ \varphi(x) = \exp\left(-\dfrac{1}{1 - r^2}\right) & \text{if} \quad r < 1 \end{cases} \quad \begin{cases} r = |x| \\ r = \sqrt{\sum_{i=1}^{n} x_i^2}. \end{cases}$$

The statement that $\mathfrak{D}$ is a vector space is valid because if φ_1 and $\varphi_2 \in \mathfrak{D}$, then $\varphi_1 + \varphi_2 \in \mathfrak{D}$ and if λ is a complex number and $\varphi \in \mathfrak{D}$, then $\lambda\varphi \in \mathfrak{D}$.

$\mathfrak{D}$ is indeed an algebra (or a ring) for ordinary multiplication, since if φ_1 and $\varphi_2 \in \mathfrak{D}$ then $\varphi_1\varphi_2 \in \mathfrak{D}$. More generally, if $\varphi \in \mathfrak{D}$ and if ψ is an infinitely differentiable function whose support is not necessarily bounded, then $\varphi\psi \in \mathfrak{D}$ and the support of $\varphi\psi$ is contained in the intersection of the supports of φ and ψ.

THEOREM 1. *Approximation Theorem. For any* $\varepsilon > 0$, *any continuous function* f, *with a bounded support* K, *can be uniformly approximated to within a distance* ε *by some function* $\varphi \in \mathfrak{D}$, *and* φ *can be required to have its support contained within an arbitrary neighbourhood of the support* K *of* f.

Proof. If d is a number > 0, the d-neighbourhood of the set K is defined to be the set K_d of points whose distance from K is $\leqslant d$. K_d is a bounded closed set containing K. We wish to show that, given f, with support K, $\varepsilon > 0$ and $d > 0$ there exists $\varphi \in \mathfrak{D}$, with support contained in K_d, such that the "distance" between f and φ, i.e., $\sup_{x \in R} |f(x) - \varphi(x)|$, is $\leqslant \varepsilon$. To do this, denote by θ_1 the function belonging to $\mathfrak{D}$ which is defined by equation (II, 1; 2). Next put

$$\text{(II, 1; 3)} \qquad \theta(x) = \frac{1}{k}\, \theta_1\left(\frac{x}{a}\right)$$

so that the support of θ is the ball $|x| \leqslant a$; k is chosen equal to

$$\iint \cdots \int_{R^n} \theta_1\left(\frac{x}{a}\right) dx = a^n \iint \cdots \int_{R^n} \theta_1(y)\, dy > 0,$$

so that

$$\text{(II, 1; 4)} \qquad \int_{R^n} \cdots \int \theta(x)\, dx = +1.$$

The function φ is then defined as follows :

$$\text{(II, 1; 5)} \quad \varphi(x) = \iint_{R^n} \cdots \int f(x - \xi)\, \theta(\xi)\, d\xi = \iint_{R^n} \cdots \int f(\xi)\, \theta(x - \xi)\, d\xi.$$

It will be shown that if the constant $a > 0$ of equation (II, 1; 3) is chosen small enough, then φ has all the required properties. First, if x does not belong to K_a, then $|x-\xi|>a$ for all $\xi \in K$. Then, since in fact the domain of integration in the second integral is K and not R^n (because of the term $f(\xi)$), and $\theta(x-\xi) \equiv 0$ for $\xi \in K$, it follows that $\varphi(x)=0$. Thus φ is zero outside K_a, so that its support is contained in K_d if $a \leqslant d$. Next, φ is infinitely differentiable. For in the second integral it is permissible to differentiate any number of times with respect to x under the sign $\int$, since the domain of integration is bounded (being the support of f) and $f(\xi)\theta(x-\xi)$ has derivatives with respect to x, continuous in x and ξ, of all orders. Lastly, consider the difference $\varphi(x)-f(x)$. As

$$\iint_{R^n} \cdots \int \theta(\xi)\, d\xi = 1,$$

we may write

$$f(x)-\varphi(x)=\iint_{R^n} \cdots \int (f(x)-f(x-\xi))\theta(\xi)\, d\xi. \tag{II, 1; 6}$$

Since f is continuous and has a bounded support, it is uniformly continuous. Given $\varepsilon>0$, it is thus possible to choose $\eta>0$ such that $|\xi| \leqslant \eta$ implies $|f(x)-f(x-\xi)| \leqslant \varepsilon$. As the domain of integration in (II, 1; 6) is in fact not R^n but the support of θ, i.e., $|\xi| \leqslant a$, it follows that in this domain $|f(x)-f(x-\xi)| \leqslant \varepsilon$ if a is chosen $\leqslant \eta$. Finally, since

$$\iint_{R^n} \cdots \int \theta(\xi)\, d\xi = 1,$$

$$|f(x)-\varphi(x)| \leqslant \varepsilon. \tag{II, 1; 7}$$

Thus, for $a \leqslant d$ and $a \leqslant \eta$, φ has all the required properties.

Topology or concept of convergence on $\mathfrak{D}$

A sequence φ_j of functions belonging to $\mathfrak{D}$ is said to converge to a function φ of $\mathfrak{D}$ as $j \to \infty$ if:

1) *The supports of the φ_j are contained in the same bounded set, independently of j.*

2) *The derivatives of any given order m of the φ_j converge uniformly as $j \to \infty$ to the corresponding derivative of φ.*

This is a type of convergence of " infinite order " since it implies uniform convergence for every derivative. Note that we do not demand that the derivatives of all orders shall simultaneously converge uniformly, but only that the derivatives of each order taken separately shall converge uniformly.

2. DISTRIBUTIONS

Definition 2. *A continuous linear functional on the vector space $\mathfrak{D}$ is called a distribution* T.

This means that to each $\varphi \in \mathfrak{D}$, T assigns a complex number $T(\varphi)$, which will also be denoted by $\langle T, \varphi \rangle$, with the properties

(II, 1; 8)
$$\begin{cases} T(\varphi_1 + \varphi_2) = T(\varphi_1) + T(\varphi_2). \\ T(\lambda\varphi) = \lambda T(\varphi), \text{ where } \lambda \text{ is any complex constant.} \\ \text{If } \varphi_j \text{ converges to } \varphi \text{ as } j \to \infty \text{ in the sense of the topology of } \mathfrak{D}, \text{ the complex numbers } T(\varphi_j) \text{ converge to the complex number } T(\varphi) \text{ as } j \to \infty. \end{cases}$$

The distributions themselves form a vector space $\mathfrak{D}'$ with the sum $T_1 + T_2$ and the product λT defined by

(II, 1; 9)
$$\langle T_1 + T_2, \varphi \rangle = \langle T_1, \varphi \rangle + \langle T_2, \varphi \rangle, \quad \langle \lambda T, \varphi \rangle = \lambda \langle T, \varphi \rangle$$

so that the scalar product $\langle T, \varphi \rangle$, for $T \in \mathfrak{D}'$ and $\varphi \in \mathfrak{D}$ is a bilinear form. The space $\mathfrak{D}'$ is part of the space $\mathfrak{D}^*$, the dual of $\mathfrak{D}$, the set of linear functionals, continuous or not, on $\mathfrak{D}$. The existence of linear functionals which are discontinuous on $\mathfrak{D}$ may be demonstrated mathematically using the axiom of choice. However, no explicit example of these can be cited and there is very little chance of ever meeting one in practice.

Examples of distributions

Example I. Let f be a locally summable function, that is to say a function summable over any bounded set. It defines a distribution T_f by

(II, 1; 10)
$$\langle T_f, \varphi \rangle = \iint_{R^n} \cdots \int f(x)\, \varphi(x)\, dx.$$

This integral indeed exists, for the domain of integration is in fact not R^n but the bounded support of φ and on this support f is summable and φ continuous so that $f\varphi$ is summable. Also the value of the integral is obviously a linear functional of φ.

It remains to show that this functional is continuous on $\mathfrak{D}$. Suppose that φ_j converges to φ in $\mathfrak{D}$ as $j \to \infty$. Let K be a bounded set containing the supports of all the functions φ_j. Then,

(II, 1; 11)
$$|\langle T_f, \varphi_j \rangle - \langle T_f, \varphi \rangle| \leqslant \left(\iint_K \cdots \int |f(x)|\, dx \right) \max |\varphi - \varphi_j|.$$

As $\max |\varphi_j - \varphi|$ tends to 0 as $j \to \infty$, the difference indeed converges to 0.

THEOREM 2. *Two functions f and g define the same functional $T_f = T_g$ if and only if they are equal almost everywhere.*

If f and g are equal almost everywhere, it is obvious that $\langle T_f, \varphi \rangle = \langle T_g, \varphi \rangle$ for any $\varphi \in \mathfrak{D}$. It is the converse which must be proved. Putting $f - g = h$, the proposition may be expressed as follows : *if h is a locally summable function and if*

$$\text{(II, 1; 12)} \qquad \iint_{\mathbf{R}^n} \cdots \int h(x)\, \varphi(x)\, dx = 0$$

for every $\varphi \in \mathcal{D}$, *then* $h(x)$ *is zero almost everywhere.*

1) It will be shown first that it is true that $\iint_{\mathbf{R}^n} \cdots \int h(x)\, \psi(x)\, dx = 0$ for any continuous (but not necessarily differentiable) function ψ with a bounded support. This results from the approximation theorem (Theorem 1) which states that for any $\varepsilon > 0$ and $d > 0$ there exists a function $\varphi \in \mathcal{D}$ such that $|\varphi - \psi| \leqslant \varepsilon$ and the support of φ is contained in K_d, where K is the support of ψ. Thus,

$$\text{(II, 1; 13)} \quad \left| \iint_{\mathbf{R}^n} \cdots \int (\varphi(x) - \psi(x)) h(x)\, dx \right| \leqslant \varepsilon \iint_{K_d} \cdots \int |h(x)|\, dx.$$

As $\iint \cdots \int_{\mathbf{R}^n} \varphi(x)\, h(x)\, dx = 0$ by hypothesis,

$$\text{(II, 1; 14)} \quad \left| \iint_{\mathbf{R}^n} \cdots \int \psi(x)\, h(x)\, dx \right| \leqslant \varepsilon \iint_{K_d} \cdots \int |h(x)|\, dx.$$

By fixing d and making ε tend to 0 it may be deduced that

$$\text{(II, 1; 15)} \qquad \iint_{\mathbf{R}^n} \cdots \int h(x)\, \psi(x)\, dx = 0.$$

2) Now, let $\chi(x)$ be a (measurable) bounded function, with a bounded support. It will be shown for this function $\chi(x)$ that

$$\text{(II, 1; 16)} \qquad \iint_{\mathbf{R}^n} \cdots \int h(x)\, \chi(x)\, dx = 0.$$

For this, we shall rely on the theory of the Lebesgue integral (Chapter I, Definition 3).

If χ *is a function measurable on* $\mathbf{R}^n$ *there exists a sequence of continuous functions* χ_j *which converge to* χ *almost everywhere as* $j \to \infty$. In the case under discussion where χ has a bounded support, it may be supposed that the supports of all the χ_j lie within a certain bounded set which is independent of j. If this is not so, the χ_j can be replaced by the functions $\alpha\chi_j$, where α is a fixed continuous function equal to 1 in the support of χ and itself has a bounded support.

Again, χ being bounded, the χ_j can be supposed bounded by $M = \sup |\chi|$. If this is not so then, putting

$$\chi_j(x) = \rho_j(x) e^{i\omega_j(x)},$$

the χ_j may be replaced by the functions $\sigma_j(x)e^{i\omega_j(x)}$ where

$$\sigma_j(x) = \min\ (M,\ \rho_j(x)).$$

These two conditions satisfied, it is true on the one hand that

$$\text{(II, 1; 17)} \qquad \iint_{R^n} \cdots \int h(x)\, \chi_j(x)\, dx = 0$$

since the χ_j are functions of type ψ, continuous and with bounded supports, while on the other hand, as $j \to \infty$,

(II, 1; 18)

$$\iint_{R^n} \cdots \int h(x)\, \chi_j(x)\, dx \quad \text{tends towards} \quad \iint_{R^n} \cdots \int h(x)\, \chi(x)\, dx.$$

For $\chi_j(x)$ converges nearly everywhere to $\chi(x)$ and $|h(x)\, \chi_j(x)|$ is summable, since its support is contained in a bounded set and it is bounded by $M|h(x)|$ which is locally summable. The result follows by Lebesgue's theorem, Chapter I, Theorem 41. Equation (II, 1; 16) can now be deduced.

3) Next, choose the function $\chi(x)$ to satisfy

$$\begin{cases} \chi(x) = 0 \ \text{ if } \ |x| > a \ \text{ or } \ h(x) = 0; \\ \chi(x) = e^{-i\omega(x)} \ \text{ if } \ |x| \leqslant a \ \text{ and } \ h(x) = r(x)e^{i\omega(x)} \neq 0. \end{cases}$$

χ is measurable with modulus equal to 1 or 0 and support contained in the ball $|x| \leqslant a$. Thus, by (II, 1; 16),

$$\text{(II, 1; 19)} \quad \iint_{R^n} \cdots \int h(x)\, \chi(x)\, dx = \iint_{|x| \leqslant a} \cdots \int |h(x)|\, dx = 0.$$

Hence $|h(x)|$ is a function $\geqslant 0$, whose integral over the ball $|x| \leqslant a$ is zero. Thus $h(x)$ is zero nearly everywhere in the ball $|x| \leqslant a$, and as this is true for arbitrary a it follows that, as we set out to prove, h is zero nearly everywhere.

COROLLARY. *If it is agreed to identify two locally summable functions f and g when they are equal almost everywhere (which is equivalent to discussing only classes of locally summable functions), then distributions constitute a generalization of the concept of a locally summable function. For this reason we henceforward identify a locally summable function f, defined almost everywhere, with the functional T_f which it defines, and write*

$$\text{(II, 1; 20)} \qquad \langle f, \varphi \rangle = \langle T_f, \varphi \rangle = \iint_{R^n} \cdots \int f(x)\, \varphi(x)\, dx.$$

In particular the functional which assigns the integral $\iint_{R^n} \cdots \int \varphi(x)\, dx$ to each function φ defines a distribution which will be identified with the function $f = 1$.

Example II. If f is a locally summable function and D a symbol denoting partial differentiation of any order with respect to $x_1, x_2, \ldots, x_n$ the functional

$$(\text{II, 1; 21}) \qquad \langle T, \varphi \rangle = \iint_{R^n} \cdots \int f(x)\, D\varphi(x)\, dx = \langle f, D\varphi \rangle$$

is a distribution.

Example III. The *Dirac distribution* δ is defined by

$$(\text{II, 1; 22}) \qquad \langle \delta, \varphi \rangle = \varphi(0).$$

The Dirac distribution $\delta_{(a)}$ at the point a of R^n is defined by

$$(\text{II, 1; 23}) \qquad \langle \delta_{(a)}, \varphi \rangle = \varphi(a).$$

In the same way the distributions

$$(\text{II, 1; 24}) \qquad \langle T, \varphi \rangle = D\varphi(a)$$

may be defined, D being any partial differential operator.

Mathematical distributions and charge distributions in physics

The distributions $T \in \mathfrak{D}'$ may be interpreted as distributions of electric or magnetic charge, mass, etc. Thus the Dirac distribution $\delta_{(a)}$ is interpreted as representing the mass $+1$ at the point a of R^n and the distribution associated with a locally summable function f as the charge distribution defined by the density f (the volume V containing the charge $\iint_V \cdots \int f(x)\, dx$). Many expressions of the type $\langle T, \varphi \rangle$ must be evaluated for physical reasons although φ does not in general belong to $\mathfrak{D}$ since each distribution T can be extended as a functional over a set of functions larger than $\mathfrak{D}$ but dependent on T; $\mathfrak{D}$ is the common set on which all functionals are defined. For example δ can be extended to all functions φ which are continuous at the origin, f to all functions φ such that $f\varphi$ is summable, etc. Thus to find the total charge, $\langle T, 1 \rangle$ is evaluated giving

$$\iint_{R^n} \cdots \int f(x)\, dx \text{ for } T = f, \quad +1 \text{ for } T = \delta_{(a)};$$

to find the moment of inertia with respect to the origin, $\langle T, r^2 \rangle$ is evaluated, giving

$$\iint \cdots \int_{R^n} f(x) r^2\, dx \quad \text{for } T = f, \quad |a^2| \text{ for } T = \delta_{(a)};$$

to find the Newtonian potential at the point $b \in R^3$ we evaluate $\left\langle T, \dfrac{1}{|x - b|} \right\rangle$ thus obtaining

$$\iiint_{R^3} \frac{f(x)\, dx}{|x - b|} \quad \text{if } T = f, \qquad \frac{1}{|a - b|} \quad \text{if } T = \delta_{(a)}, \ldots.$$

Consider the electric charge distribution on the line R, defined by a dipole of electric moment $+1$ placed at 0. We shall find the mathematical distribution to which it corresponds. The dipole is the "limit" of the system T_ε of two charges $1/\varepsilon$ and $-1/\varepsilon$ at the points of abscissa 0 and ε respectively, when $\varepsilon \to 0$. The system T_ε corresponds to the mathematical distribution defined by

$$\text{(II, 1; 25)} \qquad \langle T_\varepsilon, \varphi \rangle = \frac{1}{\varepsilon}\varphi(\varepsilon) - \frac{1}{\varepsilon}\varphi(0) = \frac{\varphi(\varepsilon) - \varphi(0)}{\varepsilon};$$

and when $\varepsilon \to 0$, $\langle T_\varepsilon, \varphi \rangle$ tends to $\varphi'(0)$. This suggests *defining* a dipole as the distribution

$$\text{(II, 1; 26)} \qquad \langle T, \varphi \rangle = \varphi'(0),$$

and this definition avoids any limiting process.

In R^n the electric dipole of moment $\mathcal{M}$, oriented in a given direction and situated at a point $a \in R^n$, is defined by the distribution

$$\text{(II, 1; 27)} \qquad \langle T, \varphi \rangle = \mathcal{M} D\varphi(a),$$

$D\varphi$ being the derivative of φ taken in the given direction. The analogy between mathematical and physical distributions is not provable; *mathematical distributions constitute the mathematically rigorous definition of physical distributions.*

The distributions encountered in physics suggest new mathematical distributions. For example, the distribution of charges over the surface S with surface density ρ is defined by

$$\text{(II, 1; 28)} \qquad \langle T, \varphi \rangle = \iint_S \rho(x)\varphi(x)\, dS.$$

This distribution must not be confused with the distribution defined by the *volume* density f, which is identified with the function f itself.

The normally oriented dipole distribution over the surface S, of surface moment density ρ, will be defined by

$$\text{(II, 1; 29)} \qquad \langle T, \varphi \rangle = \iint_S \rho(x) \frac{d\varphi}{dn}\, dS.$$

Following Dirac, quantum physicists use, instead of the distribution δ, a "Dirac function" satisfying

$$\text{(II, 1; 30)} \qquad \begin{cases} \delta(x) = 0 \quad \text{for} \quad x \neq 0, \\ \delta(0) = +\infty, \\ \displaystyle\iint_{R^n} \cdots \int \delta(x)\, dx = +1. \end{cases}$$

More generally a distribution over the space R^n of the variable x is often represented by $T(x)$, and then $\langle T, \varphi \rangle$ is written by convention as

$$\iint_{R^n} \cdots \int T(x)\varphi(x)\, dx.$$

The properties of the Dirac function are contradictory and such a function could not exist; for, if δ is nearly everywhere zero, its Lebesgue integral is zero. Besides, the function $k\,\delta(x)$ would take the same values as $\delta(x)$, namely 0 and $+\infty$, but its integral, which is equal to k, would not be the same. Naturally the physicists know very well that they are concerned with a symbolic device and not a " true " function. Nevertheless, it is more prudent to keep the concept of distribution and to write $\langle \delta, \varphi \rangle$ and not $\delta(x)$ if we wish to do more than find results intuitively which still have to be proved rigorously in terms of distributions.

In the same way $\delta_{(a)}$ is often written $\delta(x-a)$, and equation (II, 1; 23) is expressed as

(II, 1; 31) $$\iint_{R^n} \cdots \int \delta(x-a)\,\varphi(x)\, dx = \varphi(a).$$

3. THE SUPPORT OF A DISTRIBUTION

A distribution T is said to be zero in an open set Ω of R^n if $\langle T, \varphi \rangle = 0$ for any function φ of $\mathfrak{D}$ which has its support in Ω. We shall accept without proof the following.

Unification principle

Let $\{\Omega_i\}$ *be a finite or infinite family of open sets with union* Ω, *and let* $\{T_i\}$ *be a family of distributions depending on the same set of indices. The distribution** T_i *is defined on the open set* Ω_i, *and it is assumed that if* Ω_i *and* Ω_j *have a non-empty intersection, then* T_i *and* T_j *coincide in this intersection. Then there exists one and only one distribution* T, *defined on* Ω, *which coincides with* T_i *in each open set* Ω_i.

By application of the unification principle to the case where all the T are zero it can be seen that if a distribution is zero on a family of open sets then it is zero on their union. It follows that the union of all the open sets where the distribution T is zero is also an open set where T is zero and that it is the largest. The complement of this set will be called the *support* of T. Again, it may be said that the support of T is the smallest closed

*A distribution on an open set Ω is a continuous linear form on the subspace of $\mathfrak{D}$ formed by the functions φ which have their supports in Ω.

set outside which T is zero. A point belongs to the support of T if and only if T is not zero in any neighbourhood of this point.

Examples. If T is a continuous function its support as a distribution coincides with its support as a function. The support of Dirac's distribution for the point a, $\delta_{(a)}$, reduces to the single point a.

Note. If the support of T and the support of φ have no points in common then $\langle T, \varphi \rangle = 0$.

2. Differentiation of distributions

1. DEFINITION

We shall try to define $\partial T/\partial x_1$, the derivative with respect to the variable x_1 of a distribution T over R^n, in such a way that if T is a continuous function f with continuous derivatives, we recover $\partial f/\partial x_1$ in its usual sense.

Thus, if f is a continuously differentiable function, we have

(II, 2; 1)
$$\left\langle \frac{\partial f}{\partial x_1}, \varphi \right\rangle = \iint_{R^n} \cdots \int \frac{\partial f}{\partial x_1} \varphi \, dx.$$

For the simple case of a continuous function and a bounded domain of integration, Fubini's theorem gives

(II, 2; 2)
$$\iint_{x_2, x_3, \ldots, x_n} \int dx_2 \cdots dx_n \int_{-\infty}^{+\infty} \frac{\partial f}{\partial x_1} \varphi \, dx_1.$$

If the ordinary integral is integrated by parts, it becomes

(II, 2; 3)
$$[f\varphi]_{-\infty}^{+\infty} - \int_{-\infty}^{+\infty} f \frac{\partial \varphi}{\partial x_1} dx_1.$$

The integrated term disappears since φ is zero outside a bounded set. Thus (II, 2; 2) becomes

(II, 2; 4)
$$-\iint_{x_2, \ldots, x_n} \int dx_2 \cdots dx_n \int_{-\infty}^{+\infty} f \frac{\partial \varphi}{\partial x_1} dx_1$$
$$= -\iint_{R^n} \cdots \int f \frac{\partial \varphi}{\partial x_1} dx = -\left\langle f, \frac{\partial \varphi}{\partial x_1} \right\rangle$$

so that, finally,

(II, 2; 5)
$$\left\langle \frac{\partial f}{\partial x_1}, \varphi \right\rangle = -\left\langle f, \frac{\partial \varphi}{\partial x_1} \right\rangle.$$

Hence we are led to *define* the derivative $\partial T/\partial x_1$ of T by the equation

(II, 2; 6)
$$\left\langle \frac{\partial T}{\partial x_1}, \varphi \right\rangle = -\left\langle T, \frac{\partial \varphi}{\partial x_1} \right\rangle.$$

This is a valid definition of a distribution $\partial T/\partial x_1$; it is a linear functional of φ; again, if φ_j converges to φ in $\mathcal{D}$ as $j \to \infty$, $\partial\varphi_j/\partial x_1$ converges to $\partial\varphi/\partial x_1$ in $\mathcal{D}$ by the definition of this type of convergence and, since T is a distribution, $\langle T, \partial\varphi_j/x_1\rangle$ converges to $\langle T, \partial\varphi/\partial x_1\rangle$; this proves that $\langle \partial T/\partial x_1, \varphi_j\rangle$ converges to $\langle \partial T/\partial x_1, \varphi\rangle$ so that $\partial T/\partial x_1$ is indeed a distribution. All derivatives $\partial T/\partial x_i$ can be treated similarly.

Next, the second-order derivative $\partial^2 T/\partial x_i\,\partial x_j$ must be considered. We have

(II, 2; 7)
$$\left\langle \frac{\partial^2 T}{\partial x_i\,\partial x_j}, \varphi\right\rangle = -\left\langle \frac{\partial T}{\partial x_j}, \frac{\partial \varphi}{\partial x_i}\right\rangle = +\left\langle T, \frac{\partial^2 \varphi}{\partial x_j\,\partial x_i}\right\rangle,$$
$$\left\langle \frac{\partial^2 T}{\partial x_j\,\partial x_i}, \varphi\right\rangle = -\left\langle \frac{\partial T}{\partial x_i}, \frac{\partial \varphi}{\partial x_j}\right\rangle = +\left\langle T, \frac{\partial^2 \varphi}{\partial x_i\,\partial x_j}\right\rangle.$$

As φ has continuous second-order derivatives, it is known that

$$\partial^2\varphi/\partial x_i\,\partial x_j = \partial^2\varphi/\partial x_j\,\partial x_i,$$

and it follows that

(II, 2; 8)
$$\frac{\partial^2 T}{\partial x_i\,\partial x_j} = \frac{\partial^2 T}{\partial x_j\,\partial x_i}.$$

More generally, let $p = (p_1, p_2, \ldots, p_n)$ be a set of n integers $\geqslant 0$. Denote by D^p the differentiation

$$\left(\frac{\partial}{\partial x_1}\right)^{p_1}\left(\frac{\partial}{\partial x_2}\right)^{p_2}\cdots\left(\frac{\partial}{\partial x_n}\right)^{p_n},$$

and put $p_1 + p_2 + \cdots + p_n = |p|$, the order of the differentiation. Then we have the equation

$$\langle D^p T, \varphi\rangle = (-1)^{|p|}\langle T, D^p\varphi\rangle,$$

whence :

THEOREM 3. *Any distribution* T *has successive derivatives of all orders, and the order of differentiation may be changed. We have*

(II, 2; 9)
$$\langle D^p T, \varphi\rangle = (-1)^{|p|}\langle T, D^p\varphi\rangle,$$
$$D^p = \prod_i \left(\frac{\partial}{\partial x_i}\right)^{p_i}, \qquad |p| = \sum_i p_i.$$

Note. In particular, each continuous or even locally summable function f has successive derivatives of all orders in the distribution sense; but in general, these are not functions. The relation of distributions to functions is a little like that of complex numbers to real ones; any algebraic equation with real or complex coefficients has complex roots and any locally summable function or any distribution has successive derivatives of all orders which are distributions.

Generalization

Let D be a differential operator with constant coefficients

(II, 2; 10) $$D = \sum_p A_p D^p.$$

The operator ${}^tD$ defined by

(II, 2; 11) $${}^tD = \sum_p (-1)^{|p|} A_p D^p$$

is called the *transposed* (or sometimes *adjoint*) differential operator. Then

(II, 2; 12) $$\langle DT, \varphi\rangle = \langle T, {}^tD\varphi\rangle.$$

It can be seen that ${}^{tt}D = D$ and hence

(II, 2; 13) $$\langle {}^tDT, \varphi\rangle = \langle T, D\varphi\rangle.$$

For example, if D is the Laplacian $\Delta = \Sigma(\partial^2/\partial x_i^2)$,

(II, 2; 14) $$\langle \Delta T, \varphi\rangle = \langle T, \Delta\varphi\rangle.$$

2. EXAMPLES OF DERIVATIVES IN THE ONE-DIMENSIONAL CASE

If f is a continuously differentiable function its derivative in the distribution sense is, by equation (II, 2; 5), identical with its derivative in the usual sense.

Discontinuous functions

Take for T the *Heaviside function* Y, equal to 0 for $x < 0$ and to $+1$ for $x > 0$ (it is not necessary to define it for $x = 0$, which is a set of measure zero). This fundamental function of operational calculus is sometimes known as the " unit step function ".

(II, 2; 15) $$\langle Y', \varphi\rangle = -\langle Y, \varphi'\rangle = -\int_{-\infty}^{+\infty} Y(x)\varphi'(x)\,dx$$
$$= -\int_0^{\infty} \varphi'(x)\,dx = -(\varphi(x))_{\substack{x=\infty\\x=0}} = \varphi(0) = \langle \delta, \varphi\rangle.$$

Thus,

(II, 2; 16) $$Y' = \delta.$$

δ is called the " unit impulse " by electrical engineers. It can be seen that the finite discontinuity of Y appears in the form of a " point mass " in its derivative.

The following derivatives are the successive derivatives of δ.

(II, 2; 17) $$\langle \delta', \varphi \rangle = -\langle \delta, \varphi' \rangle = -\varphi'(0).$$

(II, 2; 18) $$\langle \delta^{(m)}, \varphi \rangle = (-1)^m \varphi^{(m)}(0).$$

δ' is a dipole of moment -1 at the origin.

Generalization. Let f be an infinitely differentiable function (in the usual theory-of-functions sense) for $x < 0$ and $x > 0$, such that each of its derivatives has a right-hand and a left-hand limit at $x = 0$. The "jump" in the m-th derivative $f^{(m)}$ at $x = 0$, i.e. $f^{(m)}(0+) - f^{(m)}(0-)$, will be denoted by σ_m. The derivatives of f taken in the distribution sense will be denoted by $f', f'', \ldots$, and the distributions represented by the functions equal to the derivatives of f in the usual sense for $x < 0$ and $x > 0$, and not defined for $x = 0$, will be denoted by $\{f'\}$, $\{f''\}$. (For example, if $f = \mathrm{Y}$, $f' = \delta$ and $\{f'\} = 0$.)

(II, 2; 19)
$$\langle f', \varphi \rangle = -\langle f, \varphi' \rangle = -\int_{-\infty}^{+\infty} f(x)\varphi'(x)\,dx = -\int_{-\infty}^{0} - \int_{0}^{+\infty};$$

(II, 2; 20) $$-\int_{0}^{\infty} f(x)\varphi'(x)\,dx = -\left(f(x)\varphi(x)\right)_{x=0}^{x=+\infty} + \int_{0}^{\infty} f'(x)\varphi(x)\,dx$$
$$= f(0+)\varphi(0) + \int_{0}^{\infty} f'(x)\varphi(x)\,dx.$$

Similarly,

(II, 2; 21) $$-\int_{-\infty}^{0} f(x)\varphi'(x)\,dx = -f(0-)\varphi(0) + \int_{-\infty}^{0} f'(x)\varphi(x)\,dx.$$

Adding (II, 2; 20) and (II, 2; 21),

(II, 2; 22) $$-\langle f, \varphi' \rangle = \sigma_0\varphi(0) + \int_{-\infty}^{+\infty} f'(x)\varphi(x)\,dx,$$

(II, 2; 23) $$f' = \{f'\} + \sigma_0\delta.$$

Here again the discontinuity in f appears in the derivative in the form of a point mass. Successive differentiation gives

THEOREM 4.

(II, 2; 24) $$\left\{\begin{array}{l} f' = \{f'\} + \sigma_0\delta, \\ f'' = \{f''\} + \sigma_0\delta' + \sigma_1\delta, \\ \cdots\cdots\cdots\cdots\cdots\cdots\cdots \\ f^{(m)} = \{f^{(m)}\} + \sigma_0\delta^{(m-1)} + \sigma_1\delta^{(m-2)} + \cdots + \sigma_{m-1}\delta. \end{array}\right.$$

For example, the function equal to 0 for $x < 0$ and to $\cos x$ (or $\sin x$) for $x > 0$ has for derivative the function equal to 0 for $x < 0$ and to $-\sin x$ (or cos x) for $x > 0$, with the additional term δ (or 0). This may be written

(II, 2; 25)
$$(Y(x)\cos x)' = -Y(x)\sin x + \delta,$$
$$(Y(x)\sin x)' = Y(x)\cos x.$$

The distribution pv $1/x$.

The function $1/x$ does not define a distribution because it is not summable in the neighbourhood of $x = 0$. However, it is known that the integral

(II, 2; 26)
$$\mathrm{pv}\int_{-\infty}^{+\infty}\frac{\varphi(x)}{x}\,dx = \lim_{\varepsilon\to 0}\int_{|x|\geqslant\varepsilon}$$

exists if $\varphi \in \mathscr{D}$ (and indeed if φ has a bounded support and is once differentiable at the origin); " pv " means Cauchy principal value*. A linear form in φ is thus defined and this form is continuous. To show this, assume that φ_j converges to φ in $\mathscr{D}$ as $j \to \infty$ and let $(-A, +A)$ be an interval which contains the supports of the φ_j. (It may be supposed that φ is zero since the sequence may be changed to $\varphi_j - \varphi$.) Now,

(II, 2; 27)
$$\mathrm{pv}\int_{-\infty}^{+\infty}\frac{\varphi_j}{x}\,dx = \varphi_j(0)\,\mathrm{pv}\int_{-A}^{+A}\frac{dx}{x} + \mathrm{pv}\int_{-A}^{+A}\frac{\varphi_j(x)-\varphi_j(0)}{x}\,dx.$$

The first term is zero as $\mathrm{pv}\int_{-A}^{+A} dx/x = 0$ (because $1/x$ is an odd function). In the second term the notation " pv " may be omitted because the integral is summable in the neighbourhood of $x = 0$. The mean value theorem in the form

(II, 2; 28)
$$\left|\frac{\varphi_j(x)-\varphi_j(0)}{x}\right| \leqslant \max|\varphi_j'|$$

shows that

(II, 2; 29)
$$\left|\mathrm{pv}\int_{-\infty}^{+\infty}\frac{\varphi_j}{x}\,dx\right| \leqslant 2A\max|\varphi_j'|,$$

which tends to 0 as $j \to \infty$ by virtue of the definition of the convergence of a sequence φ_j to 0 in $\mathscr{D}$. The distribution thus defined will be denoted by pv $1/x$, so that

(II, 2; 30)
$$\left\langle \mathrm{pv}\frac{1}{x}, \varphi\right\rangle = \mathrm{pv}\int_{-\infty}^{+\infty}\frac{\varphi(x)}{x}\,dx.$$

In quantum mechanics frequent use is made of the two distributions

(II, 2; 31)
$$\left.\begin{aligned}\delta^+ &= \frac{\delta}{2} + \frac{1}{2i\pi}\,\mathrm{pv}\frac{1}{x}\\ \delta^- &= \frac{\delta}{2} - \frac{1}{2i\pi}\,\mathrm{pv}\frac{1}{x}\end{aligned}\right\}\ \delta = \delta^+ + \delta^-.$$

---

*Chapter I, p. 45.

It may be noted that $\log|x|$, being locally summable, defines a distribution and that

$$(\text{II, 2; 32}) \quad \langle(\log|x|)', \varphi\rangle = -\langle\log|x|, \varphi'\rangle = -\int_{-\infty}^{+\infty}(\log|x|)\,\varphi'(x)\,dx = -\int_{-\infty}^{0} - \int_{0}^{+\infty};$$

$$(\text{II, 2; 33}) \quad -\int_{0}^{+\infty}\log x\,\varphi'(x)\,dx = -\lim_{\varepsilon\to 0}\int_{\varepsilon}^{\infty}(\log x)\varphi'(x)\,dx$$
$$= \lim_{\varepsilon\to 0}\left[(-(\log x)\varphi(x))_{\varepsilon}^{+\infty} + \int_{\varepsilon}^{\infty}\frac{\varphi(x)}{x}dx\right]$$
$$= \lim_{\varepsilon\to 0}\left[(\log\varepsilon)\,\varphi(\varepsilon) + \int_{\varepsilon}^{\infty}\frac{\varphi(x)}{x}dx\right].$$

But $(\log\varepsilon)\,(\varphi(\varepsilon) - \varphi(0))$ tends to 0 as $\varepsilon \to 0$ since

$$|\varphi(\varepsilon) - \varphi(0)| \leqslant \varepsilon \max|\varphi'|$$

and $\varepsilon\log\varepsilon \to 0$. Thus, replacing $(\log\varepsilon)\varphi(\varepsilon)$ by

$$(\log\varepsilon)\varphi(0) + (\log\varepsilon)(\varphi(\varepsilon) - \varphi(0)),$$

and then by $(\log\varepsilon)\varphi(0)$, we obtain

$$(\text{II, 2; 34}) \quad -\int_{0}^{+\infty}(\log|x|)\varphi'(x)\,dx = \lim_{\varepsilon\to 0}\left[(\log\varepsilon)\varphi(0) + \int_{\varepsilon}^{\infty}\frac{\varphi(x)}{x}dx\right].$$

Similarly

$$(\text{II, 2; 35}) \quad -\int_{-\infty}^{0}(\log|x|)\varphi'(x)\,dx = \lim_{\varepsilon\to 0}\left[-(\log\varepsilon)\,\varphi(0) + \int_{-\infty}^{-\varepsilon}\frac{\varphi(x)}{x}dx\right].$$

Adding the last two results, we have

(II, 2; 36)

$$-\int_{-\infty}^{+\infty}(\log|x|)\varphi'(x)\,dx = \lim_{\varepsilon\to 0}\left[\int_{|x|\geqslant\varepsilon}\frac{\varphi(x)}{x}dx\right] = \text{pv}\int_{-\infty}^{+\infty}\frac{\varphi(x)}{x}dx,$$

whence, finally,

$$(\text{II, 2; 37}) \qquad (\log|x|)' = \text{pv}\frac{1}{x}.$$

3. EXAMPLES OF DERIVATIVES IN THE CASE OF SEVERAL VARIABLES

Let f be a function which is infinitely differentiable in the complement of a regular hypersurface (S) and such that at every point of (S) each partial derivative has a limit on each side of (S). The difference between these limits is the "jump" or increment in the corresponding partial derivative which is uniquely defined only if the sense in which the surface is crossed is given, and changes its sign when this sense is reversed. This increment is a function defined on the surface (S). As when $n = 1$, denote by $D^p f$ a derivative of f in the distribution sense and by $\{D^p f\}$ the distribution represented by the usual derived function, defined for $x \notin (S)$ and not defined for $x \in (S)$. As (S) is a surface it is a set of measure zero.

$$\text{(II, 2; 38)} \quad \left\langle \frac{\partial f}{\partial x_1}, \varphi \right\rangle = -\left\langle f, \frac{\partial \varphi}{\partial x_1} \right\rangle = -\iint_{R^n} \cdots \int f(x) \frac{\partial \varphi}{\partial x_1} dx$$
$$= -\int_{x_2 \ldots x_n} \cdots \int dx_2 \cdots dx_n \int_{-\infty}^{+\infty} f(x) \frac{\partial \varphi}{\partial x_1} dx_1.$$

By (II, 2; 22) this is found to be equal to

$$\text{(II, 2; 39)} \quad \int_{x_2 \ldots x_n} \cdots \int dx_2 \cdots dx_n \left[\sigma_0 \varphi + \int_{-\infty}^{+\infty} \frac{\partial f}{\partial x_1} \varphi \, dx_1 \right],$$

σ_0 being the discontinuity of f when (S) is crossed in the sense of the x_1 axis, evaluated at the point of intersection of (S) with a line parallel to the x_1 axis with coordinates $x_2, \ldots, x_n$; the function φ in the product $\sigma_0 \varphi$ is evaluated at the same point. Hence

$$\text{(II, 2; 40)} \quad \left\langle \frac{\partial f}{\partial x_1}, \varphi \right\rangle = \int_{(S)} \cdots \int \sigma_0 \varphi \, dx_2 \cdots dx_n + \iint \cdots \int_{R^n} \frac{\partial f}{\partial x_1} \varphi \, dx.$$

The surface integral is equivalent to

$$\text{(II, 2; 41)} \quad \int_{(S)} \cdots \int \sigma_0 \varphi \cos \theta_1 \, dS,$$

θ_1 being the angle of the x_1 axis with the normal to (S), taken in the sense in which the surface is crossed (i.e. increasing x_1). In this form, however, the integral is seen to be independent of the sense in which the surface is crossed since $\cos \theta_1$ and σ_0 both change sign when this sense is reversed; the essential point is to use the same sense in evaluating the increment in f and the orientation of the normal. The distribution

$$\text{(II, 2; 42)} \quad \langle T, \varphi \rangle = \int_{(S)} \cdots \int \sigma_0 \varphi \cos \theta_1 \, dS$$

corresponds to masses placed on (S) with a surface density $\sigma_0 \cos \theta_1$; this distribution may be denoted by $(\sigma_0 \cos \theta_1)\delta_{(S)}$. Then for any differentiation $\partial/\partial x_i$

$$\text{(II, 2; 43)} \quad \frac{\partial f}{\partial x_i} = \left\{ \frac{\partial f}{\partial x_i} \right\} + (\sigma_0 \cos \theta_i)\delta_{(S)},$$

which is a generalization of (II, 2; 23). Differentiating again,

$$\text{(II, 2; 44)} \quad \frac{\partial^2 f}{\partial x_i^2} = \left\{ \frac{\partial^2 f}{\partial x_i^2} \right\} + \frac{\partial}{\partial x_i} [(\sigma_0 \cos \theta_i)\delta_{(S)}] + \sigma_i \cos \theta_i \delta_{(S)},$$

σ_i being the increment in $\partial f/\partial x_i$ on crossing (S). From this may be deduced

$$\Delta f = \sum_i \frac{\partial^2 f}{\partial x_i^2}$$

whose transformation will now be considered.

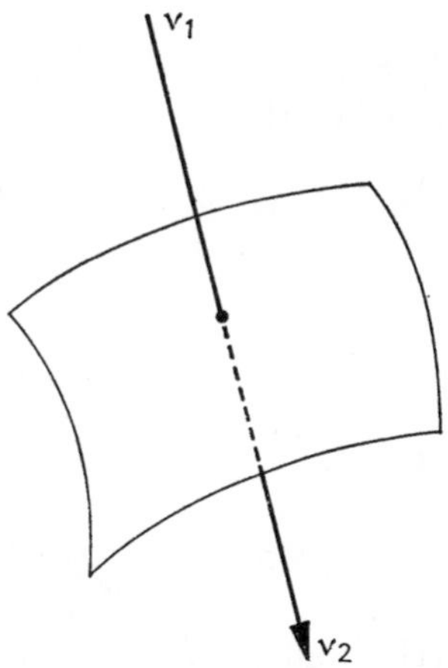

Figure (II, 1)

a) $\sum_i \sigma_i \cos \theta_i$ is the increment, denoted by σ_ν, of

$$\sum_i \cos \theta_i \frac{\partial f}{\partial x_i} = \partial f/\partial \nu,$$

the normal derivative. The increment in the normal derivative is independent of the direction of the normal; if this direction is reversed, then there is a change of sign in the evaluation of the increment and in that of the normal derivative itself, so that the increment of the normal derivative remains unchanged. Without introducing a sense of crossing the surface, this increment may be defined as $\left\{\frac{\partial f}{\partial \nu_1}\right\} + \left\{\frac{\partial f}{\partial \nu_2}\right\}$, the sum of the normal derivatives on both sides of (S).

b)

(II, 2; 45)
$$\left\langle \sum_i \frac{\partial}{\partial x_i} (\sigma_0 \cos \theta_i \delta_{(S)}), \varphi \right\rangle = -\int \cdots \int_S \sum_i \cos \theta_i \frac{\partial \varphi}{\partial x_i} \sigma_0 \, dS = -\int \cdots \int_S \frac{\partial \varphi}{\partial \nu} \sigma_0 \, dS.$$

This expression does not depend on a choice of the direction of the normal since if this direction is reversed the signs of σ_0 and $\partial \varphi/\partial \nu$ both change. The corresponding distribution is formed by dipoles directed along the normal $\vec{\nu}$, with a surface moment density $-\sigma_0$. This distribution can be represented by $\frac{\partial}{\partial \nu}(\sigma_0 \delta_{(S)})$, or by

$$\frac{\partial}{\partial \nu_1}(f_1 \delta_{(S)}) + \frac{\partial}{\partial \nu_2}(f_2 \delta_{(S)}),$$

an expression for which no choice of sense in crossing the surface is necessary.

Finally, the equation which generalizes the second equation of (II, 2; 24) is

(II, 2; 46)
$$\Delta f = \{\Delta f\} + \left(\left\{\frac{\partial f}{\partial \nu_1}\right\} + \left\{\frac{\partial f}{\partial \nu_2}\right\}\right)\delta_{(S)} + \frac{\partial}{\partial \nu_1}(f_1 \delta_{(S)}) + \frac{\partial}{\partial \nu_2}(f_2 \delta_{(S)})$$
$$= \{\Delta f\} + \sigma_\nu \delta_{(S)} + \frac{\partial}{\partial \nu}(\sigma_0 \delta_{(S)}).$$

Particular case: Green's theorem

If (S) is the boundary of the volume V and f is zero outside V then (II, 2; 46) can be written

$$\text{(II, 2; 47)}\quad \langle \Delta f, \varphi\rangle = \langle f, \Delta\varphi\rangle = \iint_V \cdots \int f\,\Delta\varphi\, dx$$
$$= \langle \{\Delta f\}, \varphi\rangle + \left\langle \frac{\partial f}{\partial \nu_i}\delta_{(S)}, \varphi\right\rangle + \left\langle \frac{\partial}{\partial \nu_i}(f\delta_{(S)}), \varphi\right\rangle$$
$$= \iint_V \cdots \int \varphi\Delta f\, dx + \int_S \cdots \int \frac{\partial f}{\partial \nu_i}\varphi\, dS - \int_S \cdots \int f\frac{\partial \varphi}{\partial \nu_i}\, dS$$

or

THEOREM 5. *Green's theorem.*

$$\text{(II, 2; 48)}\quad \iint_V \cdots \int (f\,\Delta\varphi - \varphi\Delta f)\, dx + \int_S \cdots \int \left(f\frac{\partial \varphi}{\partial \nu_i} - \varphi\frac{\partial f}{\partial \nu_i}\right) dS = 0,$$

ν_i *being the inward normal.*

Conversely, equation (II, 2; 46) could be deduced from Green's theorem (II, 2; 48).

The Laplacian of the function $\frac{1}{r^{n-2}}$, $r = \sqrt{\sum_i x_i^2}$ in R^n

The function $1/r^{n-2}$ is harmonic in the complement of the origin. For if f is a function depending only on r then (in the theory-of-function sense)

$$\text{(II, 2; 49)}\quad \frac{\partial f}{\partial x_i} = \frac{df}{dr}\frac{dr}{dx_i} = \frac{df}{dr}\frac{x_i}{r},$$
$$\frac{\partial^2 f}{\partial x_i^2} = \frac{d^2 f}{dr^2}\left(\frac{x_i}{r}\right)^2 + \frac{df}{dr}\left(\frac{1}{r} - \frac{x_i^2}{r^3}\right).$$
$$\text{(II, 2; 50)}\quad \Delta f = \frac{d^2 f}{dr^2} + \frac{n-1}{r}\frac{df}{dr}.$$

A harmonic function of r, for which $\Delta f = 0$, thus satisfies the (Euler-type) differential equation

$$\text{(II, 2; 51)}\quad \Delta f = \frac{d^2 f}{dr^2} + \frac{n-1}{r}\frac{df}{dr} = 0,$$

which has as solutions

$$\text{(II, 2; 52)}\quad \begin{cases} f = \dfrac{A}{r^{n-2}} + B, & \text{A, B constant, if } n \neq 2; \\ f = A\log\dfrac{1}{r} + B, & \text{A, B constant, if } n = 2. \end{cases}$$

A constant function is harmonic everywhere, but $1/r^{n-2}$ (for $n \neq 2$) and $\log 1/r$ (for $n = 2$) have a singularity at the origin. They are still summable in the neighbourhood of the origin (since $n-2<n$) and thus define distributions. We proceed to evaluate the Laplacians of these distributions and must expect to obtain distributions concentrated at the origin.

First method

$$\text{(II, 2; 53)} \quad \left\langle \Delta \frac{1}{r^{n-2}}, \varphi \right\rangle = \left\langle \frac{1}{r^{n-2}}, \Delta\varphi \right\rangle = \iint_{\mathrm{R}^n} \cdots \int \frac{1}{r^{n-2}} \Delta\varphi \, dx$$
$$= \lim_{\varepsilon \to 0} \iint_{r \geqslant \varepsilon} \cdots \int \frac{1}{r^{n-2}} \Delta\varphi \, dx.$$

The integral $\iint_{r\geqslant\varepsilon} \cdots \int \frac{1}{r^{n-2}} \Delta\varphi \, dx$ can be evaluated by using Green's formula (II, 2; 48) for the volume V_ε defined by the condition $r \geqslant \varepsilon$ and thus bounded by the sphere $(S_\varepsilon) : r = \varepsilon$. An alternative approach, leading to the same results, is to remark that the integral represents

$$\langle \rho_\varepsilon, \Delta\varphi \rangle = \langle \Delta\rho_\varepsilon, \varphi \rangle$$

where ρ_ε is the function equal to 0 for $r < \varepsilon$, $\frac{1}{r^{n-2}}$ for $r > \varepsilon$, which is discontinuous on the surface $r = \varepsilon$, and to apply (II, 2; 46). Remembering that the sense of crossing the surface and that of the normal are toward the interior of the domain so that $\partial/\partial\nu_i = \partial/\partial r$, and noting that

$$\left\{ \Delta \frac{1}{r^{n-2}} \right\} = 0,$$

we find that

$$\text{(II, 2; 54)} \quad \iint_{r\geqslant\varepsilon} \cdots \int \frac{1}{r^{n-2}} \Delta\varphi \, dx = -\int_{r=\varepsilon} \cdots \int \frac{1}{\varepsilon^{n-2}} \frac{\partial\varphi}{\partial r} dS$$
$$+ \int_{r=\varepsilon} \cdots \int \frac{-(n-2)}{\varepsilon^{n-1}} \varphi \, dS.$$

When $\varepsilon \to 0$, $\partial\varphi/\partial r = \sum_i \frac{x_i}{\nu} \frac{\partial\varphi}{\partial x_i}$ remains bounded and the area of the surface of integration is $S_n \varepsilon^{n-1}$, S_n being the area of the unit sphere in R^n, so that the first surface integral is of the order of ε and tends to 0 with ε. The second one can be written

$$\text{(II, 2; 55)} \quad \int_{r=\varepsilon} \cdots \int \left(\frac{-(n-2)}{\varepsilon^{n-1}} \right) \varphi(0) \, dS$$
$$+ \int_{r=\varepsilon} \cdots \int \left(\frac{-(n-2)}{\varepsilon^{n-1}} \right) (\varphi(x) - \varphi(0)) \, dS.$$

In (II, 2; 55) the value of the first integral is $-(n-2)\varphi(0)S_n$. The second integral, since

$$|\varphi(x)-\varphi(0)| \leqslant |x|\sqrt{n} \max_{\substack{|x|\leqslant r \\ i=1,2,\ldots,n}} \left|\frac{\partial\varphi}{\partial x_i}\right| = \varepsilon\sqrt{n}\max\left|\frac{\partial\varphi}{\partial x_i}\right|$$

is of the order of $\frac{\varepsilon\cdot\varepsilon^{n-1}}{\varepsilon^{n-1}} = \varepsilon$ and thus tends to 0 with ε. Thus finally

$$\left\langle \Delta\frac{1}{r^{n-2}}, \varphi\right\rangle = -(n-2)S_n\varphi(0), \tag{II, 2; 56}$$

or

$$\Delta\frac{1}{r^{n-2}} = -(n-2)S_n\delta = -(n-2)\frac{2\pi^{\frac{n}{2}}}{\Gamma\left(\frac{n}{2}\right)}\delta. \tag{II, 2; 57}$$

Second method. Denote by $\bar{\psi}(r)$ the mean of ψ on the sphere of radius r. Evaluating the multiple integral by surface integration over a sphere of radius r followed by integration with respect to* r, and using

$$\int_{|x|=r}\cdots\int \psi(x)\,dS = S_n r^{n-1}\bar{\psi}(r),$$

we find that

$$\left\langle \Delta\frac{1}{r^{n-2}}, \varphi\right\rangle = \left\langle \frac{1}{r^{n-2}}, \Delta\varphi\right\rangle = \iint_{R^n}\cdots\int \frac{1}{r^{n-2}}\Delta\varphi\,dx \tag{II, 2; 58}$$
$$= S_n\int_0^{+\infty}\frac{r^{n-1}}{r^{n-2}}\overline{\Delta\varphi}(r)\,dr.$$

It is known that the Laplacian is invariant with respect to rotation about the origin. From this it can be deduced that

$$\overline{\Delta\varphi} = \Delta\bar{\varphi} = \frac{d^2}{dr^2}\bar{\varphi}(r) + \frac{n-1}{r}\frac{d}{dr}\bar{\varphi}(r) \tag{II, 2; 59}$$

(which we shall assume without proof). Then (II, 2; 58) becomes

$$S_n\int_0^{+\infty} r\left(\bar{\varphi}'' + \frac{n-1}{r}\bar{\varphi}'\right)dr = S_n\int_0^{+\infty}[(r\bar{\varphi}')' + (n-2)\bar{\varphi}']\,dr \tag{II, 2; 60}$$
$$= S_n[r\bar{\varphi}' + (n-2)\bar{\varphi}]_0^{+\infty}$$
$$= -(n-2)S_n\bar{\varphi}(0)$$
$$= -(n-2)S_n\varphi(0),$$

which is equivalent to (II, 2; 56), from which (II, 2; 57) can again be deduced.

---

*Chapter I, Theorem 31, page 38.

For $n=2$, $1/r^{n-2}=1$ has a zero Laplacian; it is $\log 1/r$ whose Laplacian is concentrated at the origin and this may be evaluated in an analogous way to the Laplacians for $n \neq 2$. Summarizing :

THEOREM 6. $\Delta(1/r^{n-2}) = -(n-2)S_n\delta$, *for n a positive integer not equal to* 2, *and for* $n = 2$, $\Delta \log 1/r = -2\pi\delta$. *Particular cases are*

(II, 2; 61)
$$\begin{cases} n = 1 : \Delta|x| = 2\delta, \\ n = 3 : \Delta \dfrac{1}{r} = -4\pi\delta. \end{cases}$$

Applications of this equation will be met in the theory of Newtonian potentials. It is known that the potential of a charge distribution of density f satisfies Poisson's equation $\Delta U = -4\pi f$, if f is sufficiently regular. If instead of a charge distribution of density f there is a unit point charge $+1$ placed at the origin, equivalent to δ, its potential is $U = 1/r$ and this also satisfies Poisson's equation, which in this case is (II, 2; 61) for $n = 3$. In fact, it is by starting from Poisson's equation for a point charge that it will be proved in the general case.

3. Multiplication of distributions

The multiplication ST of two arbitary distributions S and T is not possible. Thus, if f is a locally summable function it defines a distribution, but there is no reason for f^2 to be also locally summable (example : for $n = 1$, $f(x) = 1/\sqrt{|x|}$, $f^2 = 1/|x|$) and thus it will not define a distribution. The more "irregular" T is then the more "regular" S must be for the product to have a meaning. We shall confine ourselves to defining the product when one of the two distributions is arbitrary and the other an infinitely differentiable function in the usual sense. Thus if α is such a function we shall define αT, $T \in \mathcal{D}'$. Suppose at first that $T = f$, a locally summable function. It will be required that in this case αf should be the product in the usual sense;

(II, 3; 1)
$$\begin{aligned} \langle \alpha f, \varphi \rangle &= \iint_{R^n} \cdots \int (\alpha(x) f(x)) \varphi(x)\, dx \\ &= \iint_{R^n} \cdots \int f(x) (\alpha(x) \varphi(x))\, dx = \langle f, \alpha\varphi \rangle. \end{aligned}$$

This suggests the following definition of αT for arbitrary T :

THEOREM 7. *If* T *is an arbitrary distribution and* α *a function which is infinitely differentiable in the usual sense, there exists a product* αT *defined by*

(II, 3; 2)
$$\langle \alpha T, \varphi \rangle = \langle T, \alpha\varphi \rangle.$$

This is a valid definition of αT as a new distribution. For the right-hand side has a meaning and depends linearly on φ, since, by Leibnitz's theorem concerning the successive derivatives of a product, it is infinitely differentiable and its support, being contained in that of φ, is bounded. Again, if φ_j tends to 0 in $\mathfrak{D}$ as $j \to \infty$ its support remains within a fixed bounded set so that the same must be true for $\alpha\varphi_j$, and since each derivative of φ_j converges uniformly to zero as $j \to \infty$ so must each derivative of $\alpha\varphi_j$, again because of Leibnitz's theorem. This, in fact, is why it is essential for α to be infinitely differentiable in the usual sense. However if T is a distribution " of order m, " which means that it can be continued to a linear functional over the space $\mathfrak{D}^m$ of m-times continuously differentiable functions with bounded support (with a topology analogous to that of $\mathfrak{D}$ but demanding uniform convergence only for any derivative of order $\leqslant m$ of φ_j), then αT has a meaning if α is only m-times continuously differentiable in the usual sense.

Examples

(II, 3; 3) $$\langle \alpha\delta, \varphi\rangle = \langle \delta, \alpha\varphi\rangle = \alpha(0)\varphi(0) = \langle \alpha(0)\delta, \varphi\rangle$$

whence

(II, 3; 4) $$\alpha\delta = \alpha(0)\delta.$$

Every product in which δ occurs is proportional to δ. Note that δ is a distribution " of order 0 " and that $\alpha\delta$ is defined as long as α is continuous in the neighbourhood of the origin. In particular, for $n = 1$,

(II, 3; 5) $$x\delta = 0.$$

On the line (i.e. for $n = 1$),

(II, 3; 6) $$\langle \alpha\delta', \varphi\rangle = \langle \delta', \alpha\varphi\rangle = -(\alpha\varphi)'_{x=0}$$
$$= -\alpha(0)\varphi'(0) - \alpha'(0)\varphi(0) = \langle \alpha(0)\delta' - \alpha'(0)\delta, \varphi\rangle$$

whence

(II, 3; 7) $$\alpha\delta' = \alpha(0)\delta' - \alpha'(0)\delta.$$

In particular,

(II, 3; 8) $$x\delta' = -\delta, \qquad x^2\delta' = 0.$$

More generally,

(II, 3; 9) $$x\delta^{(m)} = -m\delta^{(m-1)}.$$

The following proposition is of importance :

THEOREM 8. *A distribution* T *on* R ($n = 1$) *satisfies*

(II, 3; 10) $$xT = 0,$$

if and only if T *is proportional to* δ :

$$\text{(II, 3; 11)} \qquad T = C\delta.$$

Alternatively, we may say that δ is, to within a factor, the unique characteristic vector for the multiplication operator x which corresponds to the characteristic value $\lambda = 0$. An arbitrary real characteristic value λ corresponds to $\delta(x - \lambda)$. It has just been shown that $x\delta = 0$ and the converse is now required.

If T satisfies (II, 3; 10) then

$$\text{(II, 3; 12)} \qquad \langle xT, \varphi \rangle = \langle T, x\varphi \rangle = 0.$$

Thus T is zero on any function $\chi \in \mathcal{D}$ of the form $\chi = x\varphi$, $\varphi \in \mathcal{D}$. However, for a function $\chi \in \mathcal{D}$ to have the form $x\varphi$ it is necessary and sufficient that it should be zero at the origin; $\chi(0) = 0$. It is obviously necessary and it is sufficient because, if $\chi(0) = 0$, the function $\varphi(x) = \chi(x)/x$ is infinitely differentiable. This can be seen at once where $x \neq 0$ and Taylor's theorem applied to χ shows that it is in fact true everywhere. Now let θ be a given function belonging to $\mathcal{D}$ such that $\theta(0) = 1$. Then, for any $\psi \in \mathcal{D}$,

$$\text{(II, 3; 13)} \qquad \psi = \lambda\theta + \chi,$$

where

$$\lambda = \psi(0), \qquad \chi(0) = 0.$$

It has been deduced that $T(\chi) = 0$ so that

$$\text{(II, 3; 14)} \qquad T(\psi) = \lambda T(\theta) = C\psi(0)$$

with $C = T(\theta)$, giving $T = C\delta$, as required.

Note. A similar proposition is true when x is replaced by any infinitely differentiable function α having a unique single root at the origin, since then x/α is infinitely differentiable and from $\alpha T = 0$ it may be deduced that $xT = \dfrac{x}{\alpha}\,\alpha T = 0$ (and vice versa).

THEOREM 9. *The product rule for derivatives holds, giving*

$$\text{(II, 3; 15)} \qquad \frac{\partial}{\partial x_i}(\alpha T) = \frac{\partial \alpha}{\partial x_i} T + \alpha \frac{\partial T}{\partial x_i}.$$

For

$$\left\langle \frac{\partial}{\partial x_i}(\alpha T), \varphi \right\rangle = -\left\langle \alpha T, \frac{\partial \varphi}{\partial x_i} \right\rangle = -\left\langle T, \alpha \frac{\partial \varphi}{\partial x_i} \right\rangle,$$

$$\text{(II, 3; 16)} \qquad \left\langle \frac{\partial \alpha}{\partial x_i} T, \varphi \right\rangle = \left\langle T, \frac{\partial \alpha}{\partial x_i} \varphi \right\rangle,$$

$$\left\langle \alpha \frac{\partial T}{\partial x_i}, \varphi \right\rangle = \left\langle \frac{\partial T}{\partial x_i}, \alpha\varphi \right\rangle = -\left\langle T, \frac{\partial}{\partial x_i}(\alpha\varphi) \right\rangle.$$

Thus, using the linearity of T, the required proof is complete if

$$-\alpha\frac{\partial\varphi}{\partial x_i}=\frac{\partial\alpha}{\partial x_i}\varphi-\frac{\partial}{\partial x_i}(\alpha\varphi), \tag{II, 3; 17}$$

which is simply the usual equation for differentiation of $\alpha\varphi$ with the terms in a different order.

4. Topology in distribution space. Convergence of distributions. Series of distributions

The distributions T_j are said to converge to the distribution T as $j\to\infty$ if, for any $\varphi\in\mathfrak{D}$, the complex numbers $\langle T_j, \varphi\rangle$ converge to the complex number $\langle T, \varphi\rangle$ as $j\to\infty$. The convergence thus introduced for the T_j regarded as functionals on $\mathfrak{D}$ is *convergence in the ordinary sense*.

It will be said that a series $\sum_{i\in I} T_i$ is summable with sum T if, for any $\varphi\in\mathfrak{D}$, the series of numbers $\sum_{i\in I}\langle T_i, \varphi\rangle$ is summable with sum $\langle T, \varphi\rangle$. If I is the set N of integers $\geqslant 0$, a convergent series is defined in a similar way.

THEOREM 10. *If* T_j *is a sequence of distributions such that, for any* $\varphi\in\mathfrak{D}$, $\langle T_j, \varphi\rangle$ *has a limit as* $j\to\infty$, *then the sequence* T_j *has a limit in* $\mathfrak{D}'$.

If $\sum_{i\in I} T_i$ *is a series of distributions such that, for any* $\varphi\in\mathfrak{D}$, *the series* $\sum_{i\in I}\langle T_i, \varphi\rangle$ *is summable, then the series* $\sum_{i\in I} T_i$ *is summable in* $\mathfrak{D}'$.
For I = N *a convergent series is similarly defined.*

Naturally, these various statements are equivalent. Consider the first one; if $\langle T_j, \varphi\rangle$ has a limit as $j\to\infty$, this limit may be written $\langle T, \varphi\rangle$. This defines T as a functional on $\mathfrak{D}$ and this functional is evidently linear. If it can be shown to be continuous then it will be a distribution and, since T_j converges to T, the theorem will be proved. However, the continuity of T is a very special phenomenon depending on the linearity of functionals and the particular properties of the space $\mathfrak{D}$; it is not indeed usual for a *simple* limit of continuous functionals to be continuous. We shall assume the continuity of T without giving the rather subtle proof.

THEOREM 11. *If as* $j\to\infty$ *the functions* f_j *converge in the ordinary sense almost everywhere to the limit function* f *and if, independently of* j, *the* f_j *are bounded above by a locally summable function* $g\geqslant 0$, *then the distributions* f_j *converge to the distribution* f.

For, by Lebesgue's theorem on the convergence of a sequence of integrals, the $\iint_{\mathbf{R}^n} \cdots \int f_j \varphi \, dx$ converge to $\iint_{\mathbf{R}^n} \cdots \int f \varphi \, dx$ as $j \to \infty$.

Particular case. If, as $j \to \infty$, the locally summable functions f_j converge to a limit f uniformly on any bounded set, then the distributions f_j converge to the distribution f.

THEOREM 12. *Differentiation is a continuous linear operation in $\mathfrak{D}'$. If the distributions T_j converge to a distribution T as $j \to \infty$, then the derivatives T'_j converge to T'. Every summable or convergent series may be differentiated term by term under the Σ sign.*

Suppose, for example, that T_j converges to T; we show that T'_j converges to T'. Now, $\langle \mathrm{T}'_j, \varphi \rangle = -\langle \mathrm{T}_j, \varphi' \rangle$, and the latter expression tends to $-\langle \mathrm{T}, \varphi' \rangle = \langle \mathrm{T}', \varphi \rangle$ as T_j converges to T. Hence, $\lim_{j \to \infty} \langle \mathrm{T}'_j, \varphi \rangle = \langle \mathrm{T}', \varphi \rangle$, which is equivalent to $\lim_{j \to \infty} \mathrm{T}'_j = \mathrm{T}'$.

Note. In contrast it is known that, if a sequence f_j of functions which are continuous and differentiable in the usual sense converges uniformly to a limit f, f is not necessarily differentiable in the usual sense. Even if it is, the f'_j may not converge to f' in the ordinary sense. However, f' always exists in the distribution sense and the f'_j always converge to f' in $\mathfrak{D}'$.

THEOREM 13. *If f_j is a function $\geqslant 0$ for $|x| \leqslant k$, k being fixed and > 0, if as $j \to \infty$, f_j converges uniformly to 0 in any set $0 < a \leqslant |x| \leqslant 1/a < \infty$ and, finally, if, as $j \to \infty$, $\iint_{|x| \leqslant a} \cdots \int f_j(x) \, dx$ converges to $+1$ for any $a > 0$, then f_j converges to δ as $j \to \infty$.*

For letting $\varphi \in \mathfrak{D}$, we have

$$
\text{(II, 4; 1)} \qquad
\begin{aligned}
\langle f_j, \varphi \rangle &= \iint_{\mathbf{R}^n} \cdots \int f_j(x)\, \varphi(x) \, dx \\
&= \iint_{|x| \leqslant a} \cdots \int \varphi(0) f_j(x) \, dx \\
&+ \iint_{|x| \leqslant a} \cdots \int (\varphi(x) - \varphi(0)) f_j(x) \, dx \\
&+ \iint_{|x| > a} \cdots \int \varphi(x) f_j(x) \, dx.
\end{aligned}
$$

The second term has an upper bound. First, note that

$$
\text{(II, 4; 2)} \qquad |\varphi(x) - \varphi(0)| \leqslant |x| \sqrt{n} \max_{i=1, 2, \ldots, n} \left| \frac{\partial \varphi}{\partial x_i} \right|.
$$

Let M be the maximum modulus of the first partial derivatives of φ. The second integral is then bounded above by

$$a\sqrt{n}\mathrm{M}\iint_{|x|\leqslant a}\cdots\int|f_j(x)|\,dx.$$

If we choose $a \leqslant k$, the integral becomes

$$\iint_{|x|\leqslant a}\cdots\int f_j(x)\,dx \leqslant \iint_{|x|\leqslant k}\cdots\int f_j(x)\,dx.$$

Since $\iint_{|x|\leqslant k}\cdots\int f_j(x)\,dx$ converges to 1 as $j\to\infty$, it is bounded above by a constant K. Hence the second integral of the third expression of (II, 4; 1) is bounded above by $a\mathrm{KM}\sqrt{n}$. By choosing $a \leqslant \dfrac{\varepsilon}{3\mathrm{KM}\sqrt{n}}$ the second term of (II, 4; 1) is bounded above by $\varepsilon/3$, independently of j.

A *final* choice is now made for a in such a way that the above conditions $\left(a\leqslant k,\ a\leqslant\dfrac{\varepsilon}{3\mathrm{KM}\sqrt{n}}\right)$ are satisfied and a is small enough for the support of φ to be contained in $|x| \leqslant 1/a$. Then, since f_j converges uniformly to 0 as $j\to\infty$ for $a\leqslant|x|\leqslant 1/a$, the third term of (II, 4; 1) converges to 0. Thus, there exists j_1 such that for $j\geqslant j_1$ this third term is bounded above by $\varepsilon/3$.

Finally, the first term is equal to $\varphi(0)\iint_{|x|\leqslant a}\cdots\int f_j(x)\,dx$, and since $\iint_{|x|\leqslant a}\cdots\int f_j(x)\,dx$ converges to 1 as $j\to\infty$, this term converges to $\varphi(0)$ as $j\to\infty$ and there exists j_2 such that it differs from $\varphi(0)$ by not more than $\varepsilon/3$ if $j\geqslant j_2$. Then if $j_0 = \max(j_1, j_2)$, for $j\geqslant j_0$,

$$|\langle f_j,\ \varphi\rangle - \langle\delta,\ \varphi\rangle| \leqslant \varepsilon,$$

which is the required result.

Note. The condition $f_j\geqslant 0$ for $|x|\leqslant k$ is not necessary and may be replaced by the following: for a suitable fixed $k>0$, $\iint_{|x|\leqslant k}\cdots\int|f_j(x)|\,dx$ is bounded by K, independently of j.

Examples

1)

(II, 4; 3) $$f_\varepsilon = \begin{cases} 0 & \text{for } |x|>\varepsilon \\ \dfrac{n}{\varepsilon^n \mathrm{S}_n} & \text{for } |x|<\varepsilon \end{cases}$$

converges to δ as $\varepsilon\to 0$. For $f_\varepsilon\geqslant 0$ everywhere; when $|x|\geqslant a$, $f_\varepsilon = 0$ provided $\varepsilon < a$; and finally

(II, 4; 4) $$\iint_{|x|\leqslant a}\cdots\int f_\varepsilon(x)\,dx = \iint_{\substack{|x|\leqslant\varepsilon \\ \text{if } \varepsilon\leqslant a}}\cdots\int = \frac{n}{\varepsilon^n\mathrm{S}_n}\mathrm{V}_\varepsilon,$$

where V_ε is the volume of the ball of radius ε. This volume is just $\varepsilon^n S_n/n$ so that the integral of f_ε is always equal to 1 for small enough ε.

2)

$$f_\varepsilon = \frac{1}{\varepsilon^n(\sqrt{2\pi})^n} e^{-\frac{r^2}{2\varepsilon^2}} \tag{II, 4; 5}$$

converges to δ as $\varepsilon \to 0$.

First of all, note that $f_\varepsilon \geqslant 0$. For $|x| \geqslant a$, the function is bounded above by $\frac{1}{\varepsilon^n(\sqrt{2\pi})^n} e^{-a^2/2\varepsilon^2}$ which converges to 0 as $\varepsilon \to 0$. Finally,

$$\iint_{|x|\leqslant a} \cdots \int \frac{1}{\varepsilon^n(\sqrt{2\pi})^n} e^{-\frac{r^2}{2\varepsilon^2}} dx = \iint_{|x|\leqslant \frac{a}{\varepsilon}} \cdots \int \frac{1}{(\sqrt{2\pi})^n} e^{-\frac{r^2}{2}} dx \tag{II, 4; 6}$$

which tends, as $\varepsilon \to 0$, to

$$\iint_{\mathbf{R}^n} \cdots \int \frac{1}{(\sqrt{2\pi})^n} e^{-\frac{r^2}{2}} dx = +1. \tag{II, 4; 7}$$

It may be remarked that the function just considered can be replaced by

$$\frac{1}{\varepsilon^n} e^{-\pi\frac{r^2}{\varepsilon^2}} \tag{II, 4; 8}$$

since

$$\iint_{\mathbf{R}^n} \cdots \int e^{-\pi r^2} dx = +1. \tag{II, 4; 9}$$

THEOREM 14. *A trigonometric series* $\sum_{k=-\infty}^{+\infty} a_k e^{2i\pi kx}$ $(n = 1)$ *is summable in* $\mathcal{D}'$ *if, as* $|k| \to \infty$, *the moduli of the coefficients* a_k *are bounded above by a power* $A|k|^\alpha$ *of* $|k|$, $\alpha \geqslant 0$.

First, consider the series with the term a_0 omitted :

$$\sum_{\substack{k=-\infty \\ k\neq 0}}^{+\infty} \frac{a_k}{(2i\pi k)^{\beta+2}} e^{2i\pi kx}, \tag{II, 4; 10}$$

the integer β being $\geqslant \alpha$. Since the general term of this series is bounded above by A/k^2 it is uniformly summable on the real axis, and its sum is a continuous function f.

By differentiating term by term $\beta + 2$ times, in the distribution sense, the relation

$$\sum_{k=-\infty}^{+\infty} a_k e^{2i\pi kx} = a_0 + f^{(\beta+2)} \tag{II, 4; 11}$$

is obtained, which is the required result.

It also follows that the sum of the trigonometric series is a distribution of period 1, being the sum of the constant a_0 and of the derivative (in the distribution sense) of a continuous periodic function f.

It may be seen how numerous convergent trigonometric series are in $\mathcal{D}'$. Such series permit the representation of *distributions* by series of functions and, indeed, by series of functions, the exponentials, which are infinitely differentiable in the usual sense.

It is possible to prove that any periodic distribution may be expanded in $\mathcal{D}'$ as a trigonometric series and that the sufficient condition for convergence, $|a_k|$ bounded above as $|k| \to \infty$ by $|k|^\alpha$ for some suitable α, is also necessary.

For example, it may be shown that the series

$$\sum_{k=-\infty}^{+\infty} e^{2i\pi kx} \tag{II, 4; 12}$$

converges to the periodic distribution $\sum_{l=-\infty}^{\infty} \delta_{(l)}$ consisting of a mass $+1$ at each integer point of the real axis. By differentiation,

$$\sum_{k=-\infty}^{+\infty} (2i\pi k)^m e^{2i\pi kx} = \sum_{l=-\infty}^{+\infty} \delta_{(l)}^{(m)}. \tag{II, 4; 13}$$

5. Distributions with bounded supports

The space $\mathcal{E}$

This is the space of the complex functions on R^n which are infinitely differentiable and have arbitrary supports.

Let T be a distribution with bounded support K, and φ a function of $\mathcal{E}$. Let α be a function of $\mathcal{D}$ equal to 1 on a neighbourhood of the support K of T. Then $(\alpha\varphi) \in \mathcal{D}$ and $\langle T, \alpha\varphi \rangle$ is completely determined. It will be shown that this number is independent of the function α chosen. If β is another function of $\mathcal{D}$ equal to 1 on a neighbourhood of K, then $(\alpha - \beta)\varphi$ is a function of $\mathcal{D}$ which has its support in the complement of K. It follows that

$$\langle T, \alpha\varphi \rangle - \langle T, \beta\varphi \rangle = \langle T, (\alpha - \beta)\varphi \rangle = 0. \tag{II, 5; 1}$$

Q.E.D.

We may thus put

(II, 5; 2) $\langle T, \varphi \rangle = \langle T, \alpha\varphi \rangle$ for all $\varphi \in \mathcal{E}$.

Topology or concept of convergence on $\mathcal{E}$

A sequence φ_j in $\mathcal{E}$ is said to tend to 0 if the φ_j, together with each of their derivatives, converge uniformly to 0 on every compact set.

It may easily be seen that a distribution with bounded support is a continuous linear functional on $\mathcal{E}$. Conversely, a continuous linear functional $L(\varphi)$ on $\mathcal{E}$ defines a distribution with a bounded support. For :

1) Any function φ of $\mathcal{D}$ is in $\mathcal{E}$, and if the $\varphi_j \to 0$ in $\mathcal{D}$ then, *a fortiori*, they converge in $\mathcal{E}$. In consequence L is a continuous linear functional on $\mathcal{D}$ and thus defines a distribution $T \in \mathcal{D}'$, such that

(II, 5; 3) $L(\varphi) = \langle T, \varphi \rangle$ for any $\varphi \in \mathcal{D}$.

2) T has a bounded support; otherwise it would be possible to find a sequence of functions $\varphi_n \in \mathcal{D}$ each with its support in the set complementary to $|x| < n$, such that $\langle T, \varphi_n \rangle = 1$. However $\varphi_n \to 0$ in $\mathcal{E}$ and, as L is by hypothesis continuous on $\mathcal{E}$, $L(\varphi_n)$ converges to 0. It is thus impossible to have $\langle T, \varphi_n \rangle = L(\varphi_n) = 1$ for all n.

3) It will be shown that

(II, 5; 4) $L(\varphi) = \langle T, \varphi \rangle$ for all $\varphi \in \mathcal{E}$.

Let α_j be a sequence of functions belonging to $\mathcal{D}$, with $\alpha_j(x) = 1$ for $|x| < j$ and $\alpha_j(x) = 0$ for $|x| \geqslant 2j$. Then $\alpha_j\varphi \in \mathcal{D}$, $\alpha_j\varphi \to \varphi$ in $\mathcal{E}$ as $j \to \infty$ and, in consequence, $L(\alpha_j\varphi)$ converges to $L(\varphi)$. However, for large enough j, K is contained in $|x| < j$ so that, by (II, 5; 2),

(II, 5; 5) $\langle T, \varphi \rangle = \langle T, \alpha_j\varphi \rangle$.

However, by (II, 5; 3),

(II, 5; 6) $\langle T, \alpha_j\varphi \rangle = L(\alpha_j\varphi)$ for all j.

Thus, for large enough j,

(II, 5; 7) $\langle T, \varphi \rangle = L(\alpha_j\varphi)$,

from which the stated result follows by passage to the limit.

To summarize :

THEOREM 15. *The distributions with bounded supports form a vector space $\mathcal{E}'$ which is identical with the space of continuous linear functionals on the space $\mathcal{E}$.*

EXERCISES FOR CHAPTER II

Distributions on the line

Exercise II-1

If $Y(x)$ denotes Heaviside's function, evaluate, in the distribution sense, the expressions

$$\left(\frac{d}{dx} - \lambda\right) Y(x)e^{\lambda x},$$

$$\left(\frac{d^2}{dx^2} + \omega^2\right) \frac{Y(x) \sin \omega x}{\omega},$$

$$\frac{d^m}{dx^m} \frac{Y(x)x^{m-1}}{(m-1)!} \quad \text{for integer } m \geqslant 1.$$

Exercise II-2

Evaluate, in the distribution sense, the successive derivatives of $|x|$.

Exercise II-3

All derivatives are to be evaluated in the distribution sense.

(*a*) Find the successive derivatives of $|\cos x|$.

(*b*) Find the successive derivatives of the distribution $U(x)$ defined by

$$U(x) = \begin{cases} +1 & \text{for} \quad (2k-1)\dfrac{\pi}{2} < x < (2k+1)\dfrac{\pi}{2}, \\ -1 & \text{for} \quad (2k+1)\dfrac{\pi}{2} < x < (2k+3)\dfrac{\pi}{2} \end{cases}$$

where k takes all even integer values, positive, negative or zero.

(*c*) Obtain again the results of 1 by putting $|\cos x| = U(x) \cos x$ and applying equation (II, 3; 15).

Exercise II-4

Repeat Exercise *II*-3 with $|\cos x|$ replaced by $|\sin x|$ and $U(x)$ by the distribution $V(x)$ defined by

$$V(x) = \begin{cases} +1 & \text{for} \quad 2k\pi < x < (2k+1)\pi, \\ -1 & \text{for} \quad (2k+1)\pi < x < (2k+2)\pi, \end{cases}$$

where k takes all integer values, positive, negative or zero.

Exercise II-5

Evaluate in the distribution sense the derivatives of order 1, 2, 3, 4 of the two distributions :

$$T_1 = |x| \sin x, \qquad T_2 = |x| \cos x.$$

Exercise II-6
Find a distribution $F(t) = Y(t)f(t)$, where $Y(t)$ is Heaviside's function and $f(t)$ is a twice continuously differentiable function which satisfies, in the distribution sense, the equation

$$a\frac{d^2F}{dt^2} + b\frac{dF}{dt} + cF = m\delta + n\delta',$$

where a, b, c, m and n are given constants.
Particular cases:

1. $a = c = 1;\ b = 2;\ m = n = 1.$
2. $a = 1;\ b = 0;\ c = 4;\ m = 1;\ n = 0.$
3. $a = 1;\ b = 0;\ c = -4;\ m = 2;\ n = 1.$

Exercise II-7 (Examination question, Paris, 1958)
Let $G(x)$ be a function of the real variable x, with the following properties:

1. $G(x) = 0$ for $x < -1$ and for $x > +1$;
2. $G(x)$ is infinitely differentiable in each of the intervals $-1 < x < \xi$ and $\xi < x < +1$, ξ being a given number such that $-1 < \xi < 1$;
3. G and its derivatives have finite discontinuities at the points $x = -1$, $x = \xi$, $x = +1$.

(*a*) Obtain, in the distribution sense, $d^2G/dx^2 + \omega^2G$ (ω being a given real number) in terms of the derivatives of G in the usual sense.

(*b*) Is it possible to determine G and the constants α and β so as to satisfy

$$\frac{d^2G}{dx^2} + \omega^2G = \delta_{(\xi)} + \alpha\delta_{(-1)} + \beta\delta_{(+1)} \tag{1}$$

($\delta_{(a)}$ denoting the unit mass at point a)? Show that, unless ω is of the form $(k\pi/2)$, k an integer $\neq 0$, this problem has a *unique* solution, and evaluate G and the constants α and β.

(*c*) Let φ be an unknown function belonging to $\mathcal{D}$, and satisfying

$$\frac{d^2\varphi}{dx^2} + \omega^2\varphi = \psi(x), \qquad \varphi(-1) = L, \qquad \varphi(+1) = M, \tag{2}$$

the function ψ and the constants L and M being given. Show that from equation (1), $\varphi(\xi)$ may be evaluated for each ξ of the interval $(-1, +1)$, except for certain particular values of ω.

Exercise II-8

Consider the second-order differential equation

$$(1) \qquad \left(\frac{d^2}{dx^2} + k^2\right)\psi_k(x) = -4\pi\rho(x),$$

which we propose to solve by a method using the so-called Green's functions. The term *Green's function* is used for a solution of

$$(2) \qquad \left(\frac{d^2}{dx^2} + k^2\right)G_k(x, x') = -4\pi\delta(x - x')$$

where the source on the right-hand side is the Dirac distribution at the point x'.

The position of the point x is varied within the interval $(0, +\infty)$ and it is assumed that $\rho(x)$ is integrable over this domain.

(*a*) Show, by integration of (2), that the function $dG_k(x, x')/dx$ has at the point $x = x'$ a jump whose size may be obtained.

(*b*) If $u_1(x)$ and $u_2(x)$ are two linearly independent solutions of the homogeneous equation

$$\left(\frac{d^2}{dx^2} + k^2\right)u(x) = 0,$$

show that the function

$$\lambda u_1(x_<)u_2(x_>), \qquad \text{where} \qquad \begin{cases} x_< \text{ is the smallest of } x \text{ and } x' \\ x_> \text{ is the largest of } x \text{ and } x' \end{cases}$$

and where λ is a constant, is a Green's function.

(*c*) Show that, if $\psi_k(x)$ and $G_k(x, x')$ obey the same limiting conditions, then

$$\psi_k(x) = \int_0^\infty G_k(x, x')\rho(x')\, dx'.$$

(*d*) Take the following limiting conditions:

$$\text{(i)} \quad \psi_k(0) = 0; \qquad \text{(ii)} \quad \psi_k(x) \text{ tends to } e^{ikx} \text{ as } x \to \infty.$$

Find the corresponding functions $\psi_k(x)$ and $G_k(x, x')$.

Exercise II-9

(*a*) Show that, for any function ψ which is infinitely differentiable on R and any bounded interval (a, b),

$$\int_a^b (\sin\lambda x)\psi(x)\, dx \qquad \text{and} \qquad \int_a^b (\cos\lambda x)\psi(x)\, dx$$

converge to 0 when $\lambda \to \infty$.

(*b*) Deduce, using the identity

$$\varphi(x) = \varphi(0) + (\varphi(x) - \varphi(0)),$$

that the distribution $(\sin \lambda x)/x$ converges in $\mathcal{D}'$ to $\pi\delta$ when $\lambda \to \infty$.

(*c*) Show that, for real λ, the mapping

$$\varphi \in \mathcal{D} \to \text{pv} \int_{-\infty}^{+\infty} \frac{(\cos \lambda x)\, \varphi(x)}{x}\, dx$$

defines a distribution, denoted by

$$\text{pv}\cdot\frac{\cos \lambda x}{x}$$

and show that this distribution tends to 0 in $\mathcal{D}'$ as $\lambda \to \infty$.

Exercise II-10

When $n \to \infty$, find the limits in $\mathcal{D}'$ of

(*a*) $$f_n(x) = \frac{n}{\sqrt{\pi}}\, e^{-n^2x^2},$$

(*b*) $$F_n(x) = \int_{-\infty}^{x} f_n(t)\, dt.$$

Exercise II-11

Show that multiplication by a function $\alpha \in \mathcal{E}$ is a continuous linear mapping in $\mathcal{D}'$. When $a \to 0$, find the limits, in $\mathcal{D}'$, of

$$\frac{a}{x^2 + a^2} \quad \text{and} \quad \frac{ax}{x^2 + a^2}, \qquad a > 0.$$

Exercise II-12

When $h \to 0$, find the limits, in $\mathcal{D}'$, of the following distributions :

(*a*) $$\frac{\delta_{(h)} - \delta_{(-h)}}{2h},$$

(*b*) $$\frac{\delta_{(2h)} + \delta_{(-2h)} - 2\delta_{(0)}}{4h^2},$$

(*c*) $$\frac{\delta_{(nh)} - C_1^n \delta_{(n-2)h} + \cdots + (-1)^p C_p^n \delta_{(n-2p)h} + \cdots + (-1)^n \delta_{(-nh)}}{2^n h^n},$$

where

$$C_p^n = \frac{n!}{p!\,(n-p)!}.$$

Exercise II-13

For simplicity, we consider the real line R.

For any integer $m \geqslant 0$, $\mathcal{D}^m$ denotes the space of complex-valued m-times continuously differentiable functions defined on R and with bounded supports. $\mathcal{D}^m$ is endowed with the following topology : $\varphi_j \to \varphi$ in $\mathcal{D}^m$ if all the φ_j have their supports in one bounded set and if, for each integer p,

$0 < p \leqslant m$, $d^p\varphi_j/dx^p \to d^p\varphi/dx^p$ uniformly. An element of the topological dual $\mathfrak{D}'^m$ of $\mathfrak{D}^m$ (i.e., a continuous linear form on the space $\mathfrak{D}^m$) is called a *distribution of order* $\leqslant m$ on the line.

Distributions of order 0 are also called measures.

(*a*) Show that, for any integer $m \geqslant 0$, $T \in \mathfrak{D}'^m$ implies $T \in \mathfrak{D}'$ and that, if m and n are integers with $m > n$, $T \in \mathfrak{D}'^n$ implies $T \in \mathfrak{D}'^m$.

(*b*) Show that, for each integer $m \geqslant 0$, the distribution

$$\delta^{(m)}\ (\varphi \to (-1)^m \varphi^{(m)}(0))$$

belongs to $\mathfrak{D}'^m$ but not to any $\mathfrak{D}'^j$ with $j < m$.

(*c*) Show that the mapping which brings $\varphi \in \mathfrak{D}$ into correspondence with

$$T(\varphi) = \sum_{n=0}^{\infty} \frac{d^n\varphi(n)}{dx^n}$$

defines a distribution which cannot be of finite order.

Exercise II-14

(*a*) Show that the mapping which brings $\varphi \in \mathfrak{D}$ into correspondence with

$$\mathrm{Pf}\int_{-\infty}^{+\infty} \frac{\varphi(x)}{x^2}\,dx = \lim_{\varepsilon \to 0}\left[\int_{-\infty}^{-\varepsilon} \frac{\varphi(x)}{x^2}\,dx + \int_{\varepsilon}^{\infty} \frac{\varphi(x)}{x^2}\,dx - 2\,\frac{\varphi(0)}{\varepsilon}\right]$$

(the symbol Pf in front of the integral signifying "finite part") defines a distribution. This will be denoted by Pf $1/x^2$ ("pseudo-function" $1/x^2$). Obtain d/dx (pv $1/x$) in the distribution sense.

(*b*) Show that the mapping which brings $\varphi \in \mathfrak{D}$ into correspondence with

$$\mathrm{Pf}\int_{0}^{\infty} \frac{\varphi(x)}{x}\,dx = \lim_{\varepsilon \to 0}\left[\int_{\varepsilon}^{\infty} \frac{\varphi(x)}{x}\,dx + \varphi(0)\log\varepsilon\right]$$

defines a distribution. This will be denoted by Pf $Y(x)/x$. Give a similar proof for the mapping which brings $\varphi \in \mathfrak{D}$ into correspondence with

$$\mathrm{Pf}\int_{0}^{\infty} \frac{\varphi(x)}{x^2}\,dx = \lim_{\varepsilon \to 0}\left[\int_{\varepsilon}^{\infty} \frac{\varphi(x)}{x^2}\,dx - \frac{\varphi(0)}{\varepsilon} + \varphi'(0)\log\varepsilon\right].$$

The distribution thus defined will be denoted by Pf $Y(x)/x^2$. Obtain the derivative of the distribution Pf $Y(x)/x^2$.

(*c*) The distributions $\mathrm{Pf}\,\dfrac{Y(-x)}{x}$ and $\mathrm{Pf}\,\dfrac{Y(-x)}{x^2}$ are respectively defined by

$$\left\langle \mathrm{Pf}\frac{Y(-x)}{x}, \varphi\right\rangle = \mathrm{Pf}\int_{-\infty}^{0} \frac{\varphi(x)}{x}\,dx = \lim_{\varepsilon \to 0}\left[\int_{-\infty}^{-\varepsilon} \frac{\varphi(x)}{x}\,dx - \varphi(0)\log\varepsilon\right],$$

$$\left\langle \mathrm{Pf}\frac{Y(-x)}{x^2}, \varphi\right\rangle = \mathrm{Pf}\int_{-\infty}^{0} \frac{\varphi(x)}{x^2}\,dx = \lim_{\varepsilon \to 0}\left[\int_{-\infty}^{-\varepsilon} \frac{\varphi(x)}{x^2}\,dx - \frac{\varphi(0)}{\varepsilon} - \varphi'(0)\log\varepsilon\right].$$

Evaluate the derivative of the distribution $\mathrm{Pf}\dfrac{Y(-x)}{x}$. Show that

$$\mathrm{pv}\frac{1}{x} = \mathrm{Pf}\frac{Y(x)}{x} + \mathrm{Pf}\frac{Y(-x)}{x},$$

and by using the results of (*b*) and (*c*), again obtain the derivative of $\mathrm{pv}\cdot 1/x$ evaluated in (*a*).

(*d*) Evaluate the derivative of the distribution

$$\mathrm{Pf}\frac{1}{|x|} = \mathrm{Pf}\frac{Y(x)}{x} - \mathrm{Pf}\frac{Y(-x)}{x}.$$

(*e*) Find all the distributions T satisfying

$$x.\mathrm{T} = \mathrm{pv}\frac{1}{x}.$$

Exercise II-15
Evaluate $x^k\delta^{(l)}_{(0)}$. Form the linear differential equation of which $\delta_{(0)}$ is a solution.

Exercise II-16
Find all the distributions T_n satisfying

$$(E_n) \qquad x^n T_n = 1, \qquad (n \text{ a positive integer}).$$

It should be noted that if T_n is a solution of (E_n), then T'_n is, apart from a constant, a solution of (E_{n+1}).

Exercise II-17
In the plane (x, y) consider the circle (C) defined by $x^2 + y^2 = 1$. Let U be the distribution corresponding to the function equal to 0 outside (C) and to

$$\log\left[\frac{(x-a)^2 + y^2}{(ax-1)^2 + a^2y^2}\right]$$

in the interior of (C). Obtain ΔU, in the distribution sense.

Distributions in R^n

Exercise II-18
In the plane (x, y) the function $Y(x, y)$, equal to 1 for $x > 0, y > 0$, and to 0 otherwise, is called Heaviside's function. Show that in the distribution sense it is true that

$$\frac{\partial^2 Y}{\partial x\, \partial y} = \delta.$$

Exercise II-19
In the plane (x, y) consider the square ABCD, where

$$A = (1, 1); \qquad B = (2, 0); \qquad C = (3, 1); \qquad D = (2, 2).$$

Let T be the distribution corresponding to the function equal to 1 inside the square and to zero otherwise. Evaluate, in the distribution sense,

$$\frac{\partial^2 T}{\partial y^2} - \frac{\partial^2 T}{\partial x^2}.$$

Exercise II-20

In the plane (x, t) the cone

$$v^2t^2 - x^2 \geqslant 0; \qquad t \geqslant 0,$$

where v is a constant, is denoted by (C). Let $E(x, t)$ be the distribution

$$E(x, t) = \begin{cases} A \text{ in (C)}, \\ 0 \text{ otherwise}, \end{cases}$$

where A is also a constant. Evaluate, in the distribution sense,

$$\frac{1}{v^2}\frac{\partial^2 E}{\partial t^2} - \frac{\partial^2 E}{\partial x^2}.$$

If E is to be an elementary solution of the operator

$$\frac{1}{v^2}\frac{\partial^2}{\partial t^2} - \frac{\partial^2}{\partial x^2},$$

determine A as a function of v.

Exercise II-21

Show that, in the distribution sense,

$$\left(\frac{\partial}{\partial x} + i\frac{\partial}{\partial y}\right)\cdot\left(\frac{1}{x+iy}\right) = 2\pi\delta.$$

Exercise II-22

Evaluate $\Delta(\log 1/r)$ in R^2.

Exercise II-23

Show that in R^n, for $n \geqslant 3$,

$$\Delta\frac{1}{r^m} = \left\{\Delta\frac{1}{r^m}\right\} \qquad \text{for} \qquad m < n-2$$

in the distribution sense. Evaluate

$$\Delta\left(\frac{\varphi}{r^{n-2}}\right)$$

in the distribution sense for an infinitely differentiable function φ.

The result is applicable even though these functions are not differentiable at the origin.

Examples: $\varphi = e^{-kr}$, $\varphi = \sin kr$.

Exercise II-24
In R^n evaluate $\Delta^k(r^{2k-n}\log r)$ in the distribution sense, given that $2k - n$ is an even integer $\geqslant 0$.

Exercise II-25
In the plane (x, t) put

$$E(x, t) = \frac{Y(t)}{2\sqrt{\pi t}} e^{-\frac{x^2}{4t}}.$$

Evaluate in the distribution sense

$$\frac{\partial E}{\partial t} - \frac{\partial^2 E}{\partial x^2}.$$

Exercise II-26
In R^3, a function $f(r)$ depends only on $r = \sqrt{(x^2 + y^2 + z^2)}$ and is a solution for $r \neq 0$ of the equation

$$\Delta f + a^2 f = 0.$$

Write down the differential equation satisfied by $g(r) = rf(r)$. Show that, if $l = \lim_{r \to 0} [rf(r)]$, then in the distribution sense

$$(\Delta + a^2) f = kl\delta,$$

and obtain the value of the non-zero constant k.

Exercise II-27
Find an elementary solution of the operator

$$\frac{\partial^2}{\partial x\, \partial y} - a\frac{\partial}{\partial x} - b\frac{\partial}{\partial y} + ab,$$

where a and b are two given complex numbers.

Exercise II-28
Elementary solution of $\Box = \frac{1}{v^2}\frac{\partial^2}{\partial t^2} - \frac{\partial^2}{\partial x^2} - \frac{\partial^2}{\partial y^2}$.

(*a*) Set

$$\rho = v^2 t^2 - x^2 - y^2.$$

(i) Show that, for a function $f(\rho)$ depending only on ρ,

$$\Box f = Df$$

where

(1) $$D\rho = 4\rho \frac{d^2}{d\rho^2} + 6\frac{d}{d\rho}.$$

(ii) For $\varepsilon > 0$, denote by $Y_\varepsilon(\rho)$ the function

$$Y_\varepsilon(\rho) = \begin{cases} 1 \text{ for } \rho > \varepsilon, \\ 0 \text{ for } \rho < \varepsilon. \end{cases}$$

Evaluate in the distribution sense

$$D\frac{Y_\varepsilon(\rho)}{\sqrt{\rho}},$$

where D is given by (*a*).

(iii) If φ is any infinitely differentiable function of the variable ρ, with a bounded support, and T is a distribution in ρ, then

$$\langle DT, \varphi\rangle = \langle T, D^*\varphi\rangle$$

where D* is an operator to be determined.

(*b*) In the space (x, y, t) denote by (C) the cone of future waves,

$$v^2t^2 - x^2 - y^2 \geqslant 0, \qquad t \geqslant 0.$$

Let $E(x, y, t)$ be the distribution defined by

$$E(x, y, t) = \begin{cases} \dfrac{v}{2\pi}\dfrac{1}{\sqrt{v^2t^2 - x^2 - y^2}} & \text{in (C)}, \\ 0 & \text{otherwise.} \end{cases}$$

(i) Show that if $\varphi(x, y, t)$ is any infinitely differentiable function with a bounded support then

$$\langle \Box E, \varphi\rangle = \frac{1}{2\pi}\int_0^\infty \frac{d\rho}{\sqrt{\rho}}\iint_{R^2}\frac{\Box\varphi r\,dr\,d\theta}{2\sqrt{r^2+\rho}}. \tag{2}$$

Take the new set of variables ρ, r, and θ, where r and θ are polar coordinates in the (x, y) plane.

(ii) For any function $\varphi(x, y, t) \in \mathcal{D}$, let

$$\tilde{\varphi}(\rho) = v\iint_{R^2}\frac{\varphi\left(r\cos\theta, r\sin\theta, \frac{1}{v}\sqrt{r^2+\rho}\right)}{2\sqrt{r^2+\rho}}\,r\,dr\,d\theta. \tag{3}$$

Assuming that $\Box\tilde{\varphi} = D^*\tilde{\varphi}$, show that

$$\langle \Box E, \varphi\rangle = -\frac{2}{\pi v}\lim_{\varepsilon\to 0}\left[\sqrt{\varepsilon}\left(\frac{d\tilde{\varphi}}{d\rho}\right)_{\rho=\varepsilon}\right].$$

(*c*)

(i) For fixed r and t denote by $\bar{\varphi}(r, t)$ the mean of $\varphi(x, y, t)$ on the circle $x^2 + y^2 = r^2$ and show that

$$\left(\frac{d\tilde{\varphi}}{d\rho}\right)_{\rho=\varepsilon} = \frac{\pi}{2}\left[\int_0^\infty \frac{d\bar{\varphi}}{dt}\frac{r\,dr}{(r^2+\varepsilon)} - v\int_0^\infty \bar{\varphi}\left(r, \frac{1}{v}\sqrt{r^2+\varepsilon}\right)\frac{r\,dr}{(r^2+\varepsilon)^{3/2}}\right].$$

(ii) By integrating by parts the second integral on the right-hand side, show that

$$\left(\frac{d\tilde{\varphi}}{d\rho}\right)_{\rho=\varepsilon} = -\frac{\pi v}{2\sqrt{\varepsilon}}\bar{\varphi}\left(0, \frac{\sqrt{\varepsilon}}{v}\right) + A(\varepsilon; \varphi)$$

where $\sqrt{\varepsilon}\, A(\varepsilon; \varphi) \to 0$ as $\varepsilon \to 0$.

(iii) Hence deduce that $E(x, y, t)$ as defined in (*b*) is the required elementary solution.

CHAPTER III

Convolution

1. Tensor product of distributions

1. TENSOR PRODUCT OF TWO DISTRIBUTIONS

Let X^m and Y^n be two Euclidean spaces of dimensions m and n respectively, whose generic points are $x = (x_1, x_2, \ldots, x_m)$ and $y = (y_1, y_2, \ldots, y_n)$. Let the product $X^m \times Y^n$, denoted by Z^{m+n}, be the set of points $(x, y) = (x_1, x_2, \ldots, x_m, y_1, y_2, \ldots, y_n)$, which is a Euclidean space of $m + n$ dimensions.

If $f(x)$ is a function on X^m, $g(y)$ a function on Y^n, we say that their tensor product $f(x) \otimes g(y)$ is the function $h(x, y) = f(x)g(y)$ defined on Z^{m+n}.

If, in particular, f and g are locally summable on X^m and Y^n, then h is locally summable on Z^{m+n}, so that we may hope to extend the concept of tensor product to distributions.

Let $(\mathcal{D})_x$, $(\mathcal{D})_y$, $(\mathcal{D})_{x,y}$ be the $\mathcal{D}$ spaces of infinitely differentiable functions of bounded support on X^m, Y^n, $X^m \times Y^n$, respectively; and let $(\mathcal{D}')_x$, $(\mathcal{D}')_y$, $(\mathcal{D}')_{x,y}$ be the $\mathcal{D}'$ spaces of their corresponding distributions.

Let $\varphi(x, y) \in (\mathcal{D})_{x,y}$ be a function of the form $u(x)v(y)$, where $u(x) \in (\mathcal{D})_x$ and $v(y) \in (\mathcal{D})_y$. Then

(III, 1; 1)
$$\begin{aligned}\langle f(x) \otimes g(y),\ u(x)v(y)\rangle &= \int_{X^m \times Y^n} \cdots \int f(x)g(y)u(x)v(y)\,dx\,dy \\ &= \int_{X^m} f(x)u(x)\,dx \int_{Y^n} g(y)v(y)\,dy = \langle f, u\rangle\,\langle g, v\rangle.\end{aligned}$$

If now φ is not of this form, then by Fubini's rule,

(III, 1; 2)
$$\begin{aligned}\langle f(x) \otimes g(y),\ \varphi(x, y)\rangle &= \int_{X^m \times Y^n} \cdots \int f(x)g(y)\varphi(x, y)\,dx\,dy \\ &= \int_{X^m} f(x)\,dx \int_{Y^n} g(y)\varphi(x, y)\,dy = \langle f(x), \langle g(y), \varphi(x, y)\rangle\rangle\end{aligned}$$

and similarly

(III, 1; 3)
$$\langle f(x) \otimes g(y),\ \varphi(x, y)\rangle = \langle g(y), \langle f(x), \varphi(x, y)\rangle\rangle.$$

By generalizing, we can prove the following properties :

THEOREM 1. *If* $S_x \in (\mathcal{D}')_x$ *and* $T_y \in (\mathcal{D}')_y$, *there exists a unique distribution* $W_{x,y} \in (\mathcal{D}')_{x,y}$ *such that, for all* $\varphi \in (\mathcal{D})_{x,y}$ *of the form* $\varphi(x, y) = u(x)v(y)$, *where* $u \in (\mathcal{D})_x$ *and* $v \in (\mathcal{D})_y$ *we have*

$$\text{(III, 1; 4)} \qquad \langle W_{x,y}, u(x)v(y)\rangle = \langle S, u\rangle \langle T, v\rangle.$$

We call this the tensor product of the distributions S *and* T *and denote it by* $S \otimes T$. *For* $\varphi(x, y) \in (\mathcal{D})_{x,y}$, $\langle W, \varphi\rangle$ *can be found by Fubini's rule. Consider* x *as fixed: then* $\varphi(x, y)$, *regarded as a function of* y *only, belongs to* $(\mathcal{D})_y$ *and so we can form*

$$\text{(III, 1; 5)} \qquad \theta(x) = \langle T_y, \varphi(x,y)\rangle.$$

This expression depends on x; *insofar as it is a function of* x, θ *belongs to* $(\mathcal{D})_x$ *and we can form* $\langle S, \theta\rangle$. *Hence*

$$\text{(III, 1; 6)} \qquad \langle W, \varphi\rangle = \langle S, \theta\rangle = \langle S_x, \langle T_y, \varphi(x,y)\rangle\rangle,$$

and similarly

$$\text{(III, 1; 7)} \qquad \langle W, \varphi\rangle = \langle T_y, \langle S_x, \varphi(x,y)\rangle\rangle.$$

A result concerning the support of W is given by

THEOREM 2. *The support of* W *is the product* $A \times B$ *of the supports of* S *and* T; *that is to say, the set of points* (x, y) *for which* $x \in A$ *and* $y \in B$.

Proof of Theorem 2

1) Let $\varphi(x, y)$ be a function belonging to $(\mathcal{D})_{x,y}$ having its support in the complement of $A \times B$. For $x \in A$, $y \to \varphi(x, y)$ has its support in the complement of B and so $\theta(x) = 0$. Thus $\theta(x)$ has its support in the complement of A and so $\langle S_x, \theta(x)\rangle = 0$, and therefore $\langle W, \varphi\rangle = 0$. Thus the support of W is contained in $A \times B$.

2) From the formula (III, 1; 4) it is easily seen that all points of $A \times B$ belong to the support of W.

Examples and formulae

1) A function is said to be independent of x if it is of the form $g(y) = 1_x \otimes g(y)$. Similarly, we say that a distribution is independent of x if it is of the form $1_x \otimes T_y$. It will be seen that

$$\text{(III, 1; 8)} \qquad \langle 1_x \otimes T_y, \varphi\rangle = \int_{X^m} \langle T_y, \varphi(x,y)\rangle\, dx = \left\langle T_y, \int_{X^m} \varphi(x,y)\, dx\right\rangle.$$

2) If D_x^p means differentiation with respect to x and D_y^q differentiation with respect to y, then

$$\text{(III, 1; 9)} \qquad D_x^p D_y^q (S_x \otimes T_y) = (D_x^p S_x) \otimes (D_y^q T_y).$$

3)

$$\text{(III, 1; 10)} \qquad \delta_x \otimes \delta_y = \delta_{x,y}.$$

4) Let Y be Heaviside's function of n variables, equal to 1 in the "quadrant" $x_1 \geqslant 0, x_2 \geqslant 0, \ldots, x_n \geqslant 0$, and 0 elsewhere. This is the tensor product $Y(x_1) \otimes Y(x_2) \otimes \cdots \otimes Y(x_n)$. Since

$$\frac{d}{dx_i} Y(x_i) = \delta_{x_i}$$

we have

(III, 1; 11)
$$\frac{\partial^n}{\partial x_1 \partial x_2 \cdots \partial x_n} Y = \delta_{x_1, x_2, \ldots, x_n}.$$

2. TENSOR PRODUCT OF SEVERAL DISTRIBUTIONS

The theory of the tensor product of two distributions can be extended without difficulty to the tensor product of any finite number of distributions. This product is associative. If X^l, Y^m, Z^n, are three Euclidean spaces of dimensions l, m, n respectively, and R_x, S_y, T_z are three distributions on these spaces, we define $R_x \otimes S_y \otimes T_z$ by

(III, 1; 12) $\quad \langle R_x \otimes S_y \otimes T_z, u(x)v(y)w(z) \rangle = \langle R, u \rangle \langle S, v \rangle \langle T, w \rangle.$

For $\varphi \in (\mathcal{D})_{x,y,z}$, $\langle R_x \otimes S_y \otimes T_z, \varphi \rangle$ is again found by Fubini's rule, and

(III, 1; 13) $\quad R_x \otimes S_y \otimes T_z = R_x \otimes (S_y \otimes T_z) = (R_x \otimes S_y) \otimes T_z.$

2. Convolution

1. CONVOLUTION OF TWO DISTRIBUTIONS

Let S, T be two distributions on R^n. Then *convolution product* $S * T$ is a new distribution on R^n, defined by

(III, 2; 1) $\quad \langle S * T, \varphi \rangle = \langle S_\xi \otimes T_\eta, \varphi(\xi + \eta) \rangle.$

If $S * T$ exists, then obviously so does $T * S$ and the two are equal.

On the right-hand side we form $S_\xi \otimes T_\eta$, a distribution on the Euclidean space of $2n$ dimensions in the variable (ξ, η). If in $\varphi(x)$ we replace x by $\xi + \eta$ (a sum of vectors in R^n), we obtain $\varphi(\xi + \eta)$ as a function of the pair (ξ, η), and the right-hand side then has a meaning.

$S_\xi \otimes T_\eta$ always exists. If A and B are the supports of S and T in R^n, the support of $S_\xi \otimes T_\eta$ is $A \times B$, a set of pairs (ξ, η) for which $\xi \in A$, $\eta \in B$. $\varphi(\xi + \eta)$ is a well-defined function, infinitely differentiable with respect to ξ, η, but it is not of bounded support. If we take K to be the support of φ, the support of $\varphi(\xi+\eta)$ is the set of pairs (ξ, η) for which $\xi+\eta \in K$, namely the strip parallel to the "line" defined by the equation $\xi+\eta=0$, which, in the space of $2n$ dimensions, represents the n-dimensional sub-space given by the equations :

$$\xi_1+\eta_1=0, \ldots, \xi_n+\eta_n=0.$$

But the right-hand side of (III, 2; 1) will have a meaning if the support of $S_\xi \otimes T_\eta$ and that of $\varphi(\xi+\eta)$ have a bounded intersection, that is to say, if the set of points (ξ, η) satisfying $\xi \in A$, $\eta \in B$, $\xi+\eta \in K$, is bounded.

As $\langle S * T, \varphi \rangle$ must be defined for *all* $\varphi \in (\mathcal{D})_x$, the above set must be bounded *whatever* the bounded set K. Hence we deduce :

THEOREM 3. *The convolution product* $S * T$, *defined by* (III, 2; 1), *has a meaning if the supports* A *and* B *of* S *and* T *are such that, for* $\xi \in A$ *and* $\eta \in B$, $\xi+\eta$ *can only remain bounded if both* ξ *and* η *remain bounded. The product is commutative :* $S * T = T * S$.

Example 1. *If at least one of the two distributions* S, T *has a bounded support,* $S * T$ *exists.*

Let us suppose A bounded. Then $\xi \in A$ is necessarily bounded; if $\xi+\eta$ is bounded, $\eta = (\xi+\eta)-\xi$ is also bounded.

Example 2. *Convolution on the circle*

Let Γ be the circle with centre O, of radius $T/2\pi$ and perimeter T, in the Euclidean plane. The study of distributions on Γ will be made in connection with periodic distributions and Fourier series. The addition $\xi+\eta$ is here the addition of arcs (the origin being the point $s=0$ of Γ). As all the supports are bounded, the convolution always has a meaning.

Example 3. On the line R, we say that a distribution has a support " bounded from the left " if it is contained in the interval $(a, +\infty)$.

Let us suppose, then, that S and T both have their supports bounded from the left; A and B are contained in a half-line $(a, +\infty)$. Hence ξ and η are always $\geqslant a$. If $\xi+\eta$ is bounded, it remains $\leqslant C$ (constant); as $\xi \geqslant a$, it follows that $\eta \leqslant C-a$ and similarly $\xi \leqslant C-a$, hence ξ and η remain bounded, and $S * T$ exists.

Similarly $S * T$ exists if both S and T have their supports " bounded from the right ". On the other hand, if one has its support bounded from the left and the other from the right, $S * T$ does not necessarily exist.

Example 4. Suppose that in the space R^4 (space-time) A lies in the cone of future waves :

$$t = x_4 \geqslant 0, \qquad x_4^2 - x_1^2 - x_2^2 - x_3^2 \geqslant 0$$

and that B lies in the half-space $t = x_4 \geqslant 0$.

Let $\xi \in A$, $\eta \in B$ and let $\xi + \eta$ be bounded. Then $\xi_4 + \eta_4$ remains $\leqslant C$; as $\xi_4 \geqslant 0$ and $\eta_4 \geqslant 0$, ξ_4 and η_4 are both contained between 0 and C and so are bounded (Example 3).

Then, since $\xi_4^2 \geqslant \xi_1^2 + \xi_2^2 + \xi_3^2$, ξ_1, ξ_2 and ξ_3 remain bounded. Lastly, as $\xi_1 + \eta_1$ remains bounded, η_1 must also remain bounded and similarly η_2 and η_3 and so finally ξ and η remain bounded. Thus $S * T$ has a meaning.

Fundamental properties

THEOREM 4. *If* S *and* T *are two functions* f *and* g *(locally summable and only defined almost everywhere), and if their supports* A *and* B *satisfy the conditions of Theorem* 3, $S * T$ *is a locally summable function* h *defined almost everywhere by the equation*

$$\text{(III, 2; 2)} \quad h(x) = \int_{\mathbf{R}^n} f(x-t)\, g(t)\, dt = \int_{\mathbf{R}^n} f(t)\, g(x-t)\, dt.$$

Putting $W = f * g$ gives

$$\text{(III, 2; 3)} \quad \langle W, \varphi \rangle = \langle f_\xi g_\eta, \varphi(\xi + \eta) \rangle = \iint f(\xi) g(\eta) \varphi(\xi + \eta) d\xi d\eta.$$

The conditions relating to the supports A and B ensure that this double integral has a meaning. For, f and g being locally summable, $f(\xi)g(\eta)$ $\varphi(\xi + \eta)$ is locally summable but its support is bounded, and so it is summable.

Let us make the change of variable $x = \xi + \eta$, $t = \eta$; the Jacobian of this transformation is 1, and so

$$d\xi d\eta = dx\, dt$$

and

(III, 2; 4)

$$\langle W, \varphi \rangle = \iint f(x-t)\, g(t)\, \varphi(x)\, dx\, dt = \int_{\mathbf{R}^n} dx \int_{\mathbf{R}^n} \varphi(x) f(x-t)\, g(t)\, dt.$$

From Fubini's theorem, the integral on the right has a meaning for almost all values of x. Thus the integral

$$\int_{\mathbf{R}^n} f(x-t) g(t)\, dt$$

has a meaning for almost all values of x for which $\varphi(x) \neq 0$ but, as this is so for all $\varphi(x) \in (\mathcal{D})_x$, it has a meaning for almost all values of x and its value $h(x)$ is a function defined almost everywhere. Again by Fubini, $\varphi(x)h(x)$ is summable, and so h is locally summable, and

$$\text{(III, 2; 5)} \qquad \langle W, \varphi \rangle = \int_{\mathbf{R}^n} h(x)\, \varphi(x)\, dx = \langle h, \varphi \rangle$$

which proves that $W = h$.

The other expression of (III, 2; 2) is obtained by putting $\xi = t, \xi + \eta = x$, or alternatively by writing t for $x - t$ in the first expression of (III, 2; 2).

Note. If f and g are continuous functions or even if one of them is continuous or simply locally bounded, and the other is locally summable, it can be shown that $h = f * g$ is continuous.

In the case of two functions, $f * g$ can have a meaning even if the conditions relating to the supports are not satisfied. For example, if one of the two functions is *summable on* R^n, the other *bounded* on R^n, $f * g$ has a meaning and it is a continuous function bounded on R^n, the integral (III, 2; 2) then evidently having a meaning for *all* values of x. If f is summable and g is bounded, we also have the inequality

(III, 2; 6) $$|h(x)| \leqslant \left| \int_{R^n} |f(x)|\,dx \right| \sup_{x \in R^n} |g(x)|$$

or

(III, 2; 7) $$\|h\|_{L^\infty} \leqslant \|f\|_{L^1} \|g\|_{L^\infty}.$$

In the case where f and g are in L^1, $f * g$ always has a meaning, it is also in L^1 and

(III, 2; 8) $$\|f * g\|_{L^1} \leqslant \|f\|_{L^1} \|g\|_{L^1}.$$

In the case of Example 3, where we supposed the two functions f and g to have their supports in $(0, +\infty)$, h also has its support in $(0, +\infty)$ and formula (III, 2; 2) becomes

(III, 2; 9) $$h(x) = \begin{cases} 0 & \text{for } x \leqslant 0, \\ \displaystyle\int_0^x f(x - t) g(t)\,dt & \text{for } x \geqslant 0. \end{cases}$$

Examples

1) (III, 2; 10) $$Y_{(\lambda)}^{(\alpha)}(x) = Y(x) \frac{x^{\alpha-1}}{\Gamma(\alpha)} e^{\lambda x}, \quad \alpha > 0$$

for any complex λ.

(III, 2; 11) $$Y_{(\lambda)}^{(\alpha)} * Y_{(\lambda)}^{(\beta)} = Y_{(\lambda)}^{(\alpha+\beta)}.$$

For the product is zero for $x \leqslant 0$, and for $x \geqslant 0$ it has the value

(III, 2; 12) $$\left(Y_{(\lambda)}^{(\alpha)} * Y_{(\lambda)}^{(\beta)}\right)(x) = \int_0^x \frac{(x-t)^{\alpha-1}}{\Gamma(\alpha)} \frac{t^{\beta-1}}{\Gamma(\beta)} e^{\lambda(x-t)} e^{\lambda t}\,dt$$
$$= \frac{x^{\alpha+\beta-1} e^{\lambda x}}{\Gamma(\alpha)\Gamma(\beta)} \int_0^1 (1-u)^{\alpha-1} u^{\beta-1}\,du \quad (\text{putting } t = xu).$$

Now the integral is $B(\alpha, \beta) = \frac{\Gamma(\alpha)\Gamma(\beta)}{\Gamma(\alpha+\beta)}$, giving the result (III, 2; 11). In particular,

(III, 2; 13) $$Y_{(\lambda)}^{(n)}(x) = Y(x)\frac{x^{n-1}}{(n-1)!}e^{\lambda x}$$

is the n-th power of the convolution of $Y(x)e^{\lambda x}$. These convolutions play a part in the theory of differentiation of non-integral order and in the calculus of probabilities.

2) (III, 2; 14) $$G_\sigma(x) = \frac{1}{\sigma\sqrt{2\pi}}e^{-\frac{x^2}{2\sigma^2}}, \qquad \sigma > 0.$$

(III, 2; 15) $$G_\sigma * G_\tau = G_{\sqrt{\sigma^2+\tau^2}}.$$

For

(III, 2; 16) $$\begin{aligned}(G_\sigma * G_\tau)(x) &= \frac{1}{2\pi\sigma\tau}\int_{-\infty}^{+\infty} e^{-\frac{(x-t)^2}{2\sigma^2}-\frac{t^2}{2\tau^2}}dt \\ &= \frac{1}{2\pi\sigma\tau}\int_{-\infty}^{+\infty} e^{-\frac{x^2}{2\sigma^2}+\frac{xt}{\sigma^2}-\frac{t^2}{2}\left(\frac{1}{\sigma^2}+\frac{1}{\tau^2}\right)}dt \\ &= \frac{1}{2\pi\sigma\tau}e^{-\frac{x^2}{2(\sigma^2+\tau^2)}}\int_{-\infty}^{+\infty} e^{-\frac{1}{2}\frac{\sigma^2+\tau^2}{\sigma^2\tau^2}(t-\tau^2x/(\sigma^2+\tau^2))^2}dt.\end{aligned}$$

Making the change of variable

(III, 2; 17) $$\frac{\sqrt{\sigma^2+\tau^2}}{\sigma\tau}\left(t-\frac{\tau^2}{\sigma^2+\tau^2}x\right) = u,$$

the last integral becomes

(III, 2; 18) $$\frac{\sigma\tau}{\sqrt{\sigma^2+\tau^2}}\int_{-\infty}^{+\infty} e^{-\frac{u^2}{2}}du = \sqrt{2\pi}\frac{\sigma\tau}{\sqrt{\sigma^2+\tau^2}},$$

whence

(III, 2; 19) $$\begin{aligned}(G_\sigma * G_\tau)(x) &= \frac{1}{\sqrt{2\pi}\sqrt{\sigma^2+\tau^2}}e^{-\frac{x^2}{2(\sigma^2+\tau^2)}} \\ &= G_{\sqrt{\sigma^2+\tau^2}}(x).\end{aligned}$$

Formula (III, 2; 15) so obtained plays a fundamental part in the calculus of probabilities. G_σ, considered as a "probability density" is the Gauss distribution (the curve represented by $y = G_\sigma(x)$ is "bell-shaped") with mean value 0:

$$\int_{-\infty}^{+\infty} xG_\sigma(x)\,dx = 0,$$

and with mean square deviation σ:

$$\left(\int_{-\infty}^{+\infty} x^2 G\sigma(x)\, dx\right)^{1/2} = \sigma.$$

If two real independent variables obey Gauss' law with mean square deviation σ and τ, their sum obeys the law of probability defined by

$$G\sigma * G\tau$$

which is Gauss' law with deviation $\sqrt{\sigma^2 + \tau^2}$.

In particular the sum of n independent variables obeying the same law of Gauss with deviation σ obeys Gauss' law with deviation $\sigma\sqrt{n}$.

It is this property that allows us to say that the error to be expected in forming the sum of n measured quantities having the same error law is of the order of $\sqrt{n}$ times the error of each measure. On dividing by n, we see that the arithmetic mean of the resultants of n measures gives an error of the order of $1/\sqrt{n}$ times the error of each measure.

If instead of G_σ we take

$$\text{(III, 2; 20)} \qquad H_\sigma(x) = \frac{1}{\sigma} e^{-\pi\frac{x^2}{\sigma^2}}, \qquad \sigma > 0,$$

we still have

$$\text{(III, 2; 21)} \qquad H_\sigma * H_\tau = H_{\sqrt{\sigma^2+\tau^2}}.$$

3) Taking

$$\text{(III, 2; 22)} \qquad P_a(x) = \frac{1}{\pi}\frac{a}{x^2 + a^2}, \qquad a > 0,$$

then an integration shows that

$$\text{(III, 2; 23)} \qquad P_a * P_b = P_{a+b}.$$

THEOREM 5. *If α is a function infinitely differentiable (in the usual sense), and T is a distribution, then $T * \alpha$ is a function infinitely differentiable (in the usual sense) given by the formula*

$$\text{(III, 2; 24)} \qquad (T * \alpha)(x) = \langle T_t, \alpha(x - t)\rangle.$$

We say that the convolution with α is a *regularization* of T by α; α is also called the *regulator* and $T * \alpha$ is said to be *regularized* from T by α.

In the first instance $h(x) = \langle T_t, \alpha(x - t)\rangle$ has the following meaning: x is fixed, then $\alpha(x - t)$, considered as a function of t only, is in $\mathfrak{D}$, and the functional T_t may be applied to it. The result is a function of x which is well defined and infinitely differentiable by x in virtue of what has been said with regard to tensor products (see the function θ in Theorem 1). It remains to show that this function regarded as a distribution coincides with $T * \alpha$.

$$(\text{III}, 2; 25)\quad \begin{aligned}\langle T * \alpha, \varphi\rangle &= \langle T_\xi \otimes \alpha(\eta), \varphi(\xi + \eta)\rangle \\ &= \langle T_\xi, \langle \alpha(\eta), \varphi(\xi + \eta)\rangle\rangle \\ &= \left\langle T_\xi, \int_{\mathbf{R}^n} \alpha(\eta)\varphi(\xi + \eta)\, d\eta\right\rangle \\ &= \left\langle T_\xi, \int_{\mathbf{R}^n} \alpha(x - \xi)\varphi(x)\, dx\right\rangle \\ &= \langle T_\xi, \langle \varphi(x), \alpha(x - \xi)\rangle\rangle \\ &= \langle T_\xi \otimes \varphi(x), \alpha(x - \xi)\rangle \\ &= \langle \varphi(x), \langle T_\xi, \alpha(x - \xi)\rangle\rangle \\ &= \int_{\mathbf{R}^n} (\langle T_\xi, \alpha(x - \xi)\rangle)\varphi(x)\, dx \\ &= \int_{\mathbf{R}^n} h(x)\varphi(x)\, dx = \langle h, \varphi\rangle.\end{aligned}$$

Q.E.D.

Note. If T is a function f, (III, 2; 24) again gives

$$(\text{III}, 2; 26)\qquad (f * \alpha)(x) = \int_{\mathbf{R}^n} f(t)\alpha(x - t)\, dt,$$

which has been shown when α is a function which is locally summable but not necessarily differentiable.

Examples

1) $T * 1$ is the constant $\langle T, 1\rangle$.

2) If α is a polynomial of degree $\leqslant m$, $T * \alpha$ is also a polynomial of degree $\leqslant m$. For example, for one variable ($n = 1$),

$$(\text{III}, 2; 27)\quad (T * \alpha)(x) = \langle T_t, \alpha(x - t)\rangle = \sum_{k \leqslant m} \frac{x^k}{k!}\langle T_t, \alpha^{(k)}(-t)\rangle.$$

3) In the case of one dimension ($n = 1$), if T is of bounded support, the analytic function of the complex variable λ defined by

$$(\text{III}, 2; 28)\qquad \mathcal{T}(\lambda) = \langle T_x, e^{-\lambda x}\rangle$$

is called the *Laplace transform* of T.

Consider* then $S * T$ and let $U(\lambda)$ be its Laplace transform, so that

$$(\text{III}, 2; 29)\quad \begin{aligned}U(\lambda) &= \langle S * T, e^{-\lambda x}\rangle = \langle S_\xi \otimes T_\eta, e^{-\lambda(\xi+\eta)}\rangle \\ &= \langle S_\xi \otimes T_\eta, e^{-\lambda\xi}e^{-\lambda\eta}\rangle \\ &= \langle S_\xi, e^{-\lambda\xi}\rangle\, \langle T_\eta, e^{-\lambda\eta}\rangle \\ &= \mathcal{S}(\lambda)\mathcal{T}(\lambda).\end{aligned}$$

---

*If $S \in \mathcal{E}'$ and if $T \in \mathcal{E}'$ then $S * T \in \mathcal{E}'$. See the corollary to Theorem 8, p. 121.

The Laplace transform of S $*$ T *is the product (in the ordinary sense) of the Laplace transforms of* S *and* T.

THEOREM 6. *We have the three equations*

(III, 2; 30) $$\delta * T = T,$$
(III, 2; 31) $$\delta_{(a)} * T = \tau_a T, \qquad \delta_{(a)} * \delta_{(b)} = \delta_{(a+b)},$$
(III, 2; 32) $$\delta' * T = T'.$$

For

(III, 2; 33) $$\begin{aligned}\langle \delta * T, \varphi\rangle &= \langle \delta_\xi \otimes T_\eta, \varphi(\xi + \eta)\rangle \\ &= \langle T_\eta, \langle \delta_\xi, \varphi(\xi + \eta)\rangle\rangle = \langle T_\eta, \varphi(\eta)\rangle = \langle T, \varphi\rangle\end{aligned}$$

whence (III, 2; 30). This very important result shows that, in convolution, δ *is unity*. In theoretical physics the formula (III, 2; 30) is very often written (incorrectly) as

(III, 2; 34) $$\int f(x - t)\delta(t)\, dt = \int f(t)\delta(x - t)\, dt = f(x)$$

and taken as the definition of δ.

In formula (III, 2; 31) [which contains (III, 2; 30) as a particular case], $\delta_{(a)}$, which is unit mass at the point a, is also denoted by $\delta(x - a)$, and $\tau_a T$, which is T shifted by the translation a, is also denoted by T_{x-a} [or in physics $T(x - a)$]. We have seen that this shifted distribution can be defined by

(III, 2; 35) $$\langle \tau_a T, \varphi\rangle = \langle T_x, \varphi(x + a)\rangle.$$

(III, 2; 31) then follows at once since

(III, 2; 36) $$\begin{aligned}\langle \delta_{(a)} * T, \varphi\rangle &= \langle \delta_{(a)\xi} \otimes T_\eta, \varphi(\xi + \eta)\rangle \\ &= \langle T_\eta, \langle \delta_{(a)\xi}, \varphi(\xi + \eta)\rangle\rangle = \langle T_\eta, \varphi(a + \eta)\rangle \\ &= \langle \tau_a T, \varphi\rangle.\end{aligned}$$

A particular case of (III, 2; 31) is obtained by putting $T = \delta_{(b)}$, whence

$$\delta_{(a)} * \delta_{(b)} = \delta_{(a+b)}.$$

Finally, we prove (III, 2; 32):

(III, 2; 37) $$\begin{aligned}\langle \delta' * T, \varphi\rangle &= \langle \delta'_\xi \otimes T_\eta, \varphi(\xi + \eta)\rangle \\ &= \langle T_\eta, \langle \delta'_\xi, \varphi(\xi + \eta)\rangle\rangle \\ &= \langle T_\eta, -\varphi'(\eta)\rangle = -\langle T, \varphi'\rangle \\ &= \langle T', \varphi\rangle.\end{aligned}$$

Q.E.D.

In the same way we have

(III, 2; 38) $$\delta^{(m)} * T = T^{(m)}.$$

Similarly, if D is a differential operator in R^n with constant coefficients, then

(III, 2; 39) $$D\delta * T = DT.$$

In particular, if D is the Laplacian $\Delta = \sum_{i=1}^{n} \frac{\partial^2}{\partial x_i^2}$ in R^n or, in R^4 (space-time), the d'Alembertian

$$\Box = \frac{1}{v^2}\frac{\partial^2}{\partial t^2} - \frac{\partial^2}{\partial x^2} - \frac{\partial^2}{\partial y^2} - \frac{\partial^2}{\partial z^2} = \frac{1}{v^2}\frac{\partial^2}{\partial t^2} - \Delta,$$

then

(III, 2; 40) $$\Delta\delta * T = \Delta T, \qquad \Box\delta * T = \Box T.$$

In quantum physics, equation (III, 2; 32) is often written (incorrectly) as

(III, 2; 41) $$\int \delta'(x - t) f(t)\, dt = f'(x)$$

and obtained by "differentiating under the $\int$ sign" from formula (III, 2; 34) which is itself incorrect.

THEOREM 7. *If* S_j, T_j *are two sequences of distributions depending on an integer index* j, *where, as* $j \to \infty$, S_j *and* T_j *tend (in* $\mathcal{D}'$*) to* S *and* T *respectively, and if the* S_j *and* T_j *have their supports in the fixed closed sets* A *and* B *of* R^n *such that, for* $\xi \in A$ *and* $\eta \in B$, $\xi + \eta$ *can stay bounded only if both* ξ *and* η *stay bounded, then* $S_j * T_j$ *converges to* $S * T$ *as* $j \to \infty$ *(continuity of convolution).*

We state this theorem without proof.

In particular we have seen that δ can be obtained as the limit (in $\mathcal{D}'$) of infinitely differentiable functions α. Then $T = T * \delta$ is the limit of its regularizations $T * \alpha$. In particular we can show that we can find a set of *polynomials* α converging to δ. Then, if T is of bounded support, it is the limit of the polynomials $T * \alpha$, which is a form of Weierstrass' theorem:

Every distribution is the limit (in $\mathcal{D}'$*) of a set of polynomials.*

THEOREM 8. *If* S *and* T *have their supports in subsets* A *and* B *respectively in* R^n, $S * T$ *has its support contained in* $\overline{A + B}$, *the closure of the set of points* $\xi + \eta$ ($\xi \in A$, $\eta \in B$).

To see this we shall show that, if φ has its support K in the complement Ω of $\overline{A + B}$, then $\langle S * T, \varphi\rangle = 0$. We have

$$\langle S * T, \varphi\rangle = \langle S_\xi \otimes T_\eta, \varphi(\xi + \eta)\rangle.$$

The support of $S_\xi \otimes T_\eta$ is contained in $A \times B$, the set of pairs (ξ, η), $\xi \in A$, $\eta \in B$; the support of $\varphi(\xi + \eta)$ is the set of pairs (ξ, η) for which $\xi + \eta \in K$. These two supports have an empty intersection since $\xi \in A$, $\eta \in B$ implies $\xi + \eta \in A + B$ and $\overline{A + B}$ and K have an empty intersection. Thus the scalar product is zero. As Ω is open ($\overline{A + B}$ being closed) this proves that $S * T = 0$ in Ω, and so $S * T$ has its support in $\overline{A + B}$.

Corollary. *If* S *and* T *have bounded supports, the same is true of* $S * T$. *If* ($n = 1$) S *and* T *have supports bounded from the left, respectively in the half-lines* $(a, +\infty)$, $(b, +\infty)$, *then* $S * T$ *has its support in the half-line* $(a + b, +\infty)$. *If* ($n = 4$) S *and* T *have their supports in the cone of future waves* $t \geqslant 0$, $t^2 - x^2 - y^2 - z^2 \geqslant 0$, *the same is true of* $S * T$.

2. DEFINITION OF THE CONVOLUTION PRODUCT OF SEVERAL DISTRIBUTIONS. ASSOCIATIVITY OF CONVOLUTION

Let R, S, T be three distributions on R^n with supports A, B, C, respectively. We define their product of convolution $R * S * T$ by

$$\text{(III, 2; 42)} \quad \langle R * S * T, \varphi \rangle = \langle R_\xi \otimes S_\eta \otimes T_\zeta, \varphi(\xi + \eta + \zeta) \rangle.$$

It has a meaning if, for $\xi \in A$, $\eta \in B$, $\zeta \in C$, the sum $\xi + \eta + \zeta$ can only remain bounded if ξ, η and ζ all remain bounded. The associative property of the tensor product [equation (III, 1; 13)] gives the associative property of the convolution:

$$\text{(III, 2; 43)} \qquad R * S * T = (R * S) * T = R * (S * T).$$

But *the last two terms may have a meaning and not be equal* if the supports A, B, C do not have the above property; *the product of convolution is associative only if* $R * S * T$ *defined by* (III, 2; 42) *has a meaning.*

Example. $1 * \delta' * Y$, where Y is Heaviside's function.

$$\text{(III, 2; 44)} \qquad \begin{aligned} (1 * \delta') * Y &= 0 * Y = 0, \\ 1 * (\delta' * Y) &= 1 * \delta = 1. \end{aligned}$$

Equation (III, 2; 43) has the following applications.

Theorem 9. *The convolution product of several distributions has a meaning and is associative and commutative in any of the following cases:*

1) *all the distributions, except possibly for one, have their supports bounded;*

2) $n = 1$, *all the distributions have their supports bounded from the left;*

3) $n = 4$, *all the distributions have their supports in the half-space* $t \geqslant 0$, *and moreover, all, except possibly for one, in the cone of future waves*

$$t \geqslant 0, \quad t^2 - x^2 - y^2 - z^2 \geqslant 0.$$

Theorem 10. *To shift or differentiate a convolution product, we shift or differentiate any one of the factors at choice.*

This will be proved for differentiation:

$$\text{(III, 2; 45)}\quad \begin{aligned}(S * T)' &= \delta' * (S * T) = \delta' * S * T\\ &= (\delta' * S) * T = S' * T\\ &= S * \delta' * T = S * (\delta' * T) = S * T'.\end{aligned}$$

Of course, if we replace the derivative by any differential operator whatsoever with constant coefficients, it is still true that

$$\text{(III, 2; 46)}\qquad D(S * T) = DS * T = S * DT.$$

Applications to the theory of Newtonian potential ($n = 3$). The potential U of a continuous charge distribution of density $\rho(x)$ is defined by

$$\text{(III, 2; 47)}\qquad U(x) = \iiint_{R^3} \frac{\rho(t)\,dt}{|x - t|}$$

where $|\xi|$ denotes the Euclidean distance of ξ from the origin. This is just

$$\text{(III, 2; 48)}\qquad U = \rho * \frac{1}{|x|} = \rho * \frac{1}{r}, \qquad r = \sqrt{x^2 + y^2 + z^2}.$$

We are thus led to define the potential of any distribution T whatsoever by

$$\text{(III, 2; 49)}\qquad U = T * \frac{1}{r}.$$

Of course such a potential is then itself a distribution, not necessarily a function.

Let us calculate ΔU. From (III, 2; 46),

$$\text{(III, 2; 50)}\qquad \Delta U = T * \Delta\frac{1}{r} = T * (-4\pi\delta) = -4\pi T.$$

Thus we have Poisson's equation :

Theorem 11. *Poisson's equation. If* $U = T * (1/r)$ *is the potential of a distribution* T, *then*

$$\text{(III, 2; 51)}\qquad \Delta U = -4\pi T.$$

Applications to the calculus of probabilities ($n = 1$, *for simplicity*). In the calculus of probabilities the law of probability of a real variable is defined by its *probability distribution.* If this distribution is a locally summable function $f(x)$, the probability of the variable being located in (a, b) is

$$\int_a^b f(x)\,dx.$$

We have, of course, $f(x) \geqslant 0$, f summable, and

$$\int_{-\infty}^{+\infty} f(x)\, dx = +1.$$

However, a distribution such as $\delta_{(a)}$ also defines a law of probability for which the value $x = a$ is *certain*. A distribution such as

$$\tfrac{1}{3}\,\delta_{(a)} + \tfrac{2}{3}\,(\delta_{b)}$$

defines a law of probability for which x can take only the two values, $x = a$ (with probability $1/3$) and $x = b$ (with probability $2/3$).

A distribution defines a law of probability if and only if it is $\geqslant 0$, ($\langle T, \varphi\rangle \geqslant 0$ for $\varphi \geqslant 0$) *and of total mass* $+1$ ($\langle T, 1\rangle = 1$).

If T is not of bounded support, $\langle T, 1\rangle$ has no *a priori* meaning. We define it, for $T \geqslant 0$, as the upper bound of $\langle T, \varphi\rangle$ for $\varphi \in \mathcal{D}$, $0 \leqslant \varphi \leqslant 1$.

If two essentially real, independent variables have laws of probability defined by the distributions S and T, their sum has a law of probability which can be shown to be defined by the distribution $S * T$. The following theorem, obvious from the definitions, is a vital one in the calculus of probabilities.

THEOREM 12. *If* $S \geqslant 0$ *and* $T \geqslant 0$ *then* $S * T \geqslant 0$. *If* $\langle S, 1\rangle = 1$ *and* $\langle T, 1\rangle = 1$ *then* $\langle S * T, 1\rangle = 1$.

3. CONVOLUTION EQUATIONS

Let $\mathcal{A}'$ be a convolution algebra, that is to say, a vector sub-space of $\mathcal{D}'$ for which the product of convolution of two distributions or of any finite number of distributions belonging to $\mathcal{A}'$ is always defined and is in $\mathcal{A}'$. Let this product be commutative and associative and, finally, let $\delta \in \mathcal{A}'$.

Typical examples

1. $\mathcal{A}' = \mathcal{D}'(\Gamma)$ the convolution algebra of all distributions on the circle.
2. $\mathcal{A}' = \mathcal{E}'$ the convolution algebra of distributions of bounded support in R^n.
3. $\mathcal{A}' = \mathcal{D}'_+$ the convolution algebra of distributions whose support is contained in the half-line $x \geqslant 0$ ($n = 1$).
4. In R^4 (space-time) the convolution algebra of distributions whose support lies in the cone of future waves

$$t \geqslant 0, \quad t^2 - x^2 - y^2 - z^2 \geqslant 0.$$

In a convolution algebra, a convolution equation has the form

$$\text{(III, 2; 52)} \qquad A * X = B,$$

where A is the coefficient of the equation, B is the right-hand side and X is the unknown; A and B are in $\mathcal{A}'$, and likewise we seek X in $\mathcal{A}'$.

THEOREM 13. *If* A *is given, then for* (III, 2; 52) *to have at least one solution, whatever the right-hand side* B, *it is necessary and sufficient that* A *should have an inverse in the algebra* $\mathcal{A}'$, *that is to say, there exists an element* A^{-1} *satisfying*

$$\text{(III, 2; 53)} \qquad A * A^{-1} = A^{-1} * A = \delta.$$

In this case, the inverse is unique and (III, 2; 52) *always has one and only one solution given by*

$$\text{(III, 2; 54)} \qquad X = A^{-1} * B.$$

For if (III, 2; 52) always has *at least* one solution for all B, it has one for $B = \delta$, so that A has an inverse A^{-1}. Conversely, supposing that A has an inverse A^{-1}, then by convolution of the two sides with A^{-1} we deduce from (III, 2; 52)

$$\text{(III, 2; 55)} \qquad A^{-1} * A * X = A^{-1} * B,$$

that is to say (III, 2; 54). Similarly, from (III, 2; 54) we deduce (III, 2; 52) by convolution of the two sides with A.

Thus, if A^{-1} exists, (III, 2; 54) is equivalent to (III, 2; 52), that is to say the equation (III, 2; 52) has a unique solution given by (III, 2; 54). In particular the inverse of A is unique since it is the solution of (III, 2; 52) with $B = \delta$. A^{-1} *is also called the elementary solution of the convolution equation* (III, 2; 52). Finding the elementary solution is, therefore, the essential problem in solving the equation.

Notes

1) Things no longer appear the same if one is not concerned with a convolution algebra. For example, if, in R^3, $A = \Delta\delta$, there exists an inverse $A^{-1} = -1/4\pi r$. However A^{-1} has as support the entire space, and $\mathcal{D}'$ is not a convolution algebra. The two equations (III, 2; 52) and (III, 2; 54) are no longer equivalent. For example, we can no longer deduce (III, 2; 54) from (III, 2; 52), for if we compound the two sides with A^{-1}, the term $A^{-1} * A * X$ has no meaning if X is not of bounded support. So we can only say that, if X is of bounded support (in which case B is also) and satisfies (III, 2; 52), then X is given by (III, 2; 54). It is not possible to prove that the solution of (III, 2; 52) is unique, and in fact this is not so, for the homogeneous equation

$$\Delta\delta * X = 0 \qquad \text{or} \qquad \Delta X = 0$$

has an infinity of solutions, the "harmonic distributions" in particular the usual harmonic functions.

Similarly equation (III, 2; 54) has a meaning only if B is of bounded support, in which case (III, 2; 52) may be deduced. Thus $\Delta X = B$ has the particular solution

$$X = -\frac{1}{4\pi r} * B$$

when B is of bounded support. This solution is in agreement with Poisson's formula. The general solution is

$$X = -\frac{1}{4\pi r} * B + \text{a harmonic distribution.}$$

If B is not of bounded support, it is not possible to prove that a solution exists by this method.

2) If A does not have an inverse A^{-1}, the equation (III, 2; 52) has no solution for $B = \delta$. But solutions do exist for some values of B; there may be only one solution or an infinite number (the difference between any two of them being a solution of the homogeneous equation $A * X = 0$). In the algebra $\mathcal{D}'(\Gamma)$ (example 1) both cases can occur; but in $\mathcal{E}'$ and in $\mathcal{D}'_+$, there is never more than one solution. If A is an infinitely differentiable function, it cannot have an inverse, for by Theorem 5, $A * A^{-1}$ is an infinitely differentiable function, and so cannot be equal to δ.

THEOREM 14. *If* D *is a differential operator (in one variable, $n = 1$) of order m with constant coefficients, that of the m-th derivative being equal to* 1,

$$D = \frac{d^m}{dx^m} + a_1 \frac{d^{m-1}}{dx^{m-1}} + \cdots + a_{m-1}\frac{d}{dx} + a_m \qquad \text{(III, 2; 56)}$$

$D\delta$ *has an inverse in* $\mathcal{D}'_+$ *and its inverse is the product of* Y *(Heaviside's function) with the solution* Z *of the homogeneous equation* $DZ = 0$, *satisfying the "initial conditions"*

$$Z(0) = Z'(0), \ldots, Z^{(m-2)}(0) = 0, \qquad Z^{(m-1)}(0) = 1. \qquad \text{(III, 2; 57)}$$

For the function YZ has discontinuities at the origin and so its derivatives may be found by a familiar method. Alternatively, we can differentiate the product YZ by the rule for differentiating a product, and since Z is infinitely differentiable,

$$\text{(III, 2; 58)} \quad \begin{cases} (YZ)' = YZ' + \delta Z(0) \\ (YZ)'' = YZ'' + \delta Z'(0) + \delta' Z(0) \\ \cdots\cdots\cdots\cdots\cdots\cdots\cdots\cdots \\ (YZ)^{(m-1)} = YZ^{(m-1)} + \delta Z^{(m-2)}(0) + \cdots + \delta^{(m-2)}Z(0) \\ (YZ)^m = YZ^{(m)} + \delta Z^{(m-1)}(0) + \cdots + \delta^{(m-1)}Z(0). \end{cases}$$

Bearing in mind the hypotheses concerning Z, this becomes

(III, 2; 59) $$\left.\begin{aligned}(YZ)^{(k)} &= YZ^{(k)}\\ (YZ)^{(m)} &= YZ^{(m)} + \delta\end{aligned}\right\} \text{for } k \leqslant m - 1.$$

From this we deduce

(III, 2; 60) $$D\delta * (YZ) = D(YZ) = YDZ + \delta$$

and as Z is the solution of the homogeneous equation $DZ = 0$, we have

(III, 2; 61) $$D(YZ) = \delta.$$ Q.E.D.

Examples and corollaries

The elementary solution of

$$D = \frac{d}{dx} - \lambda \quad (\lambda \text{ an arbitrary complex number})$$

is

(III, 2; 62) $$(\delta' - \lambda\delta)^{-1} = Y(x)e^{\lambda x}.$$

The elementary solution of

$$D = \frac{d^2}{dx^2} + \omega^2 \quad (\omega \text{ real})$$

is

(III, 2; 63) $$(\delta'' + \omega^2\delta)^{-1} = Y(x)\,\frac{\sin \omega x}{\omega}.$$

The elementary solution of

$$D = \left(\frac{d}{dx} - \lambda\right)^m$$

is

(III, 2; 64) $$(\delta' - \lambda\delta)^{-m} = Y(x)e^{\lambda x}\frac{x^{m-1}}{(m-1)!}.$$

As an application, we seek the solution (*in the theory-of-functions sense*) of the equation

(III, 2; 65) $$Dz = f,$$

where D is given by (III, 2; 56) and f is a given function, with initial conditions

(III, 2; 66) $$z^{(k)}(0) = z_k \qquad \text{for} \qquad k \leqslant m - 1.$$

The function Yz is required. It belongs to $\mathcal{D}'_+$ and satisfies (in the distribution sense) the equation (III, 2; 58), whence

(III, 2; 67) $$D(Yz) = YDz + \sum_{k=0}^{m-1} e_k\delta^{(k)}$$

which together with

(III, 2; 68) $Dz = f, \quad e_k = z_{m-1-k} + a_1 z_{m-2-k} + \cdots + a_{m-k-1} z_0,$

yield, on putting $(D\delta)^{-1} = YZ$, the solution

$$\text{(III, 2; 69)} \qquad Yz = YZ * \left(Yf + \sum_{k=0}^{m-1} e_k \delta^{(k)} \right).$$

That is to say, for $x \geqslant 0$, we have

$$\text{(III, 2; 70)} \qquad z(x) = \int_0^x Z(x-t) f(t)\, dt + \sum_{k=0}^{m-1} e_k Z^{(k)}.$$

Hence the particular solution Z of the homogeneous equation $DZ = 0$, with the initial conditions (III, 2; 57) enables us to solve the non-homogeneous equation with right-hand side f, irrespective of the initial conditions.

To find z for $x \leqslant 0$, we must work in $\mathcal{D}'_-$, the algebra of convolution of the distributions having their support in $(-\infty, 0)$ and seek $-Y(-x)z$. The elementary solution here is $-Y(-x)Z$, where the equations (III, 2; 58) are modified by everywhere replacing $Y(x)$ by $-Y(-x)$, and similarly in (III, 2; 69) which gives finally, for $x \leqslant 0$, the same result (III, 2; 70).

Alternatively, differentiation under the $\int$ sign shows directly that $z(x)$ given by (III, 2; 70) is the solution of (III, 2; 65), with the initial conditions (III, 2; 66). For (*the derivatives having their usual meaning*)

$$\text{(III, 2; 71)} \quad \frac{d}{dx} \int_0^x Z(x-t) f(t)\, dt = \int_0^x Z'(x-t) f(t)\, dt + Z_0 f(x) = \int_0^x Z'(x-t) f(t)\, dt,$$

since $Z_0 = 0$. Similarly

$$\text{(III, 2; 72)} \quad \frac{d^k}{dx^k} \int_0^x Z(x-t) f(t)\, dt = \int_0^x Z^{(k)}(x-t) f(t)\, dt, \quad k \leqslant m-1.$$

$$\frac{d^m}{dx^m} \int_0^x Z(x-t) f(t)\, dt = \int_0^x Z^{(m)}(x-t) f(t)\, dt + f(x),$$

whence

$$\text{(III, 2; 73)} \quad D \int_0^x Z(x-t) f(t)\, dt = \int_0^x DZ(x-t) f(t)\, dt + f(x) = f(x)$$

since $DZ = 0$. Then, bearing in mind that the derivatives $Z^{(k)}$ satisfy the same equation as Z itself,

$$\text{(III, 2; 74)} \qquad D(Z^{(k)}) = 0,$$

and thus

(III, 2; 75) $$Dz = f.$$

It remains to show that z satisfies the given initial conditions, which we do not do here.

THEOREM 15. *If* A_1 *and* A_2 *have inverses in the algebra* $\mathcal{D}'_+$, $A_1 * A_2$ *has an inverse, and its inverse is*

(III, 2; 76) $$(A_1 * A_2)^{-1} = A_1^{-1} * A_2^{-1}.$$

For

(III, 2; 77) $$\begin{aligned}(A_1 * A_2) * (A_1^{-1} * A_2^{-1}) &= (A_1 * A_1^{-1}) * (A_2 * A_2^{-1}) \\ &= \delta * \delta = \delta.\end{aligned}$$

Application. Let D be a differential operator with constant coefficients (III, 2; 56). The polynomial

(III, 2; 78) $$P(z) = z^m + a_1 z^{m-1} + \cdots + a_{m-1}z + a_m$$

can be expressed as the product of factors

(III, 2; 79) $$P(z) = (z - z_1)(z - z_2) \cdots (z - z_m),$$

where the z_j may or may not be distinct. We deduce that

(III, 2; 80) $$D = \left(\frac{d}{dx} - z_1\right)\left(\frac{d}{dx} - z_2\right) \cdots \left(\frac{d}{dx} - z_m\right)$$

or

(III, 2; 81) $$D\delta = (\delta' - z_1\delta) * (\delta' - z_2\delta) \cdots * (\delta' - z_m\delta).$$

Then the inverse $(D\delta)^{-1}$ exists and is given by the following formula, bearing in mind (III, 2; 62),

(III, 2; 82) $$(D\delta)^{-1} = Y(x)e^{z_1 x} * Y(x)e^{z_2 x} * \cdots * Y(x)e^{z_m x}.$$

In particular,

(III, 2; 83) $$\begin{aligned}(\delta' - \lambda\delta)^{-m} &= Y(x)e^{\lambda x} * Y(x)e^{\lambda x} * \cdots * Y(x)e^{\lambda x} \\ &= Y(x)e^{\lambda x} \frac{x^{m-1}}{(m-1)!}\end{aligned}$$

from (III, 2; 13), which again gives (III, 2; 64).

Heaviside's operational calculus

In the algebra $\mathcal{D}'_+$, as in any algebra with a unit element, we can express a rational fraction as the sum of partial fractions. Let p be the element δ' of $\mathcal{D}'_+$ and let 1 be the unit element δ. If λ is a scalar, λ or λ_1 will be denoted by $\lambda\delta$. It is required to find $(D\delta)^{-1}$, where D is the differential operator (III, 2; 56). Use (III, 2; 79) but group the multiple roots together:

(III, 2; 84) $$P(z) = \prod_j (z - z_j)^{k_j}.$$

We have to find the inverse in $\mathcal{D}'_+$ of the element $\prod_j (p - z_j)^{k_j}$. The partial fraction expansion gives

(III, 2; 85) $$\frac{1}{\prod_j (p - z_j)^{k_j}} = \sum_j \left(\frac{c_{j,\,k_j}}{(p - z_j)^{k_j}} + \cdots + \frac{c_{j,\,1}}{p - z_j}\right).$$

Now, note that

(III, 2; 86) $$\frac{1}{(p - z_j)^{k_j}} = Y(x) e^{z_j x} \frac{x^{k_j - 1}}{(k_j - 1)!}.$$

Thus we can express $(D\delta)^{-1}$ as a sum of known quantities. For example, proceeding from (III, 2; 62) we again get (III, 2; 63). The inverse of $\delta'' + \omega^2\delta$ is written

$$\frac{1}{p^2 + \omega^2}.$$

Now,

(III, 2; 87) $$\frac{1}{p^2 + \omega^2} = \frac{1}{2\omega i}\left(\frac{-1}{p + \omega i} + \frac{+1}{p - \omega i}\right)$$

whence

(III, 2; 88) $$(\delta'' + \omega^2\delta)^{-1} = \frac{1}{2\omega i}(-Y(x)e^{-i\omega x} + Y(x)e^{i\omega x}) = Y(x)\frac{\sin \omega x}{\omega}.$$

As another example, we shall solve the integral equation

(III, 2; 89) $$\int_0^x \cos (x - t) f(t) dt = g(x)$$

for g given, f unknown, $x \geqslant 0$. This is equivalent to

(III, 2; 90) $$Y(x) \cos x * f = g.$$

The quantity $Y(x) \cos x = Y(x)\left(\dfrac{e^{ix} + e^{-ix}}{2}\right)$ as expressed in $\mathcal{D}'_+$ is

(III, 2; 91) $$Y(x) \cos x = \frac{1}{2}\left(\frac{1}{p - i} + \frac{1}{p + i}\right) = \frac{p}{p^2 + 1}.$$

The inverse in $\mathcal{D}'_+$ is

(III, 2; 92) $$\frac{p^2 + 1}{p} = p + \frac{1}{p} = \delta' + Y(x).$$

Alternatively, it can be verified that

$$\text{(III, 2; 93)}\quad Y(x)\cos(x) * (\delta' + Y) = (Y\cos x)' + Y\cos x * Y$$
$$= (-Y(x)\sin x + \delta) + Y(x)\int_0^\infty \cos t\, dt = \delta.$$

The solution of (III, 2; 90) is then

$$\text{(III, 2; 94)}\qquad f = g * (\delta' + Y) = g' + Y(x)\int_0^\infty g(t)\, dt.$$

Obviously g' must be given its meaning in the theory of distributions. It follows that the solution of (III, 2; 90) is a function only if g' is a function.

Volterra's integral equations

Consider, for $x \geqslant 0$, the equation

$$\text{(III, 2; 95)}\qquad f(x) + \int_0^\infty K(x-t) f(t)\, dt = g(x),$$

where the right-hand side g and the kernel K are locally summable functions by hypothesis, and the f is the required unknown function. If we continue f, g, K by 0 for $x < 0$ we have a convolution equation in $\mathcal{D}'_+$ of type (III, 2; 52), with $A = \delta + K$, $X = f$, $B = g$.

THEOREM 16. $A = \delta + K$ *has an inverse in* $\mathcal{D}'_+$, *for all functions* $K \in \mathcal{D}'_+$, *and its inverse is of the form* $\delta + H$, *where* H *is a function of* $\mathcal{D}'_+$.

We confine ourselves to showing this when the function K is not only locally summable but locally bounded also, that is to say bounded on every finite interval. Let us then put symbolically, as on page 128, $\delta = 1$, $K = q$. We have to invert $1 + q$; it is natural to assume an inverse in the form of the series $1 - q + q^2 + \cdots + (-1)^n q^n \cdots$ if it is convergent. This amounts to calculating

$$\text{(III, 2; 96)}\quad E = \delta - K + K^{*2} + \cdots + (-1)^n K^{*n} + \cdots$$

where K^{*n} denotes the convolution product of n functions equal to K. We show that this series is convergent in $\mathcal{D}'_+$. After the first term δ, all the terms are functions, and it will suffice to show that, on every finite interval $(0, a)$, the modulus of K^{*n} is bounded above by the general term of a convergent series of numbers. Let M_a be the maximum modulus of $K(x)$ for $0 \leqslant x \leqslant a$. Then, for $0 \leqslant x \leqslant a$,

$$\text{(III, 2; 97)}\qquad |K^{*2}(x)| = \left|\int_0^\infty K(x-t)\, K(t)\, dt\right| \leqslant x M_a^2.$$

More generally, suppose that, for $0 \leqslant x \leqslant a$,

$$|K^{*(n-1)}(x)| \leqslant \frac{x^{n-2}}{(n-2)!} M_a^{n-1}. \qquad \text{(III, 2; 98)}$$

Then

$$\begin{aligned} |K^{*n}(x)| &= \left| \int_0^x K^{*(n-1)}(t) K(x-t)\, dt \right| \\ &\leqslant M_a^n \int_0^x \frac{t^{n-2}}{(n-2)!}\, dt = \frac{x^{n-1}}{(n-1)!} M_a^n. \end{aligned} \qquad \text{(III, 2; 99)}$$

Hence it follows by induction that in the interval $0 \leqslant x \leqslant a$ there is the uniform inequality

$$|K^{*n}(x)| \leqslant M_a^n \frac{a^{n-1}}{(n-1)!}, \qquad \text{(III, 2; 100)}$$

the general term of a convergent series of numbers.

It remains to show that the distribution E thus calculated satisfies $A * E = \delta$. As the product of convolution can be calculated term by term (Theorem 7) the problem reduces, on putting $\delta = 1$, $K = q$ as usual, to proving the formula

$$(1 + q)(1 - q + q^2 + \cdots + (-1)^n q^n + \cdots) = 1 \qquad \text{(III, 2; 101)}$$

which is obvious. Finally, putting

$$-H = K - K^{*2} + K^{*3} + \cdots + (-1)^{n-1} K^{*n} + \cdots, \qquad \text{(III, 2; 102)}$$

we have

$$A^{-1} = \delta + H.$$

H is a function (zero for $x < 0$) [and if K is continuous, H is a continuous function since it is the sum of a series of continuous functions, uniformly convergent on every finite interval $(0, a)$]. The solution of (III, 2; 95) is given by

$$\left\{ \begin{aligned} &f = A^{-1} * g = (\delta + H) * g \\ &\text{or} \\ &f(x) = g(x) + \int_0^x H(x-t)\, g(t)\, dt. \end{aligned} \right. \qquad \text{(III, 2; 103)}$$

The solution f is given, in terms of the right-hand side g, by a formula analogous to that which gives g in terms of f.

Notes

1) The equation (III, 2; 95) is called an integral equation of the second kind. The equation of the first kind is

$$\int_0^x K(x-t) f(t)\, dt = g(x), \qquad \text{(III, 2; 104)}$$

which is a convolution equation, with $A = K$. The situation is then quite different. It can very well be that there is no inverse A^{-1}. (For example, if K, continued as usual by 0 for $x < 0$, is an infinitely differentiable function, then Theorem 5 shows that for all $E \in \mathcal{D}'_+$, $K * E$ is an infinitely differentiable function, and so could not be equal to δ.)

If A^{-1} exists, it is a distribution and can never be a function. For if E is a function, $K * E$ is a function and so could not be equal to δ. Thus f is never given in terms of g as g is in terms of f. For example, if $K(x)$ is the function $Y(x)\dfrac{x^{m-1}}{(m-1)!}$, the equation (III, 2; 64) shows that $A^{-1} = \delta^{(m)}$.

2) There are some of Volterra's integral equations that are not convolution equations, for example the general equation of the second kind

$$f(x) + \int_0^x K(x, \xi) f(\xi)\, d\xi = g(x), \tag{III, 2; 105}$$

where K is a continuous function of 2 variables, x, ξ, with $x \geqslant 0$, $0 \leqslant \xi \leqslant x$ We are led to put

$$\begin{cases} K^{(2)}(x, \xi) = \displaystyle\int_\xi^x K(x, \eta)\, K(\eta, \xi)\, d\eta, \\ \dots\dots\dots\dots\dots\dots\dots\dots\dots\dots \\ K^{(n)}(x, \xi) = \displaystyle\int_\xi^x K^{(n-1)}(x, \eta)\, K(\eta, \xi)\, d\eta, \end{cases} \tag{III, 2; 106}$$

(III, 2; 107)

$$-H(x, \xi) = K(x, \xi) - K^{(2)}(x, \xi) + \cdots + (-1)^{n-1} K^{(n)}(x, \xi) + \cdots$$

(a uniformly convergent series for $0 \leqslant \xi \leqslant x \leqslant a$) and the solution of equation (III, 2; 105) is given by

$$f(x) = g(x) + \int_0^x H(x, \xi) g(\xi) d\xi. \tag{III, 2; 108}$$

Systems of convolution equations

A system of n convolution equations with n unknown distributions is of the form

$$\begin{cases} A_{11} * X_1 + A_{12} * X_2 + \cdots + A_{1n} * X_n = B_1 \\ \dots\dots\dots\dots\dots\dots\dots\dots\dots\dots \\ A_{i1} * X_1 + A_{i2} * X_2 + \cdots + A_{in} * X_n = B_i \\ \dots\dots\dots\dots\dots\dots\dots\dots\dots\dots \\ A_{n1} * X_1 + A_{n2} * X_2 + \cdots + A_{nn} * X_n = B_n \end{cases} \tag{III, 2; 109}$$

where the A_{ij} are the coefficients, the B_i the right-hand-side terms and the X_i the unknowns, all the elements being in a convolution algebra $\mathcal{A}'$.

THEOREM 17. *The system* (III, 2; 109) *with given coefficients* A_{ij} *has at least one solution, whatever the right-hand side terms* B_i, *if and only if the determinant* Δ (*of convolution*) *of the* A_{ij} *has an inverse in* $\mathcal{A}'$. *This inverse is then unique and*

(III, 2; 109) *has a unique solution whatever the right-hand side terms may be, given by*

(III, 2; 110)
$$\begin{cases} X_1 = C_{11} * B_1 + C_{12} * B_2 + \cdots + C_{1n} * B_n \\ X_2 = C_{21} * B_1 + C_{22} * B_2 + \cdots + C_{2n} * B_n \\ \cdots\cdots\cdots\cdots\cdots\cdots\cdots\cdots\cdots\cdots \\ X_n = C_{n1} * B_1 + C_{n2} * B_2 + \cdots + B_{nn} * B_n \end{cases}$$

where the C_{ij} *form the matrix* (C), *which is the inverse of the matrix* (A) *with elements* A_{ij}.

Of course, the products that occur in the calculation of the determinants are convolution products; e.g., the determinant $\begin{vmatrix} L & M \\ P & Q \end{vmatrix}$ will be $L * Q - M * P$.

Equation (III, 2; 109) written in matrix form is

(III, 2; 111)
$$(A) * (X) = (B)$$

where (A) is the matrix of the A_{ij}, and (X) and (B) are matrices of n rows and one column,

(III, 2; 112)
$$(X) = \begin{pmatrix} X_1 \\ X_2 \\ \vdots \\ X_n \end{pmatrix}, \qquad (B) = \begin{pmatrix} B_1 \\ B_2 \\ \vdots \\ B_n \end{pmatrix}.$$

1) Suppose first that there is at least one solution, whatever the term on the right-hand side. Take (B) to be the system $B_j = \delta$, $B_k = 0$ for $k \neq j$, let $X_i = C_{ij}$, $i = 1, 2, \ldots, n$, be the corresponding solution, then give j the values $1, 2, \ldots, n$. Then, by definition,

(III, 2; 113)
$$\sum_k A_{ik} * C_{kj} = \begin{cases} 0 & \text{if } i \neq j \\ \delta & \text{if } i = j \end{cases}$$

where, in terms of matrices,

(III, 2; 114)
$$(A) * (C) = (\delta I)$$

where (δI) is the square matrix whose terms are δ on the principal diagonal and 0 elsewhere.

We deduce from the determinants the relation

(III, 2; 115)
$$\det(A) * \det(C) = \det(\delta I) = \delta$$

which proves that $\Delta = \det(A)$ has an inverse in the algebra $\mathcal{A}'$.

2) Conversely, suppose that Δ has an inverse Δ^{-1}. Let α_{ij} be the minor of A_{ij} in the determinant (A). Define the matrix (C) by

(III, 2; 116)
$$C_{ij} = \Delta^{-1} * \alpha_{ji}.$$

By ordinary matrix calculus,

(III, 2; 117)
$$(A) * (C) = (C) * (A) = (\delta I).$$

Then let (X) be a solution of (III, 2; 111) so that on making the convolution to the left with (C) we have

(III, 2; 118) $(C) * (B) = (C) * (A) * (X) = (\delta I) * (X) = (X)$

or

$$(X) = (C) * (B).$$

Conversely, if (X) satisfies (III, 2; 118), on making the convolution to the left with A we once more obtain (III, 2; 111).

So (III, 2; 111) and (III, 2; 118) are equivalent; or, put another way, (III, 2; 111) has a unique solution given by (III, 2; 118); or alternatively (III, 2; 109) has a unique solution given by (III, 2; 110). The matrix (C) is the inverse of the matrix (A). This must be so, for from

(III, 2; 119) $$(A^{-1}) * (A) = (\delta I)$$

we deduce, on making the product of convolution with (C),

(III, 2; 120) $$(A^{-1}) = (C).$$

These considerations have particular applications to electrical systems. This applies not only to convolution; these rules are true in any commutative algebra and are not confined to the general properties of systems of equations, matrices and determinants. Note that $\mathcal{A}'$ is commutative, but that the convolution of two matrices with coefficients in $\mathcal{A}'$ is not, in general, commutative.

3. Convolution in physics

Consider for example an electric circuit having resistance, self-inductance, and capacity. Let the emf be $e(t)$, supposed given for $t \geqslant t_0$, and let the current in the circuit be $i(t)$, supposed zero initially at time $t = t_0$. Then the function $i(t)$ will be determined for $t \geqslant t_0$.

We shall suppose $e(t)$ and $i(t)$ to be functions defined for all values of t, but zero for $t < t_0$. Then for each *excitation* defined by the function $e(t)$, we have a *response* defined by $i(t)$; this defines an *operator*, which for each $e(t)$ gives a corresponding $i(t)$. This operator has the following properties:

1) It is linear. If $i(t)$ corresponds to $e(t)$, $\lambda i(t)$ corresponds to $\lambda e(t)$; if $i_1(t)$ and $i_2(t)$ correspond respectively to $e_1(t)$ and $e_2(t)$, $i_1(t) + i_2(t)$ corresponds to $e_1(t) + e_2(t)$.

2) It commutes with translations in time, which expresses the invariability of the physical system with relation to time. That is to say that if $i(t)$ corresponds to $e(t)$, $i(t - t_1)$ obtained by a time shift corresponds to $e(t - t_1)$ obtained by a similar time shift.

3) If $e(t)$ is zero for $t < 0$, then so is $i(t)$. From 2) we then deduce that if $e(t) = 0$ for $t < t_0$, then so is $i(t)$. From these hypotheses, we usually deduce in an heuristic manner the following results.

Let the excitation $e(t)$ be $\delta(t)$, "Dirac's function". In practice this means that $e(t)$ will be very large, of the order of $1/\varepsilon$ for a very short time, $0 \leqslant t \leqslant \varepsilon$, and zero otherwise. Denote the response of the circuit by $i(t) = \mathrm{A}(t)$. This is known as the impulsive response, or the response to unit impulse. It is a "function" of t, zero for $t < 0$, from 3), but which, due to inductance and capacity, we should be wrong to suppose is zero for $t > \varepsilon$, after the subsidence of the emf.

Consider now the excitation shifted with respect to time, $e(t) = \delta(t - \tau)$. Then from 2), the corresponding response is $i(t) = \mathrm{A}(t - \tau)$. But any excitation $e(t)$ is a "linear combination" in integral form of impulsive excitations $\delta(t - \tau)$, so that

(III, 3; 1) $$e(t) = \int_0^t e(\tau)\delta(t-\tau)d\tau \quad \text{or} \quad e = e * \delta.$$

From linearity 1) we deduce that $i(t)$ is the corresponding combination of the $\mathrm{A}(t - \tau)$,

(III, 3; 2) $$i(t) = \int_0^t e(\tau)\mathrm{A}(t-\tau)d\tau$$

or

(III, 3; 3) $$i(t) = e(t) * \mathrm{A}(t).$$

The argument employed is obviously insufficient from the mathematical point of view.

In the first place we must specify that the operator is defined not only for functions zero for $t < 0$ but for distributions in $\mathcal{D}'_+$ having their support in the half-line $(0 \leqslant t \leqslant +\infty)$ and that to such a distribution corresponds a distribution i also belonging to $\mathcal{D}'_+$. Finally, we must add to the three hypotheses a hypothesis of continuity :

4) If, as $j \to \infty$, excitations e_j converge to an excitation e (in the sense of convergence of distributions), the responses i_j corresponding to the e_j converge to the response i corresponding to e.

The consideration of distributions arises naturally in the case of electromotive force. It is still more natural in the case of a current. Consider an elementary current obtained by passing a *single charge* q at the instant τ to the apparatus measuring the current magnitude. The current is the derivative with respect to time of the quantity of electricity and the quantity of electricity passed is $q\mathrm{Y}(t - \tau)$, which is zero for $t < \tau$ and q for $t \geqslant \tau$. The current is thus $q\delta(t - \tau)$.

Besides, we know from other sources that the flow of current is a discontinuous phenomenon, represented by the successive passage of a large number of elementary charges, sensibly equivalent to the flow of a continuous current.

The mathematical derivation of equation (III, 3; 3) is thus made as follows : Equation (III, 3; 3) is true for $e = \delta$ by definition of A, and so, by 2) it is true for $e(t) = \delta(t - \tau)$. Hence it is true when e is any linear combination of a finite number of impulses,

$$e(t) = \Sigma_k e_k \delta(t - \tau_k),$$

from linearity 1). Finally it is true by continuity 4) in all cases where e is the limit (in $\mathcal{D}'_+$) of finite linear combinations of impulses. If we *prove* that every distribution in $\mathcal{D}'_+$ is the limit (in $\mathcal{D}'_+$) of finite linear combinations of point-masses, then (III, 3; 3) is true for all $e \in \mathcal{D}'_+$. It is this density theorem that is really being used when we write (III, 3; 2) and when we proceed from (III, 3; 1) to (III, 3; 2). We have recourse to it again when we assert that any current is composed of a stream of charged particles or that any portion of matter is composed of a large number of point masses. *Thus the impulsive response* A *gives the response to any excitation e by the equation* (III, 3; 3).

In practice, where we have a circuit with resistance R, inductance L and capacity C, we have the equation, for $t \geqslant 0$:

(III, 3; 4) $$e(t) = \mathrm{R}i(t) + \mathrm{L}\frac{di}{dt} + \frac{1}{\mathrm{C}}\int_0^t i(\tau)\, d\tau$$

or

(III, 3; 5) $$e = \mathrm{Z} * i,$$

with

(III, 3; 6) $$\mathrm{Z} = \mathrm{R}\delta + \mathrm{L}\delta' + \frac{1}{\mathrm{C}}\mathrm{Y}.$$

The distribution $\mathrm{Z}(t)$ is the *impedance* of the circuit. It belongs to $\mathcal{D}'_+$ and gives e in terms of i by convolution. A, which is also called the *admittance*, will be the inverse of Z in the convolution algebra $\mathcal{D}'_+$:

(III, 3; 7) $$\mathrm{A} = \mathrm{Z}^{-1} \quad \text{or} \quad \mathrm{A} * \mathrm{Z} = \delta.$$

Alternatively, by differentiating both sides, we can also write (III, 3; 7) as

$$\left(\mathrm{L}\delta'' + \mathrm{R}\delta' + \frac{1}{\mathrm{C}}\delta\right) * \mathrm{A} = \delta'.$$

A is thus the function which is equal to zero for $t < 0$, and for $t \geqslant 0$ is given by the usual solution of the homogeneous equation,

$$\mathrm{L}u'' + \mathrm{R}u' + \frac{1}{\mathrm{C}}u = 0,$$

satisfying the initial conditions

$$u(0) = \frac{1}{L}, \qquad u'(0) = -\frac{R}{L^2}.$$

Equation (III, 3; 7) corresponds to (III, 3; 5) with $e = \delta$ and $i = A$. Putting as usual $\delta = 1$ (the unit element of $\mathcal{D}'_+$), $\delta' = p$, $Y = 1/p$, we have

(III, 3; 8)
$$Z = R + Lp + \frac{1}{Cp} = \frac{Lp^2 + Rp + \frac{1}{C}}{p}.$$

Hence,

(III, 3; 9)
$$A = \frac{p}{Lp^2 + Rp + \frac{1}{C}}.$$

The roots of the denominator are

(III, 3; 10)
$$p = -\frac{R}{2L} \pm \frac{\sqrt{R^2 - \frac{4L}{C}}}{2L} = -\frac{R}{2L} \pm j\sqrt{\frac{1}{LC} - \frac{R^2}{4L^2}}$$

(where $j = \sqrt{-1}$ as is the usual electrical notation). If R^2 is small compared with L/C, then

(III, 3; 11)
$$p \doteqdot -\frac{R}{2L} \pm \frac{j}{\sqrt{LC}},$$

whence the following approximation to (III, 3; 9):

(III, 3; 12)
$$A \doteqdot \frac{\frac{R}{2L} + \frac{j}{\sqrt{LC}}}{\frac{2jL}{\sqrt{LC}}\left(p + \frac{R}{2L} + \frac{j}{\sqrt{LC}}\right)} + \frac{-\frac{R}{2L} + \frac{j}{\sqrt{LC}}}{\frac{2jL}{\sqrt{LC}}\left(p + \frac{R}{2L} - \frac{j}{\sqrt{LC}}\right)}.$$

Taking into account that

$$\frac{1}{p - \lambda} = Y(t)e^{\lambda t}$$

and that R^2 is assumed to be small compared with L/C, we have the formula

(III, 3; 13)
$$A(t) = \frac{Y(t)}{L} e^{-\frac{R}{2L}t} \cos\frac{2\pi t}{T};$$

(III, 3; 14)
$$T = 2\pi\sqrt{LC}$$

is the period of natural oscillation of the circuit, and $e^{\frac{-R}{2L}t}$ the damping factor.

Exercise. Make all the calculations without neglecting R^2 compared with $\frac{L}{C}$. The example which has just been given is a very special case only. Whenever a function or a distribution i depends on any e satisfying hypotheses 1), 2), 3), and 4), there is an impulsive response A such that $i = A * e$.

Examples

1) The roles of i and e can be reversed. We can take i as the excitation, e as the response, and the impulsive response is then the impedance Z.

2) The response can be the same as the excitation. In this case $A = \delta$.

3) The excitation can be composed of several distributions or functions, for example $e_1(t)$, $e_2(t)$, and the response of several distributions or functions, for example $i_1(t)$, $i_2(t)$, $i_3(t)$. This will occur where a network involves two emf's and three unknown currents. In such a case matrices must be used.

We have the equation

(III, 3; 15)
$$\begin{pmatrix} i_1 \\ i_2 \\ i_3 \end{pmatrix} = \begin{pmatrix} A_1^1 & A_1^2 \\ A_2^1 & A_2^2 \\ A_3^1 & A_3^2 \end{pmatrix} * \begin{pmatrix} e_1 \\ e_2 \end{pmatrix},$$

that is to say,

(III, 3; 16)
$$\begin{aligned} i_1 &= A_1^1 * e_1 + A_1^2 * e_2 \\ i_2 &= A_2^1 * e_1 + A_2^2 * e_2 \\ i_3 &= A_3^1 * e_1 + A_3^2 * e_2. \end{aligned}$$

The impulsive response is thus a matrix, here of three rows and two columns. The meaning of each of these elements is obvious : for example A_3^1 is the value of i_3 for $e_1 = \delta$, $e_2 = 0$.

4) A rod AB is a conductor of heat. It is insulated except at its ends A and B. At A it receives a time-varying quantity of heat commencing at the moment $t = 0$ and such that $e_1(t)$ is the quantity of heat received per unit time at time t; the amount of heat received from the initial instant until time t is then

$$\int_0^t e_1(\tau)\,d\tau.$$

Similarly at B, with $e_2(t)$.

It is required to find the subsequent temperature in the rod, given that its temperature was zero for $t < 0$. The temperature $U(x, t)$ at the point x at time t is then, for each fixed x, a function $i(t)$ which we can take as the response, whence the impulsive responses $A_1(t)$, $A_2(t)$ (depending on x), and the equation

(III, 3; 17)
$$\begin{aligned} U(x, t) &= A_1 * e_1 + A_2 * e_2 \\ &= \int_0^t A_1(x, t-\tau)e_1(\tau)\,d\tau + \int_0^t A_2(x, t-\tau)e_2(\tau)\,d\tau. \end{aligned}$$

5) We have the same problem with the wave equation : resonance of a compartment by external sound excitation, forced vibrations of a string or organ pipe, etc.

General remarks

1) We often attempt to find $A(t)$ *experimentally*. Instead of the unit impulse $e(t) = \delta(t)$, which can be difficult to realise, we take the excitation to be Heaviside's step function $e(t) = Y(t)$. The response is then

$$i = A * Y = \int_0^t A(\tau)\, d\tau,$$

and by differentiating we obtain $A = di/dt$. However we must then take care to differentiate in the sense of distributions. On the other hand, if Y is in practice more easily realisable than δ, a derivative is more difficult to obtain with precision than the function itself.

2) Take as excitation $e(t) = e^{pt}$. Such an excitation is not automatically permissible without broadening the initial conditions, for e^{pt} is not zero for $t < 0$. However, for $\mathcal{R}p > 0$, e^{pt} tends to zero as $t \to -\infty$, and this happens to be a right choice to make. Then

$$\text{(III, 3; 18)} \qquad i(t) = A * e^{pt} = e^{pt} \int_0^\infty e^{-p\tau} A(\tau)\, d\tau = \mathcal{A}(p) e^{pt}.$$

The response is then proportional to the excitation, the factor of proportionality being $\mathcal{A}(p)$ which is termed *the isomorphic response factor.* $\mathcal{A}(p)$ is the Laplace transform of A, which thus enables us, if it is known for all values of p whose real part is sufficiently large, to determine A.

We see the mathematical nature of the restriction mentioned more clearly : we can only use this procedure if A has a Laplace transform. In the example treated previously (electric circuit) the accurate expression for $\mathcal{A}(p)$ was

$$\frac{p}{Lp^2 + Rp + \frac{1}{C}}.$$

3) Take the excitation to be $e(t) = e^{j\omega t}$, ω real. Here again $e(t)$ is not zero for $t < 0$ and, being periodic, it does not tend to zero as $t \to -\infty$. Suppose nevertheless that this excitation is permissible. The response will be

$$\text{(III, 3; 19)} \qquad i(t) = A * e^{j\omega t} = e^{j\omega t} \int_0^\infty e^{-j\omega t} A(\tau)\, d\tau = \mathcal{A}(j\omega) e^{j\omega t}.$$

The response is still proportional to the excitation, the factor of proportionality being $\mathcal{A}(j\omega)$, called *the isochronic response factor.* $\mathcal{A}(j\omega)$ is *the Fourier transform* of A, which, if known for all real values of ω, enables one to determine A.

The mathematical nature of the restriction imposed on the problem here is as follows : on the one hand A must have a Fourier transform, and this $\mathcal{A}(j\omega)$ must be a function, since it must have a value for all ω.

In the classical case of a periodic alternating current of frequency ω, it is well known that

$$z(j\omega) = R + Lj\omega + \frac{1}{Cj\omega}, \tag{III, 3; 20}$$

which is termed the impedance, and

$$\mathcal{A}(j\omega) = \frac{1}{z(j\omega)},$$

which is termed the admittance. In the latter two cases, since the excitation $e(t)$ considered is not zero for $t < 0$, the same is true of $i(t)$. There are therefore an infinite number of possible responses to the same excitation (the difference between any two of them being a response corresponding to zero excitation) and we must specify which to choose. We choose the response which is *proportional* to the excitation [the factor of proportionality being precisely $\mathcal{A}(p)$ or $\mathcal{A}(j\omega)$]. Besides, it is found in practice that all responses tend to this one as $t \to +\infty$.

EXERCISES FOR CHAPTER III

Exercise III-1

Find the inverses, in $\mathcal{D}'_+$, of the following distributions:

$$\delta'' - 5\,\delta' + 6\,\delta, \qquad Y + \delta'', \qquad Y(x)e^x + \delta'.$$

Exercise III-2

Solve the integral equation

$$\int_0^x (x - \xi) \cos (x - \xi) f(\xi)\, d\xi = g(x),$$

where g is a given function and f an unknown function with supports in $(0, +\infty)$.

Exercise III-3

$g(x)$ being a function that has its support in $(0, +\infty)$, find a function $f(x)$ having its support in $(0, +\infty)$ and satisfying for $x \geqslant 0$ the integral equation

$$\int_0^x (e^{-\xi} - \sin \xi) f(x - \xi)\, d\xi = g(x).$$

What condition must be satisfied by g for the solution to be a continuous function? Find the solution in the case where g is Heaviside's function $Y(x)$.

Exercise III-4
Solve the integral equation

$$f(x) + \int_0^x \cos(x - \xi) f(\xi)\, d\xi = g(x)$$

where g is a given function and f an unknown function with supports in $(0, +\infty)$.

Exercise III-5
Let $f(t)$ be the solution of the differential equation

$$x''' + 2x'' + x' + 2x = -10 \cos t$$

satisfying the initial conditions

$$x(0) = 0; \qquad x'(0) = 2; \qquad x''(0) = -4.$$

Let $F(t) = Y(t) f(t)$, where $Y(t)$ is Heaviside's function. Write the differential equation satisfied by $F(t)$ in the distribution sense. Then find $F(t)$ by using the operator calculus in $\mathcal{D}'_+$.

Exercise III-6
Evaluate the convolution product

$$Y(x) \sin x * Y(x) \sinh 2x.$$

Find a differential operator D with constant coefficients whose elementary solution is this convolution product. Find the solution of the differential equation

$$y^{(4)} - 3y^{(2)} - 4y = 0$$

satisfying the initial conditions

$$y(0) = 1, \qquad y'(0) = y''(0) = y'''(0) = 0.$$

Exercise III-7
Show that the following functions are in L^1:

$$e^{-|x|}; \qquad e^{-ax^2}, \quad a > 0; \qquad xe^{-ax^2}, \quad a > 0.$$

Find the convolution products

$$(a) \qquad e^{-|x|} * e^{-|x|};$$

$$(b) \qquad e^{-ax^2} * xe^{-ax^2};$$

$$(c) \qquad xe^{-ax^2} * xe^{-ax^2}.$$

Exercise III-8
Prove equation (III, 2; 23).

Exercise III-9
Find the successive powers of convolution $(f)^{*n}$ of

$$f(x) = \begin{cases} 1 \text{ for } -1 < x < 1, \\ 0 \text{ elsewhere.} \end{cases}$$

How can the result be deduced for $(f')^{*n}$? What is the support of $(f)^{*n}$?

Exercise III-10
Consider in the (x, y) plane the region $0 < x < |y|$. If f and g are two functions having their supports in this region, write $(f * g)(u, v)$. Evaluate $Y * Y$, where Y is the characteristic function of the region.

Exercise III-11
Consider the distribution

$$\sum_{n=0}^{\infty} \delta_{(n)}.$$

Let $E(x)$ be the primitive of this distribution, with its support in $(0, +\infty)$. Evaluate $E(x) * E(x)$.

Exercise III-12
Solve, in $\mathfrak{D}'_+$, the system of convolution equations

$$\begin{aligned} \delta'' * X_1 + \delta' * X_2 &= \delta \\ \delta' * X_1 + \delta'' * X_2 &= 0. \end{aligned}$$

Exercise III-13
Solve, in $\mathfrak{D}'_+$, the system of convolution equations

$$\begin{aligned} Y(x)e^{\alpha_1 x} * X_1 + Y(x)e^{\alpha_2 x} * X_2 + Y(x)e^{\alpha_3 x} * X_3 &= U_1 \\ Y(x)e^{\alpha_3 x} * X_1 + Y(x)e^{\alpha_1 x} * X_2 + Y(x)e^{\alpha_2 x} * X_3 &= U_2 \\ Y(x)e^{\alpha_2 x} * X_1 + Y(x)e^{\alpha_3 x} * X_2 + Y(x)e^{\alpha_1 x} * X_3 &= U_3 \end{aligned}$$

where U_1, U_2, U_3 are known distributions and X_1, X_2, X_3 unknown distributions in $\mathfrak{D}'_+$.

Exercise III-14 (Examination question, Paris, 1959)
Denote by $\mathfrak{J}$ the set of distributions on the real line that can be expressed as finite sums

(1) $$\sum_{m,n} a_{m,n}\delta_{(n)}^{(m)}$$

where

(2) $$\sum_{n} a_{m,n} = 0.$$

($\delta_{(n)}^{(m)}$ is the m-th derivative of the distribution defined by unit mass at the point n; m is any integer $\geqslant 0$ and n any integer of either sign.)

Denote by $\mathcal{J}_1$ the set of distributions of the type $S_1 = \delta + S$ where $S \in \mathcal{J}$. Show that $S_1 \in \mathcal{J}_1$ if and only if it is of the form (1) with a condition similar to 2), which is to be found.

(i) Show that $\mathcal{J}$ is a convolution algebra (not containing unity).

(ii) Show that, if T is a periodic distribution of period 1 (that is to say, unaltered by its translation through the distance 1), then

$$S * T = 0 \quad \text{for} \quad S \in \mathcal{J}$$

and

$$S_1 * T = T \quad \text{for} \quad S_1 \in \mathcal{J}_1.$$

Let A be a distribution with bounded support, and suppose that E is a distribution with bounded support, for which

$$A * E \in \mathcal{J}_1. \tag{3}$$

Then show that, if B is a periodic distribution of period 1, there exists one and only one distribution X which is periodic with period 1, such that

$$A * X = B. \tag{4}$$

(iii) Let D be a differential operator with constant coefficients :

$$D = \frac{d^N}{dx^N} + a_{N-1}\frac{d^{N-1}}{dx^{N-1}} + \cdots + a_2\frac{d^2}{dx^2} + a_1\frac{d}{dx} + a_0. \tag{5}$$

Show that, if $E(x)$ is a function which is zero for $x < 0$ and $x > 1$, the usual solution in the interval $0 \leqslant x \leqslant 1$ of the homogeneous equation $DE = 0$ satisfies, $DE \in \mathcal{J}_1$ in the distribution sense if and only if the quantities

$$\sigma_k = E^{(k)}(0) - E^{(k)}(1), \qquad k = 0, 1, \cdots, N - 1 \tag{6}$$

satisfy a system of N linear equations to be found; solve this system. (Solution obvious.)

(iv) Take successively

$$D = \frac{d}{dx} - \alpha \qquad \text{and} \qquad D = \left(\frac{d}{dx} - \alpha\right)\left(\frac{d}{dx} - \beta\right),$$

α and β complex, $\alpha \neq \beta$. Find the function $E(x)$ defined in (iii) by using the values found for the σ_k, and the general form of solutions of the homogeneous equation $DE = 0$. Does this function exist for all values of α and β ($\alpha \neq \beta$)?

Exercise III-15
By using the relation

$$e^{a_1x_1+a_2x_2+\cdots+a_nx_n}S * e^{a_1x_1+\cdots+a_nx_n}T = e^{a_1x_1+\cdots+a_nx_n}(S * T),$$

find a partial differential equation whose elementary solution is the distribution

$$\frac{e^{a_1x_1+\cdots+a_nx_n}}{(2-n)\,S_n r^{n-2}}$$

where n is a positive integer not equal to 2.

CHAPTER IV

Fourier series

I. Fourier series of periodic functions and distributions

I. FOURIER SERIES EXPANSION OF A PERIODIC FUNCTION

Period of a function

Any real number T such that $f(x + T) = f(x)$ is called a *period* of the function $f(x)$. The number 0 is always a period. The negative of a period of f and the sum of two periods of f are also periods of f, so that the periods of f form a sub-group of the additive group R of the real numbers. This sub-group is called the *period group* of f.

If f is continuous, its period group is a *closed* sub-group of R, for any limit T of a sequence of periods T_j of f is also a period of f. Now, there are only three categories of closed sub-groups of R :

1) The sub-group reduces to 0. A function f having no period $T \neq 0$ is termed non-periodic.

2) The whole group R. A function having every real number T as a period is constant.

3) The set of multiples lT_0 (integer $l > 0$, < 0 or $= 0$) of a number $T_0 > 0$.

If the period group of f is in one of the last two categories, f is called *periodic*. In the third case T_0 is termed the *fundamental period* of f.

Note. For a function on R^n (function of n variables) a period is a vector $\vec{T}$ such that $f(\vec{x} + \vec{T}) = f(\vec{x})$. The periods of f form an additive sub-group of the vectors of R^n, which is closed if f is continuous. However, there are then more than three categories (six in R^2, $[(n + 1)(n + 2)]/2$ in R^n).

In R^n, if $\vec{T}_1, \vec{T}_2, \ldots, \vec{T}_n$ are n independent vectors, the set of linear expressions $l_1\vec{T}_1 + l_2\vec{T}_2 + \cdots + l_n\vec{T}_n$, with positive or negative integer coefficients $l_1, l_2, \ldots, l_n$, is a very important sub-group.

A function possessing each vector of such a sub-group as a period is called *n-tuply periodic.*

Exponentials of period T. Fourier series

An exponential $e^{\lambda x}$ has the real period $T > 0$ if and only if $e^{\lambda T} = 1$, or $\lambda T = 2ki\pi$, k any integer. If T is given, the possible values of λ are

$$\lambda = ik\frac{2\pi}{T} = ik\omega,$$

ω being termed the *angular frequency* associated with period T. If we take $k = 0$, the constant 1 with R as its period group is obtained. For $k \neq 0$ an exponential of modulus 1 having the fundamental period $T/|k|$ is obtained.

The fundamental problem which we are going to consider is whether a periodic function f possessing the period $T > 0$ can be expanded in a series, called a *Fourier series*,

$$f(x) = \sum_{k=-\infty}^{+\infty} c_k e^{ik\omega x}, \tag{IV, 1; 1}$$

each term of the series being, to within a factor, one of the exponentials of period T.

Sometimes pairs of opposite value of k are grouped and the series of exponentials replaced by a series of sines and cosines,

$$f(x) = a_0 + \sum_{k=1}^{\infty} (a_k \cos k\omega x + b_k \sin k\omega x). \tag{IV, 1; 2}$$

The relations between a_k, b_k and c_k are obviously

$$\text{(IV, 1; 3)} \qquad a_0 = c_0, \qquad\qquad c_0 = a_0,$$

$$k > 0 \quad \begin{cases} a_k = c_k + c_{-k}, \\ b_k = i(c_k - c_{-k}), \end{cases} \qquad \begin{cases} c_k = \dfrac{a_k - ib_k}{2}, \\ c_{-k} = \dfrac{a_k + ib_k}{2}. \end{cases}$$

Fourier coefficients of a locally summable function

Assume that the series (IV, 1; 1) is uniformly convergent, implying that f is continuous. Then term-by-term integration enables one to obtain

$$C_k = \int_a^{a+T} f(x) e^{-ik\omega x} \frac{dx}{T}. \tag{IV, 1; 4}$$

The interval $(a, a + T)$ is a period interval of f, and the integral is obviously independent of where this interval is situated, that is to say, of a. For, if $b \neq a$, then we can write

$$\int_b^{b+T} = \int_a^{a+T} + \left(\int_{a+T}^{b+T} - \int_a^b\right).$$

The expression in parentheses is zero, since the function $g(x) = f(x)e^{-ik\omega x}$ has period T so that

$$\int_{a+T}^{b+T} g(x)\,dx = \int_a^b g(\xi + T)\,d\xi = \int_a^b g(\xi)\,d\xi,$$

by the change of variable $x = \xi + T$. The presence of dx/T shows that C_k is the *mean* of $f(x)e^{-ik\omega x}$ over a period interval. From (IV, 1; 1),

(IV, 1; 5) $$C_k = \sum_{l=-\infty}^{+\infty} c_l \int_a^{a+T} e^{i(l-k)\omega x}\,\frac{dx}{T}.$$

Now,

(IV, 1; 6) $$\int_a^{a+T} e^{im\omega x}\,\frac{dx}{T} = \begin{cases} \left[\dfrac{e^{im\omega x}}{im\omega}\right]_a^{a+T} \dfrac{1}{T} = 0 & \text{if } m \neq 0, \\ 1 & \text{if } m = 0. \end{cases}$$

Finally only one non-zero term, corresponding to $l = k$, remains in the sum Σ, so that

(IV, 1; 7) $$C_k = c_k.$$

Thus, if f is continuous and *if it possesses a uniformly convergent Fourier expansion*, this expansion is completely determined and

(IV, 1; 8) $$c_k = c_k(f) = \int_a^{a+T} f(x)e^{-ik\omega x}\,\frac{dx}{T}.$$

The quantities $c_k(f)$ are always defined, even if f is not continuous, provided that it is locally summable (summable over any finite interval or, from its periodic property, summable over any period interval). The $c_k(f)$ are always called the *Fourier coefficients* of f. They are unaltered if f is changed on a set of measure zero. It may thus be supposed that f is only defined almost everywhere.

For a series expansion in sines and cosines, it can easily be seen (either by calculating a_k and b_k from (IV, 1; 3) or directly by a method similar to that used for the c_k) that

(IV, 1; 9) $$a_0(f) = \int_a^{a+T} f(x)\,\frac{dx}{T},$$

$$a_k(f) = 2\int_a^{a+T} f(x)\cos k\omega x\,\frac{dx}{T}, \qquad b_k(f) = 2\int_a^{a+T} f(x)\sin k\omega x\,\frac{dx}{T},$$

where $k > 0$. The presence of the factor 2 originates from the fact that the mean of $\cos^2 k\omega x$ or $\sin^2 k\omega x$ (for $k > 0$) over a period interval is $\frac{1}{2}$. On the other hand, for $k = 0$, the mean of 1 is always 1.

General remarks

1) The Fourier coefficients involve the values of f *over the whole of a period interval,* unlike the coefficients of a Taylor expansion, which involve only the values of f in the neighbourhood of the point about which the expansion is made.

2) If f is an odd or even function, it is best to expand in a series of trigonometric functions. For, if f is even, only the a_k (including a_0) can be $\neq 0$, since $f(x)\sin k\omega x$ is odd and its integral [taking $(-\frac{1}{2}\mathrm{T}, +\frac{1}{2}\mathrm{T})$ as period interval] is zero. On the other hand, if f is odd only the b_k can be $\neq 0$, for $f(x)\cos k\omega x$ is odd and its integral is zero.

In each of these cases, the function whose mean is evaluated to obtain a coefficient $\neq 0$ is even, and its mean can be evaluated over a half-period, for example $(0, \mathrm{T}/2)$:

$$
\text{(IV, 1; 10)}\qquad
\begin{aligned}
&f \text{ even}: \begin{cases} a_0(f) = \dfrac{2}{\mathrm{T}}\displaystyle\int_0^{\mathrm{T}/2} f(x)\,dx, \\[2ex] a_k(f) = \dfrac{4}{\mathrm{T}}\displaystyle\int_0^{\mathrm{T}/2} f(x)\cos k\omega x\,dx, \quad k > 0, \end{cases} \\[2ex]
&f \text{ odd}: \quad b_k(f) = \dfrac{4}{\mathrm{T}}\displaystyle\int_0^{\mathrm{T}/2} f(x)\sin k\omega x\,dx.
\end{aligned}
$$

In the same way, if f satisfies other kinds of relations, certain Fourier coefficients are zero or satisfy simple relations. For example, if f possesses not only T but also the sub-multiple T/p as a period, the only c_k not to disappear correspond to values of k which are multiples of p, that is to say, for which $e^{ik\omega x}$ also has the period T/p.

3) Let f be a function given in an interval (a, b). The Fourier expansion of the periodic function f_1 of period $b - a$ which coincides with f for $a \leqslant x < b$ is called the *Fourier expansion of f in the interval (a, b)*. Note that if f is continuous at $x = a$ and $x = b$ but $f(a) \neq f(b)$, then f_1 is necessarily discontinuous at $x = a$ and $x = b$. Similarly, if f is a function given in the interval $(0, \mathrm{L})$, the cosine (or sine) series expansion of the even (or odd) function f_1 which coincides with f in the interval $(0, \mathrm{L})$ is called the cosine (or sine) series expansion of f. Note that, for a sine series, if f is continuous at $x = 0$ and $x = \mathrm{L}$ but $f(0) \neq 0$ or $f(\mathrm{L}) \neq 0$, then f_1 is necessarily discontinuous at $x = 0$ or at $x = \mathrm{L}$.

Of course, in all these cases, the Fourier coefficients are expressed as means over the values of f in the given interval (a, b) or $(0, \mathrm{L})$. Applications will be encountered in the theory of vibrating strings or resonant tubes. Note finally that we always have the inequalities

$$
\text{(IV, 1; 11)}\qquad |c_k(f)| \leqslant \int_a^{a+\mathrm{T}} |f(x)|\frac{dx}{\mathrm{T}} = \frac{1}{\mathrm{T}}\|f\|_{\mathrm{L}^1} \leqslant \sup_x |f(x)|,
$$

and, for the coefficients of the sine or cosine series,

$$\text{(IV, 1; 12)}\quad \begin{cases} |a_0(f)| \leqslant \displaystyle\int^{a+T} |f(x)| \frac{dx}{T} = \frac{1}{T} \|f\|_{L^1} \leqslant \sup_x |f(x)|, \\ \left.\begin{matrix}|a_k(f)| \\ |b_k(f)|\end{matrix}\right\} \leqslant 2 \displaystyle\int_a^{a+T} |f(x)| \frac{dx}{T} = \frac{2}{T} \|f\|_{L^1} \leqslant 2 \sup_x |f(x)| \\ \qquad\qquad\qquad\qquad\qquad\qquad\qquad \text{for } k > 0. \end{cases}$$

A theorem due to Lebesgue shows that c_k always tends to 0 as $k \to \infty$.

2. FOURIER SERIES EXPANSION OF A PERIODIC DISTRIBUTION

Periodic distribution

In this chapter we denote distributions by the letters $\mathcal{S}$, $\mathcal{C}$, etc. to avoid any confusion with the period T.

Let f be a locally summable function. The shifted function $f(x - T)$ is defined by the functional

$$\text{(IV, 1; 13)}\quad \langle f(x-T), \varphi(x)\rangle = \int_{-\infty}^{+\infty} f(x-T)\varphi(x)\,dx$$
$$= \int_{-\infty}^{+\infty} f(x)\varphi(x+T)\,dx = \langle f(x), \varphi(x+T)\rangle.$$

For a distribution $\mathcal{C}_x$ we are thus led to define the shifted distribution $\mathcal{C}_{x-T}$ by

$$\text{(IV, 1; 14)}\qquad \langle \mathcal{C}_{x-T}, \varphi(x)\rangle = \langle \mathcal{C}_x, \varphi(x+T)\rangle.$$

A distribution $\mathcal{C}$ will be called periodic with period T if $\mathcal{C}_{x-T} = \mathcal{C}_x$ or alternatively if, for any $\varphi \in \mathcal{D}$,

$$\text{(IV, 1; 15)}\qquad \langle \mathcal{C}_x, \varphi(x+T) - \varphi(x)\rangle = 0.$$

The spaces $\mathcal{D}(\Gamma)$ and $\mathcal{D}'(\Gamma)$

In the plane xOy let Γ be *the circumference of length* T *of a circle with centre* O. Any function f on Γ can be brought into correspondence with a function $\tilde{f}$ on R by putting $\tilde{f}(x) = f(M)$, M being the point whose curvilinear distance s is x; the curvilinear distance is measured in the usual sense from the point A where Γ meets Ox. $\tilde{f}$ is a periodic function with period T.

Conversely, if $\tilde{f}$ is a periodic function on R with period T, it originates by the above procedure from one and only one function f on Γ defined by $f(M) = \tilde{f}(x)$, where x is one of the curvilinear distances of M, any two of which can differ only by a multiple of T. The correspondence $f \to \tilde{f}$, between functions on Γ and functions of period T on R, is an isomorphism between these two sets of functions. For the set of functions on Γ there are definitions of continuity, differentiability d/ds with respect to the curvilinear distance, summability with respect to ds, etc., which are in correspondence with the definitions of continuity, differentiability in x, local summability, etc., of periodic functions on R.

It is essential to note that one of the most frequent reasons for introducing a periodic function is the need to consider a function on the circle, that is, a function of the angle θ, with period 2π.

The space of complex functions on Γ which are infinitely differentiable with respect to the curvilinear distance s is denoted by $\mathcal{D}(\Gamma)$. By the correspondence defined at the beginning of this section, these functions $\varphi \in \mathcal{D}(\Gamma)$ are associated with the infinitely differentiable functions $\tilde{\varphi}$ on R, which form a sub-space of $\mathcal{E}$ but have no relation to the functions of $\mathcal{D}(\mathrm{R})$. A sequence $\varphi_j \to 0$ in $\mathcal{D}(\Gamma)$ if the φ_j, as well as each of their derivatives, tend to 0 uniformly on Γ. Thus the space $\mathcal{D}(\Gamma)$ is simpler than the space $\mathcal{D}(\mathrm{R})$, which originates from the fact that Γ is a bounded set. The function equal to 1 on Γ belongs to $\mathcal{D}(\Gamma)$, while the function equal to 1 on R does not belong to $\mathcal{D}(\mathrm{R})$.

The space $\mathcal{D}'(\Gamma)$ is the space of continuous linear forms on $\mathcal{D}(\Gamma)$.

One-to-one correspondence between distributions of period T and distributions in $\mathcal{D}'(\Gamma)$

Let f be a locally summable function on Γ, $\tilde{f}$ the associated periodic function on R. For $\varphi \in \mathcal{D}(\mathrm{R})$ we construct $\tilde{\Phi}$, the sum of the functions obtained by shifting φ:

$$\tilde{\Phi}(x) = \sum_{l=-\infty}^{+\infty} \varphi(x + l\mathrm{T}). \tag{IV, 1; 16}$$

The function $\tilde{\Phi}$ is periodic with period T, infinitely differentiable and thus associated with a function $\Phi \in \mathcal{D}(\Gamma)$. Of course, for each value of x the sum Σ is in fact a finite sum. It can easily be seen that

$$\int_\Gamma f(s)\Phi(s)\, ds = \int_\mathrm{R} \tilde{f}(x)\varphi(x)\, dx, \tag{IV, 1; 17}$$

for the right-hand side here is equal to

$$\begin{aligned}\sum_{l=-\infty}^{+\infty} \int_{l\mathrm{T}}^{(l+1)\mathrm{T}} \tilde{f}(x)\varphi(x)\, dx &= \sum_{l=-\infty}^{+\infty} \int_0^\mathrm{T} \tilde{f}(x)\varphi(x + l\mathrm{T})\, dx \\ &= \int_0^\mathrm{T} \tilde{f}(x)\tilde{\Phi}(x)\, dx\end{aligned} \tag{IV, 1; 18}$$

which is equal to the left-hand side of (IV, 1; 17).

We are thus led to *define* the periodic distribution $\tilde{\mathcal{T}}$ of period T on R, associated with the distribution $\mathcal{T}$ on Γ, by

$$\text{(IV, 1; 19)} \qquad \langle \tilde{\mathcal{T}}, \varphi \rangle = \langle \mathcal{T}, \Phi \rangle.$$

This equation gives a consistent definition of a periodic distribution $\tilde{\mathcal{T}}$, for if $\varphi(x)$ is replaced by $\varphi(x + T)$, $\tilde{\Phi}$ and hence Φ are unaltered. For example, if $\mathcal{T}$ is the distribution δ on Γ, a mass $+1$ placed at the point of curvilinear distance 0, then $\tilde{\mathcal{T}}$ consists of masses $+1$ placed at each point with coordinate lT on the line:

$$\text{(IV, 1; 20)} \quad \langle \tilde{\delta}, \varphi \rangle = \sum_{l=-\infty}^{+\infty} \varphi(lT), \quad \text{or} \quad \tilde{\delta}(x) = \sum_{l=-\infty}^{+\infty} \delta(x - lT).$$

Conversely it may be shown that any periodic distribution of period T *on* R *is the distribution* $\tilde{\mathcal{T}}$ *associated with one and only one distribution* $\mathcal{T}$ *on* Γ.

Henceforth we shall consider only distributions $\mathcal{T}$ on Γ.

Convolution in $\mathcal{D}'(\Gamma)$

On Γ there is an operation of addition of arcs, to within a multiple of T. Hence we may define the convolution product of two distributions $\mathcal{S}$ and $\mathcal{T}$ by

$$\text{(IV, 1; 21)} \qquad \langle \mathcal{S} * \mathcal{T}, \varphi \rangle = \langle \mathcal{S}_\xi \otimes \mathcal{T}_\eta, \varphi(\xi + \eta) \rangle.$$

This product *always* exists since any distribution on Γ has a *bounded support.* It has properties similar to the product of convolution on R but is more simple. In particular, $\delta * \mathcal{T} = \mathcal{T}$, $\delta' * \mathcal{T} = \mathcal{T}'$, etc. For two functions f and g the product is given by the equation

$$\text{(IV, 1; 22)} \qquad h(s) = \int_\Gamma f(s - t) g(t)\, dt.$$

In terms of the associated periodic functions $\tilde{f}$ and $\tilde{g}$ on R, this product can be written

$$\text{(IV, 1; 23)} \qquad h(x) = \int_a^{a+T} \tilde{f}(x - \xi) \tilde{g}(\xi)\, d\xi,$$

$(a, a + T)$ being any period interval. This product has no relation to that of $\tilde{f}$ and $\tilde{g}$ over R, which would be an integral $\int_{-\infty}^{+\infty}$ and, for two functions which are periodic (and thus without bounded supports), would have no meaning.

Fourier series of a distribution

The Fourier coefficients of a locally summable function f on Γ are just

$$\text{(IV, 1; 24)} \qquad c_k(f) = \int_\Gamma f(s)e^{-ik\omega s}\frac{ds}{T} = \frac{1}{T}\langle f, e^{-ik\omega s}\rangle.$$

We thus *define* the Fourier coefficients of $\mathcal{T} \in \mathcal{D}'(\Gamma)$ by

$$\text{(IV, 1; 25)} \qquad c_k(\mathcal{T}) = \frac{1}{T}\langle \mathcal{T}, e^{-ik\omega s}\rangle.$$

The series

$$\text{(IV, 1; 26)} \qquad \sum_{k=-\infty}^{+\infty} c_k(\mathcal{T})e^{ik\omega s},$$

with the c_k given by (IV, 1; 25), will be called the *Fourier series* of $\mathcal{T}$.

Example. For $\mathcal{T} = \delta$, $c_k(\delta) = (1/T)\langle \delta, e^{-ik\omega s}\rangle = 1/T$ and, in consequence the Fourier series of δ is

$$\sum_{k=-\infty}^{+\infty} \frac{1}{T} e^{ik\omega s}.$$

Note that if the trigonometric series $\sum_{l=-\infty}^{\infty} C_l e^{il\omega s}$ converges to a distribution $\mathcal{T}$ in $\mathcal{D}'(\Gamma)$ [i.e., if for any $\varphi \in \mathcal{D}(\Gamma)$ the series of numbers $\sum_{l=-\infty}^{+\infty} C_l\langle e^{il\omega s}, \varphi\rangle$ converges to the number $\langle \mathcal{T}, \varphi\rangle$] then, by putting $\varphi = e^{-ik\omega s}$ and making use of (IV, 1; 6),

$$\text{(IV, 1; 27)} \qquad TC_k = \langle \mathcal{T}, e^{-ik\omega s}\rangle = Tc_k(\mathcal{T}).$$

Thus *no trigonometric series other than the Fourier series of $\mathcal{T}$ can converge to $\mathcal{T}$ in $\mathcal{D}'(\Gamma)$.*

It will be seen below that the presence of the coefficients T or $1/T$ complicates all the formulas, and that they are much simpler for $T = 1$, $\omega = 2\pi$.

2. Convergence of Fourier series in the distribution sense and in the function sense

1. CONVERGENCE OF THE FOURIER SERIES OF A DISTRIBUTION

THEOREM 1. *The Fourier series for δ converges [in the space of distributions $\mathcal{D}'(\Gamma)$] to δ:*

$$\text{(IV, 2; 1)} \qquad \sum_{k=-\infty}^{+\infty} \frac{1}{T} e^{ik\omega s} = \delta.$$

First, note that it is already known that this trigonometric series converges in $\mathcal{D}'(\Gamma)$, since the coefficients, which are all equal to $1/T$, are bounded and hence bounded above by a power of $|k|$ (see Chapter II, Theorem 14, page 97).

It remains to show that its sum $\mathcal{T}$ is in fact δ. Multiplying term by term by $e^{i\omega ks}$, we have

(IV, 2; 2) $$e^{i\omega s}\mathcal{T} = \sum_{k=-\infty}^{+\infty} \frac{1}{T} e^{i(k+1)\omega s} = \mathcal{T}$$

or

(IV, 2; 3) $$(e^{i\omega s} - 1)\,\mathcal{T} = 0.$$

$\mathcal{T}$ then satisfies the equation

(IV, 2; 4) $$\alpha\mathcal{T} = 0$$

where $\alpha = e^{i\omega s} - 1$ is a given infinitely differentiable function, everywhere $\neq 0$ except at the one point $s = 0$ of Γ, where it possesses a single root, $\alpha'(0) = i\omega \neq 0$. Hence, by Chapter II, Theorem 8 (page 92), we know that $\mathcal{T}$ is proportional to δ, i.e., or $\mathcal{T} = C\delta$.

However, by the definition of convergence of distributions,

$$\sum_{k=-\infty}^{+\infty} \frac{1}{T} e^{ik\omega s} = C\delta$$

means that

(IV, 2; 5) $$\sum_{k=-\infty}^{+\infty} \frac{1}{T} \langle e^{ik\omega s}, \varphi(s)\rangle = C\varphi(0)$$

for any $\varphi \in \mathcal{D}(\Gamma)$. Then, putting $\varphi = 1$, since

(IV, 2; 6) $$\langle e^{i\omega ks}, 1\rangle = \begin{cases} 0 & \text{for } k \neq 0, \\ T & \text{for } k = 0, \end{cases}$$

we have $C = 1$. Q.E.D.

Alternatively, from $\sum_{k=-\infty}^{+\infty} \frac{1}{T} e^{ik\omega s} = C\delta$, it follows necessarily [see (IV, 1; 27)] that

$$\frac{1}{T} = c_k(C\delta) = Cc_k(\delta) = \frac{C}{T}, \quad \text{whence} \quad C = 1.$$

Note. If instead of considering Γ we consider the real line R, we see that the series $\sum_{k=-\infty}^{+\infty} \frac{e^{ik\omega x}}{T}$ converges in $\mathcal{D}'(R)$ to the periodic distribution $\sum_{l=-\infty}^{+\infty} \delta(x - lT)$, consisting of a mass $+1$ at each point whose coordinate is an integer multiple of T:

(IV, 2; 7) $$\sum_{k=-\infty}^{+\infty} \frac{1}{T} e^{ik\omega s} = \sum_{l=-\infty}^{+\infty} \delta(x - lT).$$

THEOREM 2. *The Fourier series of any distribution* $\mathcal{T}$ *on* Γ *(or any periodic distribution* $\tilde{\mathcal{T}}$ *on* R*) converges to this distribution in* $\mathcal{D}'(\Gamma)$, *or in* $\mathcal{D}'(\mathrm{R})$.

$$\sum_{k=-\infty}^{+\infty} c_k(\mathcal{T})e^{ik\omega s} = \mathcal{T}. \tag{IV, 2; 8}$$

For, from the continuity of convolution (without any difficulty as regards the supporting sets on Γ since all of them are bounded), term-by-term convolution of the series (IV, 2; 1) with $\mathcal{T} \in \mathcal{D}'(\Gamma)$ can be performed, giving

$$\mathcal{T} = \mathcal{T} * \delta = \sum_{k=-\infty}^{+\infty} \frac{1}{\mathrm{T}}(e^{ik\omega s} * \mathcal{T}), \tag{IV, 2; 9}$$

a series which must be convergent in $\mathcal{D}'(\Gamma)$.

However, by the formula giving the product of convolution of a distribution with an infinitely differentiable function [see Chapter III (III, 2; 24)],

$$\mathcal{T} * e^{ik\omega s} = \langle \mathcal{T}_t, e^{ik\omega(s-t)}\rangle = e^{ik\omega s}\langle \mathcal{T}_t, e^{-ik\omega t}\rangle = \mathrm{T}c_k(\mathcal{T})e^{ik\omega s}, \tag{IV, 2; 10}$$

so that (IV, 2; 9) indeed gives (IV, 2; 8)*.

2. CONVERGENCE OF THE FOURIER SERIES OF A FUNCTION

If f is only supposed locally summable there is no reason to hope that the Fourier series $\sum_{k=-\infty}^{+\infty} c_k(f)e^{ik\omega x}$ converges to $f(x)$ at *all points* x, since an alteration of $f(x)$ on a set of measure zero does not change the Fourier coefficients. However it might be expected that there would be convergence *almost everywhere* (i.e. for almost all values of x) or *in the mean* (i.e. in the space L^1 of classes of functions summable over the period $(a, a + \mathrm{T})$). Unfortunately this is not so.

Suppose, then, that f is *continuous*. It might now be expected that there will be convergence at all points x, or even uniform convergence, but again this is not so. We have, however, the following theorem.

---

*It has been shown (Chapter II, Theorem 14, page 97) that if $|c_k|$ is bounded above by a power of $|k|$ as $|k| \to \infty$, then the trigonometric series $\Sigma c_k e^{ik\omega s}$ is convergent in $\mathcal{D}'(\Gamma)$. *We shall accept the converse without proof. Then, for any distribution* $\mathcal{T} \in \mathcal{D}'(\Gamma)$, $|c_k(\mathcal{T})|$ *satisfies such an inequality.*

THEOREM 3. *If f is a function of bounded variation in the period interval, then the Fourier series of f,*

$$\text{(IV, 2; 11)} \qquad \sum_{k=-\infty}^{+\infty} c_k(f)e^{ik\omega x}$$

is semi-convergent at each point $x \in \mathrm{R}$ to the arithmetic mean of the left-hand limit $f(x-0)$ and the right-hand limit $f(x+0)$:

$$\text{(IV, 2; 12)} \qquad \lim_{N \to \infty} \sum_{k=-N}^{k=+N} c_k(f)e^{ik\omega x} = \frac{1}{2}(f(x-0)+f(x+0)).$$

In particular, if f (always assumed of bounded variation) is continuous at a point x, the Fourier series at this point converges to $f(x)$. If f is continuous at each point of a closed interval (α, β), and thus also continuous from the left at α and to from right at β, then the semi-convergence is uniform in (α, β).

This theorem calls for several explanations and definitions.

1) A function f, defined in a closed interval (a, b), has a *total variation* $V(f; (a, b))$ defined as the upper bound of the sums $\sum_{i=0}^{n-1} |f(x_{i+1}) - f(x_i)|$ corresponding to the various divisions of (a, b) into a finite number of sub-intervals

$$(x_0, x_1), (x_1, x_2), \ldots, (x_{n-1}, x_n), \quad (x_0 = a, x_n = b).$$

If $V(f; (a, b))$ is *finite*, f is said to be of *bounded variation* in (a, b). Any *monotonic* bounded function is of bounded variation and

$$V(f; (a, b)) = |f(b) - f(a)|.$$

Thus any function which is the difference between two bounded increasing functions is of bounded variation, and it may be proved that conversely any function of bounded variation can be expressed as such a difference. In particular, if a bounded function has only a finite number of maxima and minima, between any two of which as well as between a and the first one and between b and the last one it is monotonic, then it is of bounded variation. This shows that the bounded functions met in practice are of bounded variation. However, it is possible to prove the existence of *continuous* functions that are not of bounded variation and even possess in *any* interval (α, β) an infinity of maxima and minima.

A continuous function with a bounded derivative (in the usual sense) is of bounded variation if the interval (a, b) is finite, and then

$$V(f; (a, b)) = \int_a^b |f'(x)|\, dx.$$

The same is true if f has only a finite number of points of discontinuity of the first kind (jumps) and if at all other points it has a derivative whose modulus is bounded by a fixed number, the interval (a, b) being still supposed finite.

2) A function f of bounded variation is not necessarily continuous, but it has at each point a *right-hand limit* and a *left-hand limit* :

$$\text{(IV, 2; 13)} \qquad \begin{cases} f(x-0) = \lim\limits_{\substack{\varepsilon \to 0 \\ \varepsilon < 0}} f(x+\varepsilon), \\ f(x+0) = \lim\limits_{\substack{\varepsilon \to 0 \\ \varepsilon > 0}} f(x+\varepsilon). \end{cases}$$

Also, these two limits are *equal* (and f is thus continuous), except at, at most, a *denumerable infinity* of points x.

3) When we say that the Fourier series is *semi-convergent*, this is to be understood as meaning that it can be summed as the formula indicates after grouping the pairs of opposite values of k. The sum $\sum_{k=-N}^{+N} c_k(f)e^{ik\omega x}$ has a limit but each sum $\sum_{k=0}^{N}, \sum_{k=-N}^{0}$ does not necessarily have a limit. On the other hand, in a series of cosines and sines the series of cosines and the series of sines are separately convergent, since they are respectively the Fourier series of

$$\frac{f(x)+f(-x)}{2} \quad \text{and} \quad \frac{f(x)-f(-x)}{2},$$

which are both of bounded variation.

4) The case where $f(x+0)$ and $f(x-0)$ are distinct reduces directly to the case where they are equal (and f is continuous) as soon as there is a proof for one particular function of bounded variation which is discontinuous at x. Now the discontinuity may always be supposed to occur at $x = 0$, and taking for f an *odd* function which is discontinuous at the origin gives $f(0+0) = -f(0-0) \neq 0$, so that the arithmetic mean is zero. Since f is odd, its Fourier series contains nothing but sine terms which are all zero for $x = 0$, so that the Fourier series does indeed converge at $x = 0$ to the arithmetic mean 0.

The proof of Theorem 3 is rather subtle. It uses the theory of the Dirichlet integral,

$$\int_a^b f(x) \frac{\sin \lambda x}{x}\, dx,$$

which is found in all treatises on differential and integral calculus.

We shall confine ourselves to a more simple case :

THEOREM 4. *If f is twice continuously differentiable, its Fourier series converges absolutely and uniformly to f.*

Assume that f possesses a continuous derivative f'. Integration by parts gives, for $k \neq 0$,

$$\text{(IV, 2; 14)} \quad c_k(f) = \left[\frac{e^{-ik\omega x}}{-ik\omega}\frac{f(x)}{\mathrm{T}}\right]_a^{a+\mathrm{T}} - \int_a^{a+\mathrm{T}} \frac{e^{-ik\omega x}}{-ik\omega} f'(x)\,\frac{dx}{\mathrm{T}}.$$

The bracketed term is zero, by periodicity, so that there remains the very important result

$$\text{(IV, 2; 15)} \qquad c_k(f) = \frac{1}{ik\omega}\, c_k(f').$$

More generally, if f is m times continuously differentiable then, for $k \neq 0$,

$$\text{(IV, 2; 16)} \qquad c_k(f) = \left(\frac{1}{ik\omega}\right)^m c_k(f^{(m)}),$$

whence

$$\text{(IV, 2; 17)} \qquad |c_k(f)| \leqslant \left|\frac{1}{k\omega}\right|^m \int_a^{a+\mathrm{T}} |f^{(m)}(x)|\,\frac{dx}{\mathrm{T}} = \frac{\mathrm{M}_m}{|k|^m}.$$

Thus the more regular f is, the more rapidly do the Fourier coefficients tend to 0 as $|k| \to \infty$.

In particular, for $m = 2$, $|c_k(f)|$ is bounded above by M_2/k^2 and, as $|e^{ik\omega x}| = 1$, the series of moduli $\sum_{k=-\infty}^{+\infty} |c_k(f)e^{ik\omega x}|$ is bounded above for all values of x by the convergent series of positive numbers $|c_0| + \sum_{\substack{k=-\infty \\ k\neq 0}}^{+\infty} \frac{\mathrm{M}_2}{k^2}$. This proves that the Fourier series of f converges absolutely and uniformly. Its sum g is a continuous function, but it remains to prove that $g = f$. It is known that the Fourier series converges to f in the distribution sense (Theorem 2), and as it converges uniformly to g it converges *a fortiori* to g in the distribution sense. Thus f and g, considered as distributions, are equal and hence, as functions, nearly everywhere equal. Being continuous, they must therefore be equal everywhere. Q.E.D.

Note. Equation (IV, 2; 16) can be extended to distributions. From (IV, 1; 25) it follows by the definition of the derivative of a distribution that

$$\text{(IV, 2; 18)} \quad \begin{cases} c_k(\mathcal{T}^{(m)}) = \dfrac{1}{\mathrm{T}}\langle \mathcal{T}^{(m)}, e^{-ik\omega s}\rangle \\ \qquad = \dfrac{(-1)^m}{\mathrm{T}}\langle \mathcal{T}, (e^{-ik\omega s})^{(m)}\rangle = (ik\omega)^m \dfrac{1}{\mathrm{T}}\langle \mathcal{T}, e^{-ik\omega s}\rangle \\ \qquad = (ik\omega)^m c_k(\mathcal{T}), \end{cases}$$

which is a generalization of (IV, 2; 16). This can be deduced in another way. From

$$\text{(IV, 2; 8)} \qquad \mathcal{T} = \sum_{k=-\infty}^{+\infty} c_k(\mathcal{T})e^{ik\omega s},$$

it follows, using term-by-term differentiation, always legitimate for distributions, that

(IV, 2; 19) $$\mathcal{T}^{(m)} = \sum_{k=-\infty}^{+\infty} c_k(\mathcal{T})\,(ik\omega)^m e^{ik\omega s}.$$

Here on the right-hand side there is a convergent trigonometric series which is the Fourier series of the left-hand side, so that

(IV, 2; 20) $$c_k(\mathcal{T}^{(m)}) = (ik\omega)^m c_k(\mathcal{T}),$$

which is again the extension (IV, 2; 18) of (IV, 2; 16) to periodic distributions.

3. Hilbert bases for Hilbert space. Mean-square convergence of a Fourier series

1. DEFINITION OF A HILBERT SPACE

A *Hilbert space* is defined as a vector space $\mathcal{H}$ over the field of complex numbers C, endowed with a form $(\vec{x}, \vec{y})$, called a scalar product, with values in C and possessing the following properties :

1) $(\vec{x}, \vec{y})$ is a *Hermitian form* :

(IV, 3; 1) $$\begin{cases} (\vec{x}_1 + \vec{x}_2, \vec{y}) = (\vec{x}_1, \vec{y}) + (\vec{x}_2, \vec{y}), \\ (\vec{x}, \vec{y}_1 + \vec{y}_2) = (\vec{x}, \vec{y}_1) + (\vec{x}, \vec{y}_2), \end{cases}$$

(IV, 3; 2) $$(\lambda\vec{x}, \mu\vec{y}) = \lambda\bar{\mu}(\vec{x}, \vec{y}).$$

The scalar product is thus said to be linear in $\vec{x}$ and semi-linear in $\vec{y}$.

(IV, 3; 3) $$(\vec{x}, \vec{y}) = \overline{(\vec{y}, \vec{x})}$$ (Hermitian property).

The Hermitian property is equivalent to

(IV, 3; 4) $(\vec{x}, \vec{x})$ is real for all $\vec{x}$.

2) $(\vec{x}, \vec{x})$ is *positive definite* or, alternatively,

(IV, 3; 5) $$\begin{cases} (\vec{x}, \vec{x}) \geqslant 0 \text{ for all } \vec{x}, \\ (\vec{x}, \vec{x}) = 0 \text{ is equivalent to } \vec{x} = \vec{0}. \end{cases}$$

Hence, by writing

$$(\vec{x} + \lambda\vec{y}, \vec{x} + \lambda\vec{y}) \geqslant 0$$

we derive *Schwarz' inequality*

$$(\text{IV, 3; 6}) \qquad |(\vec{x}, \vec{y})| \leqslant \sqrt{(\vec{x}, \vec{x})}\ \sqrt{(\vec{y}, \vec{y})},$$

whence one can deduce *Minkowski's inequality*

$$(\text{IV, 3; 7}) \qquad \sqrt{(\vec{x} + \vec{y}, \vec{x} + \vec{y})} \leqslant \sqrt{(\vec{x}, \vec{x})} + \sqrt{(\vec{y}, \vec{y})}.$$

It follows from this that $\|\vec{x}\| = \sqrt{(\vec{x}, \vec{x})}$ is a *norm* on $\mathcal{H}$. $\mathcal{H}$ is thus a *normed space.*

3) $\mathcal{H}$ is *complete* in the metric defined by this norm and is thus a *Banach space**.

2. HILBERT BASIS

If $\mathcal{H}$ is a Hilbert space and if $(\vec{e}_i)_{i\in I}$ is a family of vectors $\vec{e}_i$ belonging to $\mathcal{H}$, then we say that $(\vec{e}_i)_{i\in I}$ is *a Hilbert basis of* $\mathcal{H}$ *if:*

1) The $\vec{e}_i$ are orthogonal in pairs and are all of norm 1, so that

$$(\text{IV, 3; 8}) \qquad (\vec{e}_i, \vec{e}_j) = \delta_{ij} = \begin{cases} 0, & i \neq j, \\ 1, & i = j. \end{cases}$$

This condition implies that the $\vec{e}_i$ are linearly independent.

2) The system of the $(\vec{e}_i)_{i\in I}$ is *total* or *topologically generating* (the term *complete* is often found in the literature but is incorrect and leads to confusion), which means that the set of finite linear combinations of the $\vec{e}_i$ is *dense in* $\mathcal{H}$. It can be proved that this is so *if and only if there exists no vector* $\vec{x} \neq \vec{0}$ *orthogonal to all the* $\vec{e}_i$.

Now let $(\vec{e}_i)_{i\in I}$ be a Hilbert basis of a Hilbert space $\mathcal{H}$. For each $\vec{x} \in \mathcal{H}$ it is possible to obtain complex numbers x_i which are called the *coordinates* of $\vec{x}$ with respect to the Hilbert basis :

$$(\text{IV, 3; 9}) \qquad x_i = (\vec{x}, \vec{e}_i).$$

The following theorem can then be proved :

THEOREM 5 (on Hilbert bases).

1) *Let* $\vec{x} \in \mathcal{H}$, $x_i = (\vec{x}, \vec{e}_i)$.

The series $\sum_{i\in I} x_i\vec{e}_i$ *is summable in* $\mathcal{H}$ *and its sum is* $\vec{x}$.

The series $\sum_{i\in I} |x_i|^2$ *is summable and its sum is* $\|\vec{x}\|^2$.

---

*A Banach space is a complete normed vector space.

Also, if $\vec{y} \in \mathcal{H}$ *and* $y_i = (\vec{y}, \vec{e_i})$ *the series* $\sum_{i \in I} x_i \bar{y}_i$ *is summable and its sum is* $(\vec{x}, \vec{y})$.

2) *Conversely, let* $(x_i)_{i \in I}$ *be a family of complex numbers such that the series* $\sum_{i \in I} |x_i|^2$ *is summable. Then the series* $\sum_{i \in I} x_i \vec{e_i}$ *is summable in* $\mathcal{H}$ *and its sum* $\vec{x}$ *is the only vector of* $\mathcal{H}$ *such that* $(\vec{x}, \vec{e_i}) = x_i$ *for all* i.

The convergence of $\sum_{i \in I} x_i \vec{e_i}$ means that, for any $\varepsilon > 0$, there exists a finite set $J \subset I$ such that, for any finite set $K \supset J$,

$$\left\| \left(\sum_{i \in K} x_i \vec{e_i} \right) - \vec{x} \right\| \leqslant \varepsilon. \tag{IV, 3; 10}$$

3. THE SPACE $L^2(T)$

Let $L^2(a, b)$ be the space of the *classes* of functions f which are square summable over (a, b) (*classes* meaning that any two of these functions which are equal almost everywhere are to be identified), that is to say, measurable and such that $\int_a^b |f(x)|^2 \, dx$ is finite.

$L^2(a, b)$ is a vector space, for if $f \in L^2$ and $g \in L^2$, then

$$|f + g|^2 \leqslant 2(|f|^2 + |g|^2) \quad \text{so that} \quad f + g \in L^2. \tag{IV, 3; 11}$$

Again, it follows from $2|fg| \leqslant |f|^2 + |g|^2$ that, if $f \in L^2$ and $g \in L^2$, then the product fg is *summable* in (a, b). In particular, by taking $g = 1$ and the interval (a, b) as *finite* it can be seen that any function whose square is summable over a finite interval is *a fortiori* summable. The converse is false, as the example of $f(x) = 1/\sqrt{x}$ in the interval $(0, 1)$ at once shows. On the other hand, the proposition is not true in any sense for an infinite interval :

$f(x) = 1/x$ has a summable square but is not itself summable in $(1, +\infty)$;

$f(x) = \dfrac{1}{\sqrt{x}(x+1)}$ is summable but its square is not summable in $(0, +\infty)$.

On $L^2(a, b)$ consider the scalar product

$$(f, g) = \int_a^b f(x)\overline{g(x)} \, dx. \tag{IV, 3; 12}$$

This is obviously a Hermitian form positive definite, $\left(\int_a^b |f(x)|^2 \, dx = 0 \right.$ implies that f is nearly everywhere zero).

It may be proved (*Riesz-Fischer theorem*) that $L^2(a, b)$ is complete in the metric defined by the norm

$$\|f\| = \sqrt{\int_a^b |f(x)|^2\, dx} \tag{IV, 3; 13}$$

so that $L^2\,(a, b)$ is a *Hilbert space.*

If $b = a + T$, the space $L^2(a, a + T)$ is obviously isomorphic to the space of the classes of periodic functions of period T, with locally summable squares and with the scalar product defined by the integral of (IV, 3; 12) over *any* period interval. We shall also adopt here a more convenient scalar product

$$(f, g) = \int_a^{a+T} f(x)\,\overline{g(x)}\,\frac{dx}{T} \tag{IV, 3; 14}$$

and the *Hilbert space thus defined will be denoted by* $L^2(T)$.

Convergence in this space is also called *mean square convergence* : functions $f_j \in L^2(T)$ converge to $f \in L^2(T)$ as $j \to \infty$ if

$$\lim_{j\to\infty} \left(\int_a^{a+T} |f_j(x) - f(x)|^2 \frac{dx}{T}\right) = 0. \tag{IV, 3; 15}$$

Then consider the system

$$\vec{e}_k = e^{ik\omega x}, \qquad k = \cdots -2, -1, 0, +1, +2, \ldots \tag{IV, 3; 16}$$

These vectors are orthogonal in pairs and of norm 1 in $L^2(T)$ by (IV, 1; 6), so that

$$\int_a^{a+T} e^{i(k-l)\omega x} \frac{dx}{T} = \delta_{kl}. \tag{IV, 3; 17}$$

It will be assumed provisionally that this system is *total* in $L^2(T)$. It is then a Hilbert basis. Thus any $f \in L^2(T)$ has coordinates which are identical with its Fourier coefficients :

$$f_k = \int_a^{a+T} f(x) e^{-ik\omega x} \frac{dx}{T} = c_k(f). \tag{IV, 3; 18}$$

The general Theorem 5 on Hilbert bases then gives :

THEOREM 6. 1) *If $f \in L^2(T)$, the series formed by the squares of its Fourier coefficients is summable and*

$$\sum_{k=-\infty}^{+\infty} |c_k(f)|^2 = \int_a^{a+T} |f(x)|^2 \frac{dx}{T} \tag{IV, 3; 19}$$

(*Bessel-Parseval equation*).

Also, if $g \in L^2(T)$,

(IV, 3; 20) $$\sum_{k=-\infty}^{+\infty} c_k(f)\,\overline{c_k(g)} = \int_a^{a+T} f(x)\,\overline{g(x)}\,\frac{dx}{T}.$$

The Fourier series of f converges to f in mean square :

(IV, 3; 21) $$\sum_{k=-\infty}^{+\infty} c_k(f)e^{ik\omega x} = f(x) \quad \textit{in the following sense :}$$

(IV, 3; 22) $$\lim_{\substack{M\to\infty \\ N\to\infty}} \left[\int_a^{a+T} \left|\left(\sum_{k=-M}^{+N} c_k(f)e^{ik\omega x}\right) - f(x)\right|^2 \frac{dx}{T}\right] = 0.$$

2) *Conversely, if* $(c_k)_{k=-\infty}^{+\infty}$ *is a system of complex numbers such that the series* $\sum_{k=-\infty}^{+\infty} |c_k|^2$ *converges, then the series* $\sum_{k=-\infty}^{+\infty} c_k e^{ik\omega x}$ *converges in the mean square to a function* $f \in L^2(T)$, *and f is the only function of* $L^2(T)$ *with the* c_k *as Fourier coefficients.*

For Theorem 6 to be proved it remains to show that the $(\vec{e}_k)$ form a topologically generating system. For this it is sufficient (page 159) to prove that any function $f \in L^2(T)$ having all its Fourier coefficients zero is itself almost everywhere zero. Now, this is obvious since, by Theorem 2, f is the limit in $\mathcal{D}'$ of the Fourier series and hence is zero as a distribution and almost everywhere zero as a function.

General remarks

1) In the case of series of cosines and sines the Bessel-Parseval equation is written

(IV, 3; 23) $$|a_0|^2 + \sum_{k=1}^{\infty} \frac{|a_k|^2 + |b_k|^2}{2} = \int_a^{a+T} |f(x)|^2 \frac{dx}{T}.$$

This originates from the fact that although any two of the functions $\cos k\omega x$, $\sin k\omega x$ are orthogonal, they are not of norm 1 for $k > 0$, since

$$\|\cos k\omega x\|^2 = \|\sin k\omega x\|^2 = \tfrac{1}{2}.$$

The Bessel-Parseval equation means that, because of the orthogonality property, the mean of $|f(x)|^2$ is the sum of the means of the squares of the moduli of the terms of the Fourier series.

2) There is no analogous result for $L^1(T)$, the space of locally summable periodic functions.

However, in this case, it is obvious that the $|c_k(f)|$ are bounded by $\frac{1}{T}\|f\|_{L^1} = \int_a^{a+T} |f(x)| \frac{dx}{T}$ and it can be proved (Lebesgue's theorem) that $c_k(f)$ converges to zero as $|k| \to \infty$. However, there is no theorem of convergence for the series !

3) If $\sum_{k=-\infty}^{+\infty} |c_k|$ converges then c_k tends to 0 as $|k| \to \infty$ so that, *a fortiori*, $\sum_{k=-\infty}^{+\infty} |c_k|^2$ also converges. The converse is false, as is shown by the example $c_k = 1/k$ (for $k \neq 0$). If $\sum_{k=-\infty}^{+\infty} |c_k|$ converges, then the Fourier series converges absolutely and uniformly, and its sum is a continuous function.

4) The fact that the Fourier series for f converges to f in $L^2(T)$ does not imply that it converges to f nearly everywhere.

5) Let J be a finite set of integers (of either sign). Among the "trigonometric polynomials"

$$P(x) = \sum_{k \in J} C_k e^{ik\omega x},$$

the one giving the best mean square approximation to f, that is to say the one for which the integral

$$\int_a^{a+T} |P(x) - f(x)|^2 \frac{dx}{T}$$

is a minimum, is

$$P(x) = \sum_{k \in J} c_k(f) e^{ik\omega x}.$$

For the Bessel-Parseval equation gives

$$\text{(IV, 3; 24)} \quad \int_a^{a+T} |P(x) - f(x)|^2 \frac{dx}{T} = \sum_{k \in J} |C_k - c_k(f)|^2 + \sum_{k \notin J} |c_k(f)|^2.$$

Q.E.D.

The value of the minimum is thus

$$\text{(IV, 3; 25)} \quad \sum_{k \notin J} |c_k(f)|^2$$

which can be made as small as we like by making J large enough.

4. The convolution algebra $\mathcal{D}'(\Gamma)$

Let $\mathcal{S}$ and $\mathcal{T}$ be two distributions on Γ. Then

$$
\text{(IV, 4; 1)} \qquad
\begin{aligned}
c_k(\mathcal{S} * \mathcal{T}) &= \frac{1}{\mathrm{T}} \langle \mathcal{S} * \mathcal{T}, e^{-ik\omega s} \rangle \\
&= \frac{1}{\mathrm{T}} \langle \mathcal{S}_\xi \otimes \mathcal{T}_\eta, e^{-ik\omega(\xi+\eta)} \rangle \\
&= \frac{1}{\mathrm{T}} \langle \mathcal{S}_\xi, e^{-ik\omega\xi} \rangle \langle \mathcal{T}_\eta, e^{-ik\omega\eta} \rangle \\
&= \mathrm{T}\, c_k(\mathcal{S}) c_k(\mathcal{T}).
\end{aligned}
$$

THEOREM 7. *The Fourier coefficients of $\mathcal{S} * \mathcal{T}$ are, to within a factor* T, *the products of the corresponding Fourier coefficients of $\mathcal{S}$ and $\mathcal{T}$,*

$$
\text{(IV, 4; 2)} \qquad c_k(\mathcal{S} * \mathcal{T}) = \mathrm{T} c_k(\mathcal{S}) c_k(\mathcal{T}).
$$

As $\mathcal{S} * \mathcal{T}$ is the sum of its Fourier series, a practical means of dealing with convolutions on Γ may be derived. Put

$$
\text{(IV, 4; 3)} \qquad \mathcal{S} = \sum_{k=-\infty}^{+\infty} c_k(\mathcal{S}) e^{ik\omega s}, \quad \mathcal{T} = \sum_{k=-\infty}^{+\infty} c_k(\mathcal{T}) e^{ik\omega s},
$$

whence

$$
\mathcal{S} * \mathcal{T} = \mathrm{T} \sum_{k=-\infty}^{+\infty} c_k(\mathcal{S}) c(\mathcal{T}) e^{ik\omega s}.
$$

Several results follow:

1) The fact that δ is the unit element in convolution is related to the fact that its Fourier coefficients are equal to $1/\mathrm{T}$, and that 1 is the unit element in multiplication.

2) The equation $\delta' * \mathcal{T} = \mathcal{T}'$ is easily verified by using term-by-term differentiation of the Fourier series:

$$
\text{(IV, 4; 4)} \qquad \delta' = \frac{d\delta}{ds} = \frac{1}{\mathrm{T}} \sum_{k=-\infty}^{+\infty} ik\omega\, e^{ik\omega s} = \sum_{k=-\infty}^{+\infty} \frac{2i\pi k}{\mathrm{T}^2} e^{ik\omega s},
$$

$$
\text{(IV, 4; 5)} \qquad
\begin{cases}
\mathcal{T} = \displaystyle\sum_{k=-\infty}^{+\infty} c_k(\mathcal{T}) e^{ik\omega s}, \\
\mathcal{T}' = \dfrac{d\mathcal{T}}{ds} = \displaystyle\sum_{k=-\infty}^{+\infty} ik\omega c_k(\mathcal{T})\, e^{ik\omega s} = \sum_{k=-\infty}^{+\infty} \frac{2i\pi k}{\mathrm{T}} c_k(\mathcal{T}) e^{ik\omega s},
\end{cases}
$$

whence it follows that:

$$c_k(\mathcal{C}') = \mathrm{T}c_k(\delta')c_k(\mathcal{C}) \quad \text{or} \quad \mathcal{C}' = \delta' * \mathcal{C}. \tag{IV, 4; 6}$$

3) Convolution equations on Γ are very easily solved by Fourier series. Consider the equation

$$\mathcal{A} * \mathcal{H} = \mathcal{B}. \tag{IV, 4; 7}$$

Denoting the Fourier coefficients of $\mathcal{A}$, $\mathcal{B}$ and $\mathcal{H}$ by a_k, b_k and x_k respectively, equation (IV, 4; 7) becomes the denumerable infinity of equations

$$\mathrm{T}a_k x_k = b_k; \tag{IV, 4; 8}$$

$\mathcal{H}$ is thus given by the series

$$\mathcal{H} = \frac{1}{\mathrm{T}} \sum_{k=-\infty}^{+\infty} \frac{b_k}{a_k} e^{ik\omega s}. \tag{IV, 4; 9}$$

It is still necessary to show that this series has a meaning. If all the a_k are $\neq 0$, the $x_k = (1/\mathrm{T})(b_k/a_k)$ are well-defined and it is sufficient to check that the series on the right-hand side of (IV, 4; 9) converges in $\mathcal{D}'(\Gamma)$, that is to say that the b_k/a_k are bounded above by a power of $|k|$ as $|k| \to \infty$. If some of the a_k are zero, then $\mathcal{B}$ must satisfy the compatibility conditions that the corresponding $b_k = 0$. If these conditions are satisfied, the values of the corresponding x_k are indeterminate. This is similar to a system of linear algebraic equations where there are as many indeterminate elements in the solution as there are compatibility conditions for the right-hand side. It is still necessary to show that the series (IV, 4; 9) converges.

In particular, the inverse $\mathcal{A}^{-1}$ of $\mathcal{A}$ in the convolution algebra, satisfying $\mathcal{A} * \mathcal{A}^{-1} = \delta$, is given by

$$\mathcal{A}^{-1} = \frac{1}{\mathrm{T}^2} \sum_{k=-\infty}^{+\infty} \frac{1}{a_k} e^{ik\omega s}. \tag{IV, 4; 10}$$

This exists if all the $a_k \neq 0$ and if the series (IV, 4; 10) is the Fourier series of a distribution, that is to say, if the $|a_k|$ are bounded below by a power $\geqslant 0$ of $1/|k|$ as $|k| \to \infty$.

4) Divisors of 0 exist in $\mathcal{D}'(\Gamma)$, $\mathcal{A}$ being a divisor of zero if and only if certain of its coefficients a_k are zero. For if $a_k = 0$, then

$$\mathcal{A} * e^{ik\omega s} = \mathrm{T}a_k e^{ik\omega s} = 0. \tag{IV, 4; 11}$$

5) It is possible to solve much more difficult problems. For example, let it be required to find all distributions $\mathcal{H} \in \mathcal{D}'(\Gamma)$ satisfying

$$\mathcal{H} * \mathcal{H} = \delta \tag{IV, 4; 12}$$

(a convolution equation of the second degree).

The Fourier coefficients x_k satisfy

$$\text{(IV, 4; 13)} \qquad Tx_k^2 = \frac{1}{T} \quad \text{or} \quad x_k = \pm\frac{1}{T}$$

the choices between $+1$ and -1 for the various values of k being independent. All the series $\sum x_k e^{ik\omega s}$ thus formed are convergent, since $|x_k|$ is bounded by $1/T$, independently of k.

Thus this second-degree equation has an infinity of solutions!

It is only in an algebra without divisors of 0 *that an equation of degree m has not more than m roots.*

6) It will be recalled that differential equations occur among the convolution equations. Let it be required to solve in $\mathcal{D}'(\Gamma)$ the equation

$$\text{(IV, 4; 14)} \qquad \left(\frac{d}{ds} - \lambda\right)\mathcal{H} = \mathcal{B}$$

or

$$\text{(IV, 4; 15)} \qquad (\delta' - \lambda\delta) * \mathcal{H} = \mathcal{B}.$$

It is necessary to solve the infinity of equations,

$$\text{(IV, 4; 16)} \qquad T\left(\frac{2i\pi k}{T^2} - \frac{\lambda}{T}\right)x_k = b_k.$$

A. First suppose that λ is not equal to any of the values $2i\pi k/T = ik\omega$. Then the x_k are determined uniquely by

$$\text{(IV, 4; 17)} \qquad x_k = \frac{b_k}{ik\omega - \lambda}.$$

As $|k| \to \infty$, $|b_k| \leqslant A|k|^\alpha$, giving $|x_k| \leqslant \frac{A|k|^{\alpha-1}}{\omega}(1 + \varepsilon)$, so that the series $\sum_{k=-\infty}^{+\infty} x_k e^{ik\omega s}$ converges to the unique solution $\mathcal{H}$. In particular, there is a unique distribution $\mathcal{A}^{-1}$ such that

$$\text{(IV, 4; 18)} \qquad \left(\frac{d}{ds} - \lambda\right)\mathcal{A}^{-1} = \delta.$$

Here $\mathcal{A}^{-1}$ is the sum of the series given by

$$\text{(IV, 4; 19)} \qquad \mathcal{A}^{-1} = \frac{1}{T}\sum_{k=-\infty}^{+\infty} \frac{1}{ik\omega - \lambda} e^{ik\omega s} \quad \text{in } \mathcal{D}'(\Gamma).$$

The unique solution of (IV, 4; 14) is then

$$\mathcal{H} = \mathcal{A}^{-1} * \mathcal{B}, \tag{IV, 4; 20}$$

which is equivalent to (IV, 4; 17).

The distribution $\mathcal{A}^{-1}$ is quite simple to calculate in an elementary way. For in the complement of the point $s = 0$ on Γ, it satisfies the equation

$$\left(\frac{d}{ds} - \lambda\right)\mathcal{A}^{-1} = 0. \tag{IV, 4; 21}$$

The classical solution in R of this equation is of the form

$$\mathcal{A}^{-1} = Ce^{\lambda x}. \tag{IV, 4; 22}$$

Hence consider the *periodic function* $\tilde{f}$ of period T on R, which is equal to $e^{\lambda x}$ in $0 < x < T$.

The associated function f on Γ is equal to $e^{\lambda x}$, if care is taken to choose for each point of Γ its curvilinear distance which satisfies $0 < s < T$. Thus f is discontinuous at the point $s = 0$, so that

$$f' = \{f'\} + (1 - e^{\lambda T})\,\delta. \tag{IV, 4; 23}$$

Then f satisfies the differential equation

$$f' - \lambda f = [\{f'\} - \lambda f] + (1 - e^{\lambda T})\,\delta = (1 - e^{\lambda T})\,\delta, \tag{IV, 4; 24}$$

the expression in brackets being zero by the properties of f. As λ is not equal to any of the numbers $ik\omega$, $1 - e^{\lambda T}$ is not equal to 0 and the function

$$\frac{f}{1 - e^{\lambda T}} \tag{IV, 4; 25}$$

is the solution of (IV, 4; 18). Since this solution is unique it is thus $\mathcal{A}^{-1}$. Again, the evaluation of the Fourier coefficients of this function is immediate:

$$c_k\left(\frac{f}{1 - e^{\lambda T}}\right) = \frac{1}{1 - e^{\lambda T}} \int_0^T \tilde{f}(x)\, e^{-ik\omega x} \frac{dx}{T} \tag{IV, 4; 26}$$

$$= \frac{1}{1 - e^{\lambda T}} \frac{1}{T} \left[\frac{e^{(\lambda - ik\omega)x}}{\lambda - ik\omega}\right]_{x=0}^{x=T} = \frac{1}{T} \frac{1}{ik\omega - \lambda} = c_k(\mathcal{A}^{-1}).$$

Note that, except at its point of finite discontinuity, f has a bounded derivative. Thus f is of bounded variation and, by Theorem 3, its Fourier series is semi-convergent. In particular, at the point of discontinuity,

$$\text{(IV, 4; 27)} \qquad \frac{1}{2}\frac{1+e^{\lambda T}}{1-e^{\lambda T}} = \frac{1}{T}\sum_{k=-\infty}^{+\infty} \frac{1}{ik\omega - \lambda} = \frac{1}{T}\lim_{N\to\infty}\sum_{k=-N}^{+N} \frac{1}{ik\omega - \lambda}.$$

Here there is convergence after grouping pairs of terms corresponding to opposite values of k.

However the series $\sum_{k=-\infty}^{0}$ and $\sum_{k=0}^{+\infty}$ are divergent. This can be seen immediately for pure imaginary λ since, to within a factor $1/i$, the general term keeps the same sign and is $\sim 1/k\omega$ as $|k| \to \infty$. On the other hand, grouping the terms in pairs yields

$$\text{(IV, 4; 28)} \qquad \frac{1}{ik\omega - \lambda} + \frac{1}{-ik\omega - \lambda} = \frac{-2\lambda}{\lambda^2 + k^2\omega^2} \sim \frac{-2\lambda}{k^2\omega^2}$$

as $k \to \infty$. Equation (IV, 4; 27) gives a kind of "partial-fraction" expansion for the function of λ on the left-hand side, which has poles at $\lambda = ik\omega$ and residues $-1/T$.

Note finally that $f \in L^2(T)$ and that the series $\sum \left|\frac{1}{ik\omega - \lambda}\right|^2$ is in fact convergent.

B. Now suppose that $\lambda = ik_0\omega$. Equation (IV, 4; 14) has a solution only if $b_{k_0} = 0$. If this is so it has an infinity of solutions since x_{k_0} is indeterminate, as can be seen from the fact that $e^{ik_0\omega s}$ is the solution of the corresponding homogeneous equation. Note that there is no inverse $\mathcal{A}^{-1}$ of $\mathcal{A}$ because, for any k_0, $c_{k_0}(\delta) \neq 0$.

Consider as a particular case $k_0 = 0$. Equation (IV, 4; 14) becomes

$$\text{(IV, 4; 29)} \qquad \frac{d\mathcal{H}}{ds} = \mathcal{B}.$$

The problem reduces to that of finding a primitive of $\mathcal{B}$. Such a primitive exists if and only if $b_0 = 0$. In essence, the point is that if $\tilde{f}$ is a periodic function of period T on R it has primitives $\tilde{F}$, but if $\int_a^{a+T} \tilde{f}(x)\,dx \neq 0$, none of these primitives is periodic since then $\tilde{F}(a+T) - \tilde{F}(a) \neq 0$. In this case all the primitives differ from one another by the value of the constant x_0, but one and only one of them, corresponding to $x_0 = 0$, will in its turn have primitives, and so on. For example,

$$\text{(IV, 4; 30)} \qquad P_0 = T\delta - 1$$

has primitives. One of these, P_1, has the coefficient $c_0(P_1) = 0$, and P_1 is then the periodic function of period T equal to $\frac{1}{2}\,T/-x$ in the interval (0, T). By successive integration a sequence of periodic functions $\tilde{P}_m$ may thus be obtained, satisfying

$$\text{(IV, 4; 31)} \qquad \frac{d}{dx}\tilde{P}_m = \tilde{P}_{m-1}, \quad \int_0^T \tilde{P}_m(x)\,dx = 0.$$

In (0, T), $\tilde{P}_m$ (which is a function for $m \geqslant 1$) is equal to a polynomial of degree m, called a *Bernoulli polynomial*. Its Fourier expansion is

$$\text{(IV, 4; 32)} \qquad \tilde{P}_m = \sum_{\substack{k=-\infty \\ k\neq 0}}^{+\infty} \left(\frac{1}{ik\omega}\right)^m e^{ik\omega x}.$$

For $m \geqslant 2$, the Fourier series is absolutely and uniformly convergent, so that $\tilde{P}_m$ is continuous and hence the corresponding Bernoulli polynomial takes the same values for $x = 0$ and $x = T$. For even m, putting $x = 0$,

$$\text{(IV, 4; 33)} \qquad \tilde{P}_m(0) = 2(-1)^{m/2}\sum_{k=1}^{\infty}\left(\frac{1}{k\omega}\right)^m,$$

which enables us to obtain successively

$$\text{(IV, 4; 34)} \qquad \sum_{k=1}^{\infty}\frac{1}{k^2} = \frac{\pi^2}{6}, \qquad \sum_{k=1}^{\infty}\frac{1}{k^4} = \frac{\pi^4}{90}, \qquad \text{etc.}$$

The example just discussed can be interpreted in a different way. Consider the operator

$$\text{(IV, 4; 35)} \qquad A = \frac{d}{ds}$$

on the vector space $\mathcal{D}'(\Gamma)$. Its *eigenvalues* λ, i.e., the complex numbers λ for which there exists a distribution $\mathcal{T} \neq 0$ such that $(d/ds - \lambda)\,\mathcal{T} = 0$, are the numbers $\lambda = ik\omega$, $k = \cdots -2, -1, 0, 1, 2 \ldots.$ For $\lambda = ik\omega$, there is an eigenvector, determined to within a scalar factor, namely

$$\text{(IV, 4; 36)} \qquad \vec{e}_k = e^{ik\omega s}.$$

Now, as in the case of vector spaces of finite dimension, these $\vec{e}_k$ form, if not a basis of $\mathcal{D}'(\Gamma)$, at least a "topological basis" in the sense that any vector $\mathcal{T} \in \mathcal{D}'(\Gamma)$ admits of one and only one expansion with respect to the $\vec{e}_k$, which is precisely the *Fourier expansion* of $\mathcal{T}$. The operator A is then expressed, in terms of this basis, by the relations

(IV, 4; 37)
$$\begin{cases} \mathcal{T} = \sum_{k=-\infty}^{+\infty} c_k(\mathcal{T})e^{ik\omega s}, \\ A\mathcal{T} = \frac{d\mathcal{T}}{ds} = \sum_{k=-\infty}^{+\infty} ik\omega c_k(\mathcal{T})e^{ik\omega s}. \end{cases}$$

Thus the Fourier expansion is just the spectral decomposition of the space $\mathcal{D}'(\Gamma)$ *relative to the operator* $A = d/ds$.

We might try to replace the space $\mathcal{D}'(\Gamma)$ by the space $L^2(T)$, but d/ds does not operate in this space. On the other hand, the operator d/ds may be replaced by any convolution operator in $\mathcal{D}'(\Gamma)$. The operator $\mathcal{T} \to \mathcal{A} * \mathcal{T}$ possesses eigenvectors $e^{ik\omega s}$ for the eigenvalues $Tc_k(\mathcal{A}) = Ta_k$.

EXERCISES FOR CHAPTER IV

Exercise IV-1

Calculate the Fourier series of the function, of period 2π, which is equal to x in $-\pi < x < \pi$.

Evaluate

$$\sum_{n=1}^{\infty} \frac{1}{n^2} \quad \text{and} \quad \sum_{n=0}^{\infty} \frac{1}{(2n+1)^2}.$$

Exercise IV-2

Expand the functions $u_0(x)$ of Exercises VII-1, VII-2, VII-3 and VII-4 in sine series.

Exercise IV-3

Calculate the Fourier series of $f(x) = e^{\alpha e^{ix}}$, α being any complex number. Prove the relation

$$\frac{1}{2\pi}\int_0 e^{2\alpha \cos x}\, dx = \sum_{n=0}^{\infty} \frac{\alpha^{2n}}{(n\,!)^2}.$$

Show that it is equivalent to equation (IX, 1; 73) for $\nu = 0$.

Exercise IV-4

(*a*) Let $\mathcal{T}$ be a distribution of period T. Show that its Fourier series may be put in the form

$$\sum_{n=0}^{\infty} a_n \cos\frac{2\pi nx}{T} + \sum_{n=1}^{\infty} b_n \sin\frac{2\pi nx}{T}$$

and give expressions for the coefficients a_n and b_n. In the two following questions put $\check{\varphi}(x) = \varphi(-x)$, for a function $\varphi(x)$.

(*b*) A distribution $\mathfrak{T}$ is said to be *odd* if

$$\langle \mathfrak{T}, \check{\varphi} \rangle = -\langle \mathfrak{T}, \varphi \rangle \text{ for all } \varphi \in \mathfrak{D}.$$

Show that the Fourier series of an odd distribution of period T reduces to a sine series.

(*c*) A distribution $\mathfrak{T}$ is said to be *even* if

$$\langle \mathfrak{T}, \check{\varphi} \rangle = \langle \mathfrak{T}, \varphi \rangle \text{ for all } \varphi \in \mathfrak{D}.$$

Show that the Fourier series of an even distribution of period T reduces to a cosine series.

Exercise IV-5

Expand in a Fourier series the distribution of period 1 which is equal to $-\delta_{(-b)} + \delta_{(b)}$, where $0 < b < \frac{1}{2}$, in the interval $(-\frac{1}{2}, +\frac{1}{2})$.

Exercise IV-6

Let

$$U(x) = \begin{cases} +1 & \text{for } \ 2k < x < 2k + 1, \\ -1 & \text{for } \ (2k+1) < x < (2k+2), \qquad k \text{ any integer.} \end{cases}$$

(*a*) Expand $U(x) \cos \pi x$ in Fourier series : (i) of exponentials; (ii) of sines.

(*b*) Expand the distribution

$$2 \sum_{n=-\infty}^{+\infty} \delta'_{(-n)} - \pi^2 U(x) \cos \pi x$$

in a series of exponentials.

(*c*) Solve in $\mathfrak{D}'(\Gamma)$, Γ being a circle with circumference of length 1, the equation

$$X * \cos \pi s = 2\delta' - \pi^2 \cos \pi s.$$

Verify the results obtained.

Exercise IV-7

Define the distribution

$$\text{pv } \cot x = (\log |\sin x|)'.$$

Show that $\cos x$ (pv $\cot x$) $= \sin x$ and that pv $\cot x$ is an odd distribution of period π. Hence find its Fourier series expansion.

Exercise IV-8 (Proof of equations (IV, 3; 19) and (IV, 3; 20).

Let f and g be two periodic functions, of period T, in the space $L^2(T)$. Show that

$$h(x) = \frac{1}{T} \int_0^T f(x+t) g(t) \, dt$$

is also periodic with period T and belongs to the space $L^2(T)$. Obtain the Fourier coefficients $c_n(h)$ of h in terms of the Fourier coefficients $c_n(f)$ and $c_n(g)$ of f and g respectively.

Assuming that the Fourier series of $h(x)$ is convergent and represents $h(x)$ for all x, prove equation (IV, 3; 20) for f and g and equation (IV, 3; 19) for f.

Exercise IV-9

(*a*) Expand the function $f(x) = |\sin^3 x|$ in a cosine series, giving the general coefficient a_n in the form of a rational fraction $P(n)/Q(n)$. Is this Fourier series convergent? Are the differentiated Fourier series convergent? Write down Parseval's equation in this case.

(*b*) Calculate directly, in the distribution sense,

$$\left(\frac{d^2}{dx^2} + 1\right)\left(\frac{d^2}{dx^2} + 9\right)f.$$

Show that the result permits the Fourier expansion of f to be recovered, without any evaluation of integrals, by regarding f as the solution of a differential equation.

Exercise IV-10

Expand the functions x and x^3 for $-\pi < x < \pi$ in Fourier series and hence evaluate

$$\sin x - \frac{\sin 2x}{2^3} + \frac{\sin 3x}{3^3} + \cdots + (-1)^{n+1}\frac{\sin nx}{n^3} + \cdots$$

for all values of x. Show that the expansions found may be deduced from that of $\delta_{(\pi)}$.

Exercise IV-11

Let

$$f(x) = \begin{cases} \dfrac{\pi}{3} & \text{for } 0 < x < \dfrac{\pi}{3}, \\ 0 & \text{for } \dfrac{\pi}{3} < x < \dfrac{2\pi}{3}, \\ -\dfrac{\pi}{3} & \text{for } \dfrac{2\pi}{3} < x < \pi. \end{cases}$$

Show that $f(x)$ can be written over the three specified intervals as the sum of a series of sines and a series of cosines. Find the sum of these two series for $x = 0$, $x = \pi/3$, $x = 2\pi/3$, $x = \pi$ and for an arbitrary value of x.

Verify the expansions found by differentiating $f(x)$, considered as a distribution.

Exercise IV-12
Expand sin mx, cos mx, $\dfrac{e^{mx} + e^{-mx}}{e^{m\pi} - e^{-m\pi}}$ in Fourier series for $-\pi < x < \pi$.

Exercise IV-13
Construct the curve represented by the equation

$$\sum_{1}^{\infty} (-1)^{n-1} \frac{\sin ny \cos nx}{n^3} = 0.$$

Exercise IV-14
Let

$$S_n(x) = 2 \sum_{1}^{n} (-1)^{p-1} \frac{\sin px}{p}.$$

Determine the coordinates of the maxima and minima in the range $0 < x < \pi$. $S_n(y)$ attains its maximum y_n for $x = \dfrac{n\pi}{n+1}$. Show that

$$\lim_{n \to \infty} y_n = 2 \int_0^{\pi} \frac{\sin x}{x} dx$$

and that this number is $> \pi$. What conclusion can be drawn?

Exercise IV-15
Note. All functions and distributions occurring below are periodic and of period 1. $\mathfrak{D}$ denotes the space of infinitely differentiable functions φ on (0, 1) such that $\varphi^{(n)}(0) = \varphi^{(n)}(1)$ for all n. Thus $\mathfrak{D}$ may be regarded as the space of functions on the circle of unit circumference in the plane. By our choice of period the Fourier series expansion of $T \in \mathfrak{D}'$ may be expressed as

$$T = \sum_{n=-\infty}^{+\infty} \langle T_\xi, e^{2\pi i n(x-\xi)} \rangle.$$

(*a*) It will be said that $T \in \mathfrak{D}'$ is of ***positive type*** (which can be written $T \gg 0$) if all the Fourier coefficients of T are positive. For all φ of $\mathfrak{D}$ put

$$(\varphi * \tilde{\varphi})(x) = \int_0^1 \varphi(t)\, \overline{\varphi(t - x)}\, dt.$$

Prove the following result : $T \gg 0$ if and only if $\langle T, \varphi * \tilde{\varphi} \rangle \geqslant 0$ for every $\varphi \in \mathfrak{D}$.

Before the main result show rapidly that $\langle T, \varphi * \tilde{\varphi} \rangle$ has a meaning when $\varphi \in \mathfrak{D}$, that is to say $\varphi \in \mathfrak{D}$ implies $\varphi * \tilde{\varphi} \in \mathfrak{D}$. Give several simple examples of distributions $\gg 0$. Is every distribution $T \gg 0$ a positive distribution, i.e. is $\langle T, \varphi \rangle \gg 0$ if φ is a positive function?

Conversely, is every positive distribution a distribution of positive type? Justify all your answers by examples. Give an example of a distribution which is at once positive and of positive type.

(*b*) Consider a continuous function $f(x)$ of the real variable x which is periodic of period 1 and a distribution of positive type. Prove the following result: For any real numbers $x_1 \dots x_l$, the quadratic form whose coefficients are the $a_{ij} = f(x_i - x_j)$ is positive definite Hermitian; i.e. $\sum_{i,j} a_{ij}\xi_i\bar{\xi}_j \geqslant 0$ for any l complex numbers $\xi_1, \dots, \xi_l$. To prove this result start from the definition $\langle f, \varphi * \tilde{\varphi}\rangle \geqslant 0$ and approximate φ by suitably chosen sequences.

Prove the following properties of f:

$$f(-x) = \overline{f(x)};\ |f(x)| \leqslant f(0) \text{ (and thus } f(0) \geqslant 0\text{) for all } x.$$

(*c*) (i) Establish a necessary and sufficient condition for a periodic distribution T to be the derivative of a periodic distribution S. Does the periodic distribution S possess a primitive?

(ii) In the following, X denotes an unknown periodic distribution. Consider the differential equation

$$\frac{d^m X}{dx^m} + a_1 \frac{d^{m-1} X}{dx^{m-1}} + \cdots + a_m X = S, \quad \text{where S is periodic.}$$

In choosing the a_i and S give examples of the following cases:

1. S possesses a primitive and there is no solution.

2. S does not possess a primitive and there is a solution of the equation.

3. $$\frac{d^m \delta}{dx^m} + a_1 \frac{d^{m-1} \delta}{dx^{m-1}} + \cdots + a_m \delta$$

is of positive type and there is no solution.

4. $$\frac{d^m \delta}{dx^m} + a_1 \frac{d^{m-1} \delta}{dx^{m-1}} + \cdots + a_m \delta$$

is not of positive type and there is a solution.

Exercise IV-16 (Examination questions, Paris, 1958).

(*a*) Denote by $f(x)$ the periodic function of period 2π which is equal to $\cosh \lambda x$ (real $\lambda > 0$) in the interval $(-\pi, +\pi)$. Find its Fourier series expansion, evaluating the coefficients by the usual method.

(*b*) Show that f satisfies, in the distribution sense, a differential equation $f'' - \lambda^2 f = A$. By writing f as a Fourier series with unknown coefficients, show that it is possible to recover from the differential equation the Fourier coefficients obtained in (*a*).

(*c*) What result is given by the Fourier expansion found at $x = \pi$? Hence deduce a series expansion for $\pi \dfrac{\cosh \lambda\pi}{\sinh \lambda\pi}$. By finding a suitable integral with respect to λ and then taking an exponential of this, show that an infinite product expansion of $\sinh \lambda\pi$ is obtained.

Exercise IV-17 (Fourier series in R^2).

Let $f(\vec{x}) = f(x_1, x_2)$ be a function defined on R^2, with periods $(T_1, 0)$, $(0, T_2)$.

(*a*) Show that if there exists a uniformly convergent double series of the form

$$(1) \qquad \sum_{p,\, q=-\infty}^{+\infty} c_{pq} e^{2\pi i \left(\frac{px_1}{T_1} + \frac{qx_2}{T_2}\right)}$$

with sum $f(\vec{x})$, then necessarily

$$(2) \qquad c_{mn} = \frac{1}{T_1 T_2} \iint_{\substack{\alpha_1 \leqslant x_1 \leqslant \alpha_1 + T_1 \\ \alpha_2 \leqslant x_2 \leqslant \alpha_2 + T_2}} f(x_1, x_2) e^{-2\pi i \left(\frac{mx_1}{T_1} + \frac{nx_2}{T_2}\right)} dx_1\, dx_2$$

and that the result is independent of α_1 and α_2.

(*b*) If $f(\vec{x})$ is a sufficiently regular function of period $\vec{T}$ the series (1) with coefficients c_{pq} given by (2) is termed the Fourier series of $f(\vec{x})$. Show that the series (1) can also be written in the form

$$\sum_{m,n=0}^{\infty} a_{mn} \cos \frac{2\pi m x_1}{T_1} \cos \frac{2\pi n x_2}{T_2} + \sum_{m,n=1}^{\infty} b_{mn} \sin \frac{2\pi m x_1}{T_1} \sin \frac{2\pi n x_2}{T_2}$$
$$+ \sum_{m,n=0}^{\infty} d_{mn} \cos \frac{2\pi m x_1}{T_1} \sin \frac{2\pi n x_2}{T_2} + \sum_{m,n=0}^{\infty} e_{mn} \sin \frac{2\pi m x_1}{T_1} \cos \frac{2\pi n x_2}{T_2}$$

and give expressions for the coefficients a_{mn}, b_{mn}, d_{mn}, and e_{mn}. Consider the particular case when $f(x_1, x_2)$ is of the form $g(x_1) \otimes h(x_2)$.

(*c*) Show that if $x_1 \to f(x_1, x_2)$ is an odd function for fixed x_2 and if $x_2 \to f(x_1, x_2)$ is also an odd function for fixed x_1, then the Fourier series for $f(\vec{x})$ reduces to

$$\sum_{m,\, n=1}^{\infty} b_{mn} \sin \frac{2\pi m x_1}{T_1} \sin \frac{2\pi n x_2}{T_2}$$

with $$\frac{b_{mn}}{4} = \frac{4}{T_1 T_2} \int_0^{T_1/2} dx_1 \int_0^{T_2/2} dx_2 f(x_1, x_2) \sin \frac{2\pi m x_1}{T_1} \sin \frac{2\pi n x_2}{T_2}.$$

(*d*) Give the general form of the Fourier series of a function $f(x_1, x_2)$ which is an odd function of one variable when the other is fixed.

(*e*) Consider a function $f(x_1, x_2)$ defined in the rectangle

$$0 \leqslant x_1 \leqslant A_1, \qquad 0 \leqslant x_2 \leqslant A_2.$$

State how this function must be continued to other values of x_1 and x_2 for the Fourier series to have the form

$$\sum_{m,\,n=1}^{\infty} b_{mn} \sin \frac{\pi m x_1}{A_1} \sin \frac{\pi n x_2}{A_2}. \tag{3}$$

Application. Expand in a Fourier series of the form (3) the function $f_0(x_1, x_2)$ defined in the rectangle $0 \leqslant x_1 \leqslant A_1$, $0 \leqslant x_2 \leqslant A_2$ by the relations :

$$f_0(x_1, x_2) = \begin{cases} \dfrac{2h}{A_1}(A_1 - x_1) & \text{in the region (I)} \\ \dfrac{2h}{A_1} x_1 & \text{in the region (II)} \\ \dfrac{2h}{A_2}(A_2 - x_2) & \text{in the region (III)} \\ \dfrac{2h}{A_2} x_2 & \text{in the region (IV),} \end{cases}$$

h being a given positive constant and the regions I, II, III and IV being bounded by the diagonals of the rectangle as indicated by the figure below. By using the results of Chapter VII, § 2, nº 3 it is possible to write down the solution $u(x_1, x_2, t)$ of the vibrating membrane equation for the above rectangle, given that it is zero on the boundary of the rectangle and satisfies the initial conditions

$$u_0(x_1, x_2) = f_0(x_1, x_2), \qquad u_1(x_1, x_2) = 0.$$

(The variables x_1 and x_2 are respectively the variables x and y of Chapter VII.)

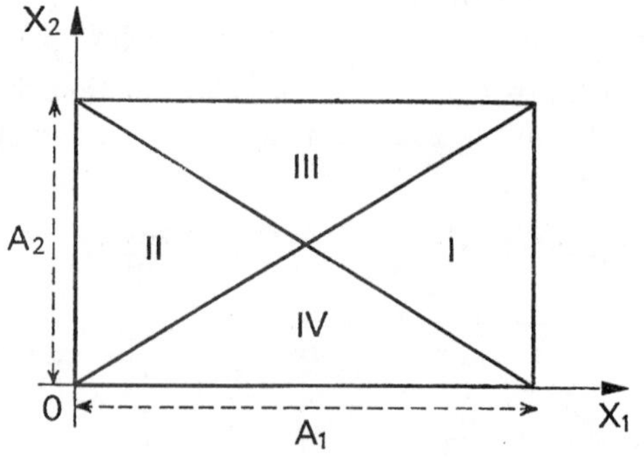

Figure (IV, 1)

Exercise IV-18 (Periodic distribution on R).
Let $\mathcal{W}_{x_1x_2} = \mathcal{S}_{x_1} \otimes \mathcal{T}_{x_2}$ be the distribution on R^2 which is the tensor product of a distribution $\mathcal{S}_{x_1}$ of period T_1 and a distribution $\mathcal{T}_{x_2}$ of period T_2. Let $c_m(\mathcal{S})$ and $c_n(\mathcal{T})$ be the Fourier coefficients of $\mathcal{S}$ and $\mathcal{T}$. Then the series

$$\sum_{m,\, n=-\infty}^{+\infty} c_m(\mathcal{S}) c_n(\mathcal{T}) e^{2\pi i \left(\frac{mx_1}{T_1} + \frac{nx_2}{T_2}\right)}$$

is called the Fourier series of $\mathcal{W}$.

Consider in R^2 the rectangle $-A_i \leqslant x_i \leqslant A_i$, $i = 1, 2$. Let $\vec{\alpha} = (\alpha_1, \alpha_2)$ be a point of the rectangle with $0 < \alpha_i < A_i$, $i = 1, 2$. Let

$$-\vec{\alpha} = (-\alpha_1, -\alpha_2), \qquad \vec{\beta} = (-\alpha_1, \alpha_2), \qquad \vec{\gamma} = (\alpha_1, -\alpha_2).$$

Expand in a Fourier series the distribution of period $(2A_1, 2A_2)$ which is equal in the rectangle to

$$\delta(\vec{\alpha}) + \delta(-\vec{\alpha}) - \delta(\vec{\beta}) - \delta(\vec{\gamma}).$$

For the rectangle $0 \leqslant x_1 \leqslant A_i$, $i = 1, 2$, find the solution of the vibrating membrane equation which is zero on the boundary and satisfies the initial conditions

$$u_0(x_1, x_2) = 0,$$
$$u_1(x_1, x_2) = \delta_{(\vec{\alpha})}.$$

CHAPTER V

The Fourier transforms

1. Fourier transforms of functions of one variable

1. INTRODUCTION

Let $f(x)$ be a locally summable non-periodic function. Let $f_T(x)$ denote the function equal to $f(x)$ in the interval $(-T/2, +T/2)$ and continued outside this interval so as to be periodic with period T. For $T \to \infty$, f_T tends to f uniformly on any finite interval. f_T may be expanded in the Fourier series

(V, 1; 1)
$$\begin{cases} f_T(x) = \sum_{n=-\infty}^{+\infty} c_{n,T}\, e^{ni\omega x}, \qquad \omega = \frac{2\pi}{T}, \\ c_{n,T} = \frac{1}{T}\int_{-\frac{T}{2}}^{+\frac{T}{2}} f(x) e^{-ni\omega x}\, dx. \end{cases}$$

Consider an interval $(\lambda, \lambda + \Delta\lambda)$. The values of n for which $2\pi\lambda \leqslant n\omega \leqslant 2\pi(\lambda + \Delta\lambda)$ are $2\pi\Delta\lambda/\omega$ in number (with an error $\leqslant 2$). The corresponding group of terms

$$\sum_{2\pi\lambda \leqslant n\omega \leqslant 2\pi(\lambda+\Delta\lambda)} c_{n,T}\, e^{ni\omega x}$$

can be written approximately as

(V, 1; 2)
$$c_\lambda 2\pi \frac{\Delta\lambda}{\omega} e^{2i\pi\lambda x} = c_\lambda T \Delta\lambda e^{2i\pi\lambda x},$$

with

(V, 1; 3)
$$c_\lambda = \frac{1}{T}\int_{-\frac{T}{2}}^{+\frac{T}{2}} f(x) e^{-2i\pi\lambda x}\, dx.$$

Thus this group of terms may be written approximately as

$$\text{(V, 1; 4)} \qquad \begin{cases} \displaystyle\sum_{2\pi\lambda \leqslant n\omega \leqslant 2\pi(\lambda+\Delta\lambda)} c_{n,\,\mathrm{T}} e^{ni\omega x} = \mathrm{C}(\lambda) e^{2i\pi\lambda x}\,\Delta\lambda, \\ \mathrm{C}(\lambda) = \displaystyle\int_{-\frac{\mathrm{T}}{2}}^{+\frac{\mathrm{T}}{2}} f(x)\, e^{-2i\pi\lambda x}\, dx = \mathrm{T} c_{\lambda}. \end{cases}$$

As $\mathrm{T} \to \infty$, it is possible to replace $\int_{-\mathrm{T}/2}^{+\mathrm{T}/2}$ by $\int_{-\infty}^{+\infty}$. Also, the sum $\sum_{n=-\infty}^{+\infty}$ may be expressed as the sum of all the groups of terms corresponding to the intervals

$$(0, \Delta\lambda),\quad (\Delta\lambda, 2\Delta\lambda),\quad \ldots,\quad (k\Delta\lambda, (k+1)\Delta\lambda),\quad \ldots,$$
$$(-\Delta\lambda, 0)\ (-2\Delta\lambda, -\Delta\lambda),\quad \ldots,\quad (-(k+1)\,\Delta\lambda, -k\Delta\lambda),\quad \ldots,$$

of magnitude $\Delta\lambda$, which can be written in the form (V, 1; 4). Thus we arrive intuitively at the equations

$$\text{(V, 1; 5)} \qquad f(x) = \int_{-\infty}^{+\infty} \mathrm{C}(\lambda) e^{2i\pi\lambda x}\, d\lambda,$$

$$\text{(V, 1; 6)} \qquad \mathrm{C}(\lambda) = \int_{-\infty}^{+\infty} f(x) e^{-2i\pi\lambda x}\, dx.$$

The Fourier series with sum f has been replaced by the Fourier integral (V, 1; 5), the Fourier coefficient $\mathrm{C}(\lambda)$ being given by (V, 1; 6). Obviously, what has been done here is not a rigorous proof. We shall proceed to look more closely into these matters.

2. FOURIER TRANSFORMS

Definition. *If $f(x)$ is a complex-valued function of the real variable x, the complex-valued function $\mathrm{C}(\lambda)$ of the real variable λ defined by*

$$\text{(V, 1; 7)} \qquad \mathrm{C}(\lambda) = \int_{-\infty}^{+\infty} e^{-2i\pi\lambda x} f(x)\, dx$$

is called the Fourier transform of f. In particular,

$$\text{(V, 1; 8)} \qquad \mathrm{C}(0) = \int_{-\infty}^{+\infty} f(x)\, dx.$$

We write

$$\text{(V, 1; 9)} \qquad \mathrm{C} = \mathscr{F} f \quad \text{or} \quad \mathrm{C}(\lambda) = \mathscr{F}[f(x)].$$

Similarly, we may define the transform

(V, 1; 10) $$C_1 = \overline{\mathcal{F}} f, \qquad C_1(\lambda) = \int_{-\infty}^{+\infty} e^{2\pi i\lambda x} f(x)\, dx.$$

$\mathcal{F} f$ and $\overline{\mathcal{F}} f$ obviously exist if f is summable ($f \in L^1$) and in this case they are, by Lebesgue's theorem, continuous functions of λ since $e^{\pm 2i\pi\lambda x} f(x)$ is continuous in λ at any fixed x and has its modulus bounded above by $|f(x)|$, which is a summable function independent of λ. They are bounded since

(V, 1; 11) $$|C(\lambda)| = \int_{-\infty}^{+\infty} |f(x)|\, dx = \|f\|_{L^1}.$$

A theorem due to Lebesgue shows that $C(\lambda)$ tends to 0 as $\lambda \to \pm\infty$. It is still possible to define $\mathcal{F} f$ in certain other cases : for example, if $f(x)$ is a *monotonic* function for $x > 0$ and $x < 0$ separately, or again if f is of bounded variation and thus a finite sum of such functions, and $f(x)$ tends to 0 as $|x| \to \infty$. It may be verified by Abel's theorem that $C(\lambda)$ exists and is continuous for $\lambda \neq 0$. Also there is the inequality

(V, 1; 12) $$|C(\lambda)| \leqslant \frac{\left(\int_{-\infty}^{+\infty} |df| = \text{total variation of} f\right)}{2\pi|\lambda|}$$

so that here again $C(\lambda) \to 0$ as $|\lambda| \to \infty$.

This inequality, used in the proof of Abel's theorem, follows at once on integrating by parts if f has a continuous derivative. It will be encountered again below: see equation (V, 1; 16).

3. FUNDAMENTAL RELATIONS AND INEQUALITIES

Suppose that $f \in L^1$, f is continuous and differentiable, and $f' \in L^1$. Integrating by parts for $\lambda \neq 0$,

(V, 1; 13)
$$\int_{-A}^{+A} e^{-2i\pi\lambda x} f(x)\, dx = \left[\frac{e^{-2i\lambda\pi x}}{-2i\pi\lambda} f(x)\right]_{x=-A}^{x=+A} - \int_{-A}^{+A} \frac{e^{-2i\pi\lambda x}}{-2i\pi\lambda} f'(x)\, dx.$$

Let A tend to $+\infty$. The term in brackets then tends to zero, since, as we shall show, $f(x)$ tends to zero as $|x| \to \infty$. For,

$$f(x) - f(0) = \int_0^x f'(t)\, dt,$$

so that, f' being summable, $f(x)$ indeed has a finite limit as $x \to \pm\infty$, since $f(\pm\infty) = f(0) + \int_0^{\pm\infty} f'(t)\, dt$. However, this limit cannot be

$\neq 0$ so long as f is summable. Since f' is summable, the integral on the right-hand side of (V, 1; 13) has the infinite integral as a limit so that, finally,

$$\text{(V, 1; 14)} \qquad C(\lambda) = \int_{-\infty}^{+\infty} \frac{e^{-2\pi i\lambda x}}{2\pi i\lambda} f'(x)\, dx \qquad \text{for} \qquad \lambda \neq 0.$$

This result may also be written

$$\text{(V, 1; 15)} \qquad 2i\pi\lambda C(\lambda) = \int_{-\infty}^{+\infty} e^{-2i\pi\lambda x} f'(x)\, dx = \mathscr{F} f';$$

it is still valid by continuity when $\lambda = 0$ (both sides are then zero, the left-hand side because it contains λ as a factor, the right-hand side because it is equal to $\left.\int_{-\infty}^{+\infty} f'(x)\, dx = f(+\infty) - f(-\infty) = 0\right)$. Hence there follows the important inequality

$$\text{(V, 1; 16)} \qquad |2\pi\lambda|\,|C(\lambda)| \leqslant \int_{-\infty}^{+\infty} |f'(x)|\, dx = \|f'\|_{L^1}.$$

More generally, if f is m times differentiable and if its derivatives of order $\leqslant m$ are summable then

$$\text{(V, 1; 17)} \qquad (2i\pi\lambda)^m C(\lambda) = \mathscr{F} f^{(m)},$$

$$\text{(V, 1; 18)} \qquad |2\pi\lambda|^m |C(\lambda)| \leqslant \|f^{(m)}\|_{L^1}.$$

The same results apply for the operation $\overline{\mathscr{F}}$, so long as $2i\pi\lambda$ is replaced by $-2i\pi\lambda$.

We now investigate whether $C(\lambda)$ is differentiable and whether

$$\text{(V, 1; 19)} \qquad C'(\lambda) = \int_{-\infty}^{+\infty} (-2i\pi x) e^{-2i\pi\lambda x} f(x)\, dx.$$

This differentiation under the $\int$ sign is legitimate if the expression thus obtained is an integral which is uniformly convergent with respect to λ for any finite interval in λ. This will be the case *if $xf(x)$ is summable.* Then

$$\text{(V, 1; 20)} \qquad C'(\lambda) = \mathscr{F}[-2i\pi x f(x)].$$

More generally, if $x^m f(x)$ is also summable, $C(\lambda)$ is m times continuously differentiable and then

$$\text{(V, 1; 21)} \qquad C^{(m)}(\lambda) = \mathscr{F}[(-2i\pi x)^m f(x)],$$

$$\text{(V, 1; 22)} \qquad |C^{(m)}(\lambda)| \leqslant \|(2\pi x)^m f(x)\|_{L^1}.$$

SUMMARY : *The more times that $f(x)$ is differentiable (with continuous derivatives) the faster does $C(\lambda)$ decrease at infinity; the faster that $f(x)$ decreases at infinity the more times is $C(\lambda)$ differentiable (with bounded derivatives).*

The same results are valid for the operator $\overline{\mathscr{F}}$ so long as $-2i\pi x$ is replaced by $2i\pi x$.

We shall sum up all these results in a theorem :

THEOREM 1. *Any summable function $f(x)$ has a Fourier transform*

(V, 1; 23) $$\mathscr{F}f = \mathrm{C}, \qquad \mathrm{C}(\lambda) = \int_{-\infty}^{+\infty} e^{-2i\pi\lambda x} f(x)\, dx,$$

which is continuous and bounded and tends to 0 as $|\lambda| \to \infty$:

(V, 1; 24) $$|\mathrm{C}(\lambda)| \leqslant \|f\|_{\mathrm{L}^1}.$$

If f is m times continuously differentiable and if its derivatives of order $\leqslant m$ are summable, then

(V, 1; 25) $$\mathscr{F}[f^{(m)}] = (2i\pi\lambda)^m \mathrm{C}(\lambda),$$ $$|2\pi\lambda|^m |\mathrm{C}(\lambda)| \leqslant \|f^{(m)}\|_{\mathrm{L}^1}.$$

If $x^m f(x)$ is summable, $\mathrm{C}(\lambda)$ is m times continuously differentiable and

(V, 1; 26) $$\mathscr{F}[(-2i\pi x)^m f(x)] = \mathrm{C}^{(m)}(\lambda),$$ $$|\mathrm{C}^{(m)}(\lambda)| \leqslant \|(2\pi x)^m f(x)\|_{\mathrm{L}^1}.$$

We also have

THEOREM 2. *If $\mathrm{C}(\lambda) = \mathscr{F}[f(x)]$ then*

(V, 1; 27) $$\frac{1}{|k|}\mathrm{C}\left(\frac{\lambda}{k}\right) = \mathscr{F}[f(kx)], \qquad k \text{ real} \neq 0.$$

For, by the change of variable $kx = \xi$,

$$\mathscr{F}[f(kx)] = \int_{-\infty}^{+\infty} e^{-2i\pi\lambda x} f(kx)\, dx = \int_{-\infty}^{+\infty} e^{-2i\pi\lambda\frac{\xi}{k}} f(\xi)\,\frac{d\xi}{|k|} = \frac{1}{|k|}\mathrm{C}\left(\frac{\lambda}{k}\right).$$

In particular

(V, 1; 28) $$\mathrm{C}(-\lambda) = \mathscr{F}[f(-x)],$$

and it follows that if f is an odd or even function then so is C.

4. SPACES $\mathscr{S}$ OF INFINITELY DIFFERENTIABLE FUNCTIONS WITH ALL DERIVATIVES DECREASING RAPIDLY

The space $\mathscr{S}$ is the space of complex functions on R which are infinitely differentiable and which, together with all their derivatives, decrease more rapidly than any power of $1/|x|$ as $|x| \to \infty$.

Results. If $\varphi \in \mathcal{S}$ then :

(1) For any integers l and m both $\geqslant 0$, the function $x^l\varphi^{(m)}(x)$ is bounded and summable, for from the fact that $x^{l+2}\varphi^{(m)}(x)$ is bounded by a constant M it follows that $x^l\varphi^{(m)}(x)$ is bounded by M/x^2 and hence summable.

(2) For any integers l and m both $\geqslant 0$, $(x^l\varphi(x))^{(m)}$ is bounded and summable by Leibnitz's theorem for the derivatives of a product.

Topology in the space $\mathcal{S}$

A sequence φ_j of functions belonging to $\mathcal{S}$ converges to 0 in $\mathcal{S}$ as $j \to \infty$ if for any integers l and m, both $\geqslant 0$, the sequence $x^l\varphi_j^{(m)}(x)$ converges uniformly to 0 in R. With regard to the Fourier transforms in the space $\mathcal{S}$ we then have :

THEOREM 3. *If $\varphi(x) \in (\mathcal{S})_x$, its Fourier transform $C(\lambda)$ is in $(\mathcal{S})_\lambda$. Also, if the $\varphi_j(x)$ converge to 0 in $(\mathcal{S})_x$ as $j \to \infty$, then their Fourier transforms $C_j(\lambda)$ converge to 0 in $(\mathcal{S})_\lambda$.*

Proof

1) Since $\varphi \in \mathcal{S}$, $x^l\varphi(x)$ is summable for any integer $l \geqslant 0$ so that, by the second part of Theorem 1, $C(\lambda)$ is infinitely differentiable.

2) For any integer $l \geqslant 0$ each derivative of $C(\lambda)$ decreases faster as $|\lambda| \to \infty$ than an arbitrary power of $1/|\lambda|$. We have, in fact, the inequality

$$(V, 1; 29) \qquad |(2\pi\lambda)^m C^{(l)}(\lambda)| \leqslant \|((2i\pi x)^l \varphi(x))^{(m)}\|_{L^1}.$$

3) Consider now a sequence φ_j, which converges to 0 in the sense of $\mathcal{S}$. Then, by (V, 1; 29),

$$(V, 1; 30) \qquad |(2\pi\lambda)^m C_j^{(l)}(\lambda)| \leqslant \|((2i\pi x)^l \varphi_j(x))^{(m)}\|_{L^1}$$

and on the right-hand side of this inequality there is a sequence of numbers which converges to 0 as $j \to \infty$. In consequence $\lambda^m C_j^{(l)}(\lambda)$ converges uniformly to 0 on the line as $j \to \infty$. Q.E.D.

Examples

1) Let
$$f(x) = \begin{cases} 1 & \text{for } |x| \leqslant A, \\ 0 & \text{for } |x| > A. \end{cases}$$

Its Fourier transform is

$$(V, 1; 31) \qquad C(\lambda) = \int_{-A}^{+A} e^{-2i\pi\lambda x}\, dx = \frac{\sin 2\pi\lambda A}{\pi\lambda}.$$

Note that f is not continuously differentiable. However $|\lambda C(\lambda)|$ is bounded

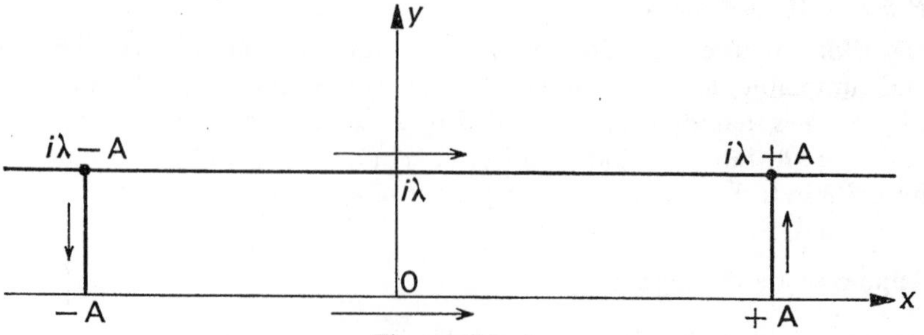

Figure (V, 1)

although $|\lambda^2 C(\lambda)|$ is not. As f has a bounded support, $x^m f(x)$ is summable for any m. It may be verified that $C(\lambda)$ is infinitely differentiable and that each of its derivatives is bounded. Since $\|f\|_{L^1} = 2A$, we have

$$|C(\lambda)| \leqslant 2A = C(0).$$

2) $f(x) = e^{-\pi x^2}$. Then

$$C(\lambda) = \int_{-\infty}^{+\infty} e^{-\pi x^2 - 2i\pi\lambda x}\, dx = \int_{-\infty}^{+\infty} e^{-\pi(x+i\lambda)^2 - \pi\lambda^2}\, dx. \tag{V, 1; 32}$$

To evaluate this integral in x (for fixed λ), introduce the complex variable $Z = x + i\lambda$ so that

$$C(\lambda) = e^{-\pi\lambda^2} \int_{i\lambda-\infty}^{i\lambda+\infty} e^{-\pi Z^2}\, dZ. \tag{V, 1; 33}$$

Now $e^{-\pi Z^2}$ is an integral function and it is thus permissible to change the contour of integration [see Fig. (V, 1]:

$$\int_{i\lambda-\infty}^{i\lambda+\infty} = \lim_{A\to\infty} \int_{i\lambda-A}^{i\lambda+A} = \lim_{A\to\infty} \left[\int_{i\lambda-A}^{-A} + \int_{-A}^{+A} + \int_{+A}^{i\lambda+A} \right]. \tag{V, 1; 34}$$

The first and third integrals of the last expression tend to 0 as $A \to \infty$. For, taking the third as an example,

$$\left| \int_{A}^{i\lambda+A} e^{-\pi Z^2}\, dZ \right| = \left| \int_0^{\lambda} e^{-\pi(A^2-y^2) - 2i\pi A y} i\, dy \right| \leqslant |\lambda| e^{-\pi(A^2-\lambda^2)}, \tag{V, 1; 35}$$

which for fixed λ certainly tends to 0 as $A \to \infty$. Thus

$$C(\lambda) = e^{-\pi\lambda^2} \lim_{A\to\infty} \int_{-A}^{+A} e^{-\pi Z^2}\, dZ = e^{-\pi\lambda^2} \int_{-\infty}^{+\infty} e^{-\pi x^2}\, dx = e^{-\pi\lambda^2}, \tag{V, 1; 36}$$

the last integral (Gauss' integral) being equal to 1.

Hence, finally

(V, 1; 37) $$\mathcal{F}[e^{-\pi x^2}] = e^{-\pi\lambda^2}.$$

Applying Theorem 2, with $k > 0$,

(V, 1; 38) $$\mathcal{F}[e^{-kx^2}] = \mathcal{F}\left[e^{-\pi\left(\sqrt{\frac{k}{\pi}}x\right)^2}\right]$$

$$= \sqrt{\frac{\pi}{k}}\,e^{-\pi\left(\sqrt{\frac{\pi}{k}}\lambda\right)^2} = \sqrt{\frac{\pi}{k}}\,e^{-\frac{\pi^2}{k}\lambda^2}.$$

Note that with $e^{-\pi x^2} \in \mathcal{S}_x$ we obtain a transform $e^{-\pi\lambda^2} \in \mathcal{S}_\lambda$.

3) $$f(x) = Y(x)e^{-ax}\frac{x^{\alpha-1}}{\Gamma(\alpha)} \quad \text{for } \alpha > 0,\ a > 0.$$

Then

(V, 1; 39) $$C(\lambda) = \int_0^\infty e^{-ax-2i\pi\lambda x}\frac{x^{\alpha-1}}{\Gamma(\alpha)}\,dx.$$

We make the change of variable

(V, 1; 40) $$Z = (a + 2i\pi\lambda)x.$$

As x varies over the positive real axis $(0, +\infty)$, Z varies over the complex half-line $(0, (a + 2i\pi\lambda)\infty)$. Then

(V, 1; 41) $$C(\lambda) = \int_0^{(a+2i\pi\lambda)\infty} e^{-Z}\frac{\left(\dfrac{Z}{a+2i\pi\lambda}\right)^{\alpha-1}}{\Gamma(\alpha)}\,\frac{dZ}{a+2i\pi\lambda}$$

$$= \left(\frac{1}{a+2i\pi\lambda}\right)^{\alpha}\int_0^{(a+2i\pi\lambda)\infty} e^{-Z}\frac{Z^{\alpha-1}}{\Gamma(\alpha)}\,dZ,$$

the branch of Z^α for $\mathcal{R}Z > 0$ being chosen to be continuous with the real positive branch for real $Z > 0$. Here again the function $e^{-Z}Z^{\alpha-1}$ is holomorphic for $\mathcal{R}Z > 0$ and by considering a contour of the form of Fig. (V, 2),

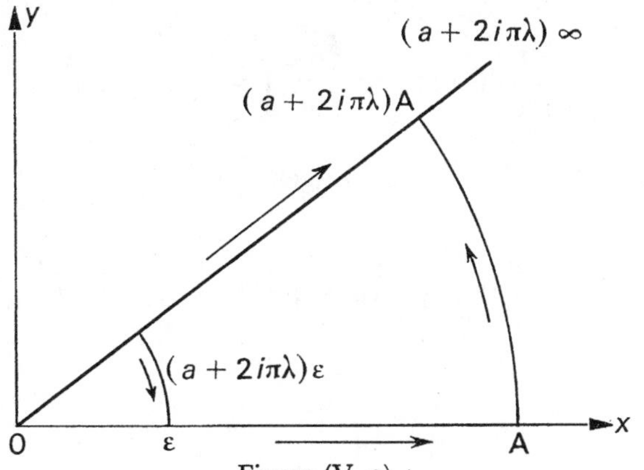

Figure (V, 2)

it is possible to write

$$\text{(V, 1; 42)} \quad \int_0^{(a+2i\pi\lambda)\infty} e^{-Z}Z^{\alpha-1}dZ = \lim_{\substack{\varepsilon \to 0 \\ A \to \infty}} \int_{(a+2i\pi\lambda)\varepsilon}^{(a+2i\pi\lambda)A}$$

$$= \lim_{\substack{\varepsilon \to 0 \\ A \to \infty}} \left[\int_{(a+2i\pi\lambda)\varepsilon}^{\varepsilon} + \int_{\varepsilon}^{A} + \int_{A}^{(a+2i\pi\lambda)A} \right].$$

Here again the first and third integrals of the last expression converge to 0 as $\varepsilon \to 0$, $A \to \infty$ (as an exercise, prove this rigorously) and $\int_\varepsilon^A$ tends to $\int_0^\infty$. Then

$$\text{(V, 1; 43)} \qquad \int_0^{(a+2i\pi\lambda)\infty} e^{-Z}Z^{\alpha-1}\, dZ = \int_0^\infty e^{-x}x^{\alpha-1}\, dx = \Gamma(\alpha),$$

giving

$$\text{(V, 1; 44)} \qquad \mathcal{F}\left[Y(x)e^{-ax}\frac{x^{\alpha-1}}{\Gamma(\alpha)}\right] = \left(\frac{1}{a + 2i\pi\lambda}\right)^\alpha,$$

the branch being chosen to be continuous with the real positive branch of $1/a^\alpha$ for $\lambda = 0$. Note that for $\lambda = 0$ the value

$$\text{(V, 1; 45)} \qquad C(0) = \frac{1}{a^\alpha}$$

follows immediately, the change of variables involving no complex quantity. It is known that, $f(x)$ being summable, $C(\lambda)$ must be continuous. Similarly by (V, 1; 28) ($-x$ being equal to $|x|$ for $x < 0$),

$$\text{(V, 1; 46)} \qquad \mathcal{F}\left[Y(-x)e^{-a|x|}\frac{|x|^{\alpha-1}}{\Gamma(\alpha)}\right] = \left(\frac{1}{a - 2i\pi\lambda}\right)^\alpha,$$

whence, on adding (V, 1 ; 44), we get

$$\text{(V, 1; 47)} \quad \mathcal{F}\left[e^{-a|x|}\frac{|x|^{\alpha-1}}{\Gamma(\alpha)}\right] = \left(\frac{1}{a + 2i\pi\lambda}\right)^\alpha + \left(\frac{1}{a - 2i\pi\lambda}\right)^\alpha.$$

In particular, for $\alpha = 1$,

$$\text{(V, 1; 48)} \qquad \mathcal{F}[e^{-a|x|}] = \frac{2a}{a^2 + 4\pi^2\lambda^2}.$$

2. Fourier transforms of distributions in one variable

1. DEFINITION

We shall endeavour to find an expression for $\mathcal{F}f$ by considering f and $\mathcal{F}f$ no longer as functions but as distributions. Then

$$\text{(V, 2; 1)} \qquad \langle \mathcal{F}f, \varphi \rangle = \int_{-\infty}^{+\infty} \varphi(\lambda) d\lambda \int_{-\infty}^{+\infty} e^{-2i\pi\lambda x} f(x)\, dx$$
$$= \iint e^{-2i\pi\lambda x} f(x) \varphi(\lambda) dx\, d\lambda,$$

a result which is valid since the double integral does exist. In fact,

$$|e^{-2i\pi\lambda x} f(x)\varphi(\lambda)| = |f(x)||\varphi(\lambda)|$$

is summable as a product of a summable function of x with a summable function of λ. It is thus possible to write

$$\text{(V, 2; 2)} \qquad \langle \mathcal{F}f, \varphi \rangle = \int_{-\infty}^{+\infty} f(x)\, dx \int_{-\infty}^{+\infty} e^{-2i\pi x\lambda} \varphi(\lambda) d\lambda$$
$$= \langle f, \mathcal{F}\varphi \rangle.$$

This relation is valid even when $\varphi \notin \mathcal{D}$, so long as $\varphi \in L^1$. In this case, $\mathcal{F}f = C(\lambda)$ is bounded, as is $\mathcal{F}\varphi = \gamma(x)$. The two quantities

$$\text{(V, 2; 3)} \qquad \langle \mathcal{F}f, \varphi \rangle = \int_{-\infty}^{+\infty} C(\lambda)\varphi(\lambda)\, d\lambda,$$

$$\text{(V, 2; 4)} \qquad \langle f, \mathcal{F}\varphi \rangle = \int_{-\infty}^{+\infty} f(x)\gamma(x)\, dx$$

then both have a meaning and are equal.

Equation (V, 2; 2) suggests that, for a distribution U_x, the Fourier transform $\mathcal{F}U = V_\lambda$ should be defined by the relation

$$\text{(V, 2; 5)} \qquad \langle \mathcal{F}U, \varphi \rangle = \langle U, \mathcal{F}\varphi \rangle \text{ for all } \varphi(\lambda) \in (\mathcal{D})_\lambda.$$

However this relation has no meaning for an arbitrary $U \in \mathcal{D}'_x$. For, if $\varphi \in (\mathcal{D})_\lambda$, there is no reason for $\mathcal{F}\varphi$ to belong to $(\mathcal{D})_x$ [it will even be seen in Theorem 5 that $\mathcal{F}\varphi$ belongs only to $(\mathcal{D})_x$ if $\varphi = 0$] so that the right-hand side of (V, 2; 5) has no meaning. The Fourier transform of an arbitrary distribution does not exist. We shall instead consider a particular type of distribution.

2. TEMPERED DISTRIBUTIONS : THE SPACE $\mathcal{S}'$

From the definition of the space $\mathcal{S}$ and its topology it can be seen that the space $\mathcal{D}$ is contained in $\mathcal{S}$ and that if the φ_j converge to 0 in the sense of $\mathcal{D}$ then they converge *a fortiori* to 0 in the space $\mathcal{S}$.

A distribution T (that is to say a continuous linear form on $\mathcal{D}$) is termed a *tempered distribution* if it may be extended to a continuous linear form on $\mathcal{S}$.

Examples

1) A summable function f is tempered. For it is possible to put

(V, 2; 6) $$\langle f, \varphi\rangle = \int_{-\infty}^{+\infty} f(x)\varphi(x)\, dx,$$

the integral existing if φ is bounded. Now any function φ belonging to $\mathcal{S}$ is in fact bounded.

2) A bounded function is tempered. For (V, 2; 6) then has a meaning if $\varphi \in L^1$. Now a function φ belonging to $\mathcal{S}$ is in fact summable. More generally, any locally summable function which is slowly increasing, that is to say satisfying

(V, 2; 7) $$|f(x)| \leqslant A|x|^k \quad \text{for} \quad |x| \to \infty,$$

is tempered. For $\varphi \in \mathcal{S}$ satisfies for $|x| \to \infty$ an inequality

(V, 2; 8) $$|\varphi(x)| < \frac{B}{|x|^{k+2}},$$

so that $|f(x)\,\varphi(x)| \leqslant \dfrac{AB}{|x|^2}$ is summable. The term *tempered* clearly evokes this slow increase at infinity.

3) A distribution T with a bounded support is tempered. For $\langle T, \varphi\rangle$ has a meaning even for *any* infinitely differentiable function φ, with no restriction as to rate of decrease at infinity.

4) If T is tempered, so are its successive derivatives. For T′ is defined by

(V, 2; 9) $$\langle T', \varphi\rangle = -\langle T, \varphi'\rangle.$$

Now if $\varphi \in \mathcal{S}$ then $\varphi' \in \mathcal{S}$ also, so that if T is tempered, $\langle T, \varphi'\rangle$ has a meaning for $\varphi \in \mathcal{S}$. This shows that $\langle T', \varphi\rangle$ has a meaning for $\varphi \in \mathcal{S}$ so that T′ is tempered.

5) If T is tempered and if α is a polynomial in x, then αT is also tempered. For α T is defined by

(V, 2; 10) $$\langle \alpha T, \varphi\rangle = \langle T, \alpha\varphi\rangle.$$

If $\varphi \in \mathcal{S}$ and α is a polynomial then $\alpha\varphi$ is also in $\mathcal{S}$. As T is tempered, $\langle T, \alpha\varphi \rangle$ has a meaning so that $\langle \alpha T, \varphi \rangle$ has a meaning and αT is also tempered.

6) It can be easily shown that a function f, such as e^x, which is not slowly increasing for $x \to \infty$ *is not tempered.*

The space of tempered distributions, which is the dual of the space $\mathcal{S}$, *will be denoted by* $\mathcal{S}'$.

3. FOURIER TRANSFORMS OF TEMPERED DISTRIBUTIONS

Let $U_x \in (\mathcal{S}')_x$. For $\varphi(\lambda) \in (\mathcal{S})_\lambda$ the right-hand side of (V, 2; 5) has a meaning since $\mathcal{F}\varphi = \gamma(x)$ is in $(\mathcal{S})_x$ (Theorem 3). Also if $\varphi_j(\lambda)$ converges to 0 in the sense of $(\mathcal{S})_\lambda$, then, by the same Theorem 3, $\mathcal{F}\varphi_j = \gamma_j(x)$ converges to 0 in the sense of $(\mathcal{S})_x$ and consequently $\langle U_x, \mathcal{F}\varphi_j \rangle$ converges to 0. The right-hand side of (V, 2; 5) thus defines a continuous linear form on $(\mathcal{S})_\lambda$ and hence a tempered distribution V_λ which is by definition $\mathcal{F}U$.

The operation $\overline{\mathcal{F}}$, of course, has similar properties. Hence

THEOREM 4. *If* U_x *is a tempered distribution,* $\mathcal{F}U$ *and* $\overline{\mathcal{F}}U$ *may be defined by the equations*

$$\text{(V, 2; 11)} \qquad \langle \mathcal{F}U, \varphi \rangle = \langle U, \mathcal{F}\varphi \rangle, \quad \langle \overline{\mathcal{F}}U, \varphi \rangle = \langle U, \overline{\mathcal{F}}\varphi \rangle$$

for any function $\varphi(\lambda) \in (\mathcal{S})_\lambda$. *Then* $\mathcal{F}U$ *and* $\overline{\mathcal{F}}U$ *are tempered.*

Fourier transforms of distributions with bounded supports

Let $U_x \in \mathcal{E}'$ (and thus $U_x \in \mathcal{S}'$). It is then possible to evaluate, for fixed λ,

$$\text{(V, 2; 12)} \qquad V(\lambda) = \langle U_x, e^{-2i\pi\lambda x} \rangle.$$

Also, as $e^{-2i\pi\lambda x}$ is infinitely differentiable in λ and x, it follows (Chapter III, Theorem 1) that $V(\lambda)$ is an infinitely differentiable function of λ. Again, even for a *complex* fixed λ, $e^{-2i\pi\lambda x} \in (\mathcal{E})_x$ so that $V(\lambda)$ exists for complex λ, is infinitely differentiable and hence is an integral function. It will now be shown that this function $V(\lambda)$, with λ considered real, is the Fourier transform of U_x and that equation (V, 2; 12) is thus an extension of the definition (V, 1; 7) when $f(x)$ is replaced by a distribution U_x

with a bounded support. For, when $\varphi(\lambda) \in (\mathcal{S})_\lambda$,

$$\text{(V, 2; 13)} \quad \begin{aligned} \langle \mathcal{F}U, \varphi \rangle &= \langle U, \mathcal{F}\varphi \rangle \\ &= \left\langle U_x, \int_{-\infty}^{+\infty} e^{-2i\pi\lambda x}\, \varphi(\lambda)\, d\lambda \right\rangle \\ &= \int_{-\infty}^{\infty} \langle U_x, e^{-2i\pi\lambda x}\, \varphi(\lambda) \rangle d\lambda \qquad \text{by (III, 1; 8)} \\ &= \int_{-\infty}^{\infty} \varphi(\lambda)\, V(\lambda)\, d\lambda = \langle V, \varphi \rangle, \end{aligned}$$

whence it follows at once that $\mathcal{F}U = V$. This may be expressed by

THEOREM 5. *If* U_x *is a distribution having a bounded support, then its Fourier transform is a function* $V(\lambda)$, *which may be continued over complex values of* λ *as an integral function, given by the equation*

$$\text{(V, 2; 12)} \qquad V(\lambda) = \langle U_x, e^{-2i\pi\lambda x} \rangle.$$

Further examples where the Fourier transform of a distribution or even of a function is a distribution will be seen below [equation (V, 2; 26)]. As a particular case of Theorem 5 it can be seen that if f is a function with bounded support, then its Fourier transform, given by (V, 1; 7), is a holomorphic function of λ. It can thus only have a bounded support if it is identically zero. As it will be seen that $\mathcal{F}\varphi$ can be zero only if φ is zero it follows that if $\varphi(x) \in (\mathcal{D})_x$ is not $\equiv 0$, its Fourier transform is not in $(\mathcal{D})_\lambda$.

Fourier transforms of distributions with point supports

Equation (V, 2; 12) gives

$$\text{(V, 2; 14)} \qquad \mathcal{F}\delta = \langle \delta_x, e^{-2i\pi\lambda x} \rangle = 1.$$

Thus the Fourier transform of δ is the constant function 1,

$$\text{(V, 2; 15)} \qquad \mathcal{F}\delta' = \langle \delta'_x, e^{-2i\pi\lambda x} \rangle = 2i\pi\lambda,$$

$$\text{(V, 2; 16)} \qquad \mathcal{F}\delta^{(m)} = \langle \delta^{(m)}, e^{-2i\pi\lambda x} \rangle = (2i\pi\lambda)^m.$$

Direct application of (V, 2; 11) gives the same result:

$$\text{(V, 2; 17)} \quad \begin{aligned} \langle \mathcal{F}\delta, \varphi \rangle = \langle \delta, \mathcal{F}\varphi \rangle &= \left\langle \delta_x, \int_{-\infty}^{+\infty} e^{-2i\pi\pi\lambda}\varphi(\lambda)\, d\lambda \right\rangle \\ &= \int_{-\infty}^{+\infty} \varphi(\lambda)\, d\lambda = \langle 1, \varphi \rangle, \text{ whence } \mathcal{F}\delta = 1. \end{aligned}$$

Finally,

$$\text{(V, 2; 18)} \qquad \mathcal{F}[\delta(x - a)] = \langle \delta_{x-a}, e^{-2i\pi\lambda x} \rangle = e^{-2i\pi a\lambda}.$$

Proof of the equation $\mathscr{F}1 = \delta$.

Note first that equations (V, 1; 25) and (V, 1; 26) (without the inequalities) can be extended to tempered distributions. If $\mathscr{F}U_x = V_\lambda$ then

(V, 2; 19) $$\mathscr{F}[U_x^{(m)}] = (2i\pi\lambda)^m V_\lambda,$$

(V, 2; 20) $$\mathscr{F}[(-2i\pi x)^m U_x] = V_\lambda^{(m)}.$$

As an example we prove the first of these results. Let $\varphi(\lambda) \in \mathscr{S}_\lambda$ and $\gamma(x) = \mathscr{F}\varphi$. Using (V, 2; 26),

(V, 2; 21)

$$\begin{aligned}\langle \mathscr{F}U_x^{(m)}, \varphi(\lambda)\rangle &= \langle U^{(m)}, \mathscr{F}\varphi\rangle = \langle U^{(m)}, \gamma\rangle \\ &= \langle U, (-1)^m \gamma^{(m)}\rangle \\ &= \langle U, (-1)^m \mathscr{F}[(-2i\pi\lambda)^m \varphi(\lambda)]\rangle = \langle \mathscr{F}U, (2i\pi\lambda)^m \varphi(\lambda)\rangle \\ &= \langle V_\lambda, (2i\pi\lambda)^m \varphi(\lambda)\rangle = \langle (2i\pi\lambda)^m V_\lambda, \varphi(\lambda)\rangle,\end{aligned}$$

whence $\mathscr{F}U_x^{(m)} = (2i\pi\lambda)^m V_\lambda$. The second result may be proved using (V, 1; 25).

Now apply (V, 2; 19) with $U = 1$ (which is a bounded function and thus a tempered distribution) and $m = 1$. Then, as the derivative of 1 is zero,

(V, 2; 22) $$0 = \mathscr{F}[0] = 2i\pi\lambda V_\lambda,$$

(V, 2; 23) $$\lambda V_\lambda = 0.$$

However it was seen in Chapter II that a distribution V_λ satisfying (V, 2; 23) is necessarily proportional to δ. Thus,

(V, 2; 24) $$\mathscr{F}1 = C\delta.$$

Apply definition (V, 2; 11), taking as the function φ the simplest function of $\mathscr{S}$ whose Fourier transform is known explicitly :

$$\varphi(\lambda) = e^{-\pi\lambda^2}, \qquad \mathscr{F}\varphi = e^{-\pi x^2}, \qquad \text{[Eq. (V, 1; 37)]}$$

(V, 2; 25) $$\begin{aligned}\langle \mathscr{F}1, e^{-\pi\lambda^2}\rangle &= \langle 1, \mathscr{F}e^{-\pi\lambda^2}\rangle = \langle 1, e^{-\pi x^2}\rangle \\ &= \int_{-\infty}^{+\infty} e^{-\pi\lambda^2}\, dx = 1 \qquad \text{(Gauss integral).}\end{aligned}$$

Now $\langle \delta, e^{-\pi\lambda^2}\rangle = 1$ so that $C = 1$ and therefore

(V, 2; 26) $$\mathscr{F}1 = \delta.$$

Naturally, as the right-hand side is real, the equation is invariant under change of i into $-i$ and we have

(V, 2; 27) $$\overline{\mathscr{F}}1 = \delta.$$

The Fourier inversion theorem

The relation (V, 1; 5) is associated with the relation (V, 1; 6). However, it does not necessarily have a meaning in the theory of functions sense. For it follows from $f(x) \in L^1$ that $C(\lambda)$ is bounded but not necessarily summable [example: equation (V, 1; 44) for $\alpha < 1$] so that (V, 1; 5) may not have a meaning.

Note, moreover, that a relation such as (V, 1; 5) giving $f(x)$ for all values of x would be absurd, since a change of values of f on a set of measure zero would not alter $C(\lambda)$.

On the other hand, $C(\lambda)$ as a bounded function is a tempered distribution and hence has a transform under $\overline{\mathcal{F}}$. Is this transform $f(x)$? This is, in fact, the case and we shall see more generally that if $\mathcal{F}U = V$ then $\overline{\mathcal{F}}V = U$ [a result of which (V, 2; 27) is, by (V, 2; 14), a particular case].

First suppose that $\varphi(x)$ is a function belonging to $\mathcal{S}_x$ and $\mathcal{F}\varphi = \gamma(\lambda)$. We wish to show that $\overline{\mathcal{F}}\gamma = \varphi$. At the point a,

$$\text{(V, 2; 28)} \qquad \overline{\mathcal{F}}[\gamma](a) = \int_{-\infty}^{+\infty} e^{+2i\pi\lambda a}\gamma(\lambda)\, d\lambda.$$

Note also that

$$\text{(V, 2; 29)} \qquad e^{2i\pi\lambda a}\gamma(\lambda) = \int_{-\infty}^{+\infty} e^{-2i\pi\lambda(x-a)}\varphi(x)\, dx,$$

which by a change of variable becomes

$$= \int_{-\infty}^{+\infty} e^{-2\pi i\lambda x}\varphi(x + a)\, dx = \mathcal{F}[\varphi(x + a)].$$

Hence (V, 2; 28) may be written

$$\begin{aligned} \text{(V, 2; 30)} \quad \overline{\mathcal{F}}[\gamma](a) &= \int_{-\infty}^{+\infty} \mathcal{F}[\varphi(x + a)]\, d\lambda = \langle 1, \mathcal{F}[\varphi(x + a)]\rangle \\ &= \langle \mathcal{F}1, \varphi(x + a)\rangle \qquad \text{by (V, 2; 26)} \\ &= \langle \delta, \varphi(x + a)\rangle = \varphi(a). \end{aligned}$$

Thus (V, 1; 5) has been proved when $f = \varphi \in \mathcal{S}$. Then, for $\varphi \in \mathcal{S}$,

$$\text{(V, 2; 31)} \qquad \overline{\mathcal{F}}\mathcal{F}\varphi = \varphi \qquad \text{and, similarly,} \qquad \mathcal{F}\overline{\mathcal{F}}\varphi = \varphi.$$

The same relation follows immediately for $U \in \mathcal{S}'$.

$$\begin{aligned} \text{(V, 2; 32)} \qquad \langle \overline{\mathcal{F}}\mathcal{F}U, \varphi\rangle = \langle \mathcal{F}U, \overline{\mathcal{F}}\varphi\rangle &= \langle U, \mathcal{F}\overline{\mathcal{F}}\varphi\rangle \\ &= \langle U, \varphi\rangle, \end{aligned}$$

and thus

$$\text{(V, 2; 33)} \qquad \overline{\mathcal{F}}\mathcal{F}U = U \qquad \text{and, similarly,} \qquad \mathcal{F}\overline{\mathcal{F}}U = U.$$

Hence :

THEOREM 6. *In the space of tempered distributions, $\overline{\mathcal{F}}$ and $\mathcal{F}$ are inverse operations :*

(V, 2; 33) $$\overline{\mathcal{F}}\mathcal{F}\mathrm{U} = \mathcal{F}\overline{\mathcal{F}}\mathrm{U} = \mathrm{U}.$$

Thus if $\mathcal{F}\mathrm{U} = \mathrm{V}$, then $\overline{\mathcal{F}}\mathrm{V} = \mathrm{U}$.

COROLLARY. $\mathcal{F}\mathrm{U} = 0$ *only if* $\mathrm{U} = 0$.

For if $\mathrm{V} = \mathcal{F}\mathrm{U} = 0$, then $\mathrm{U} = \overline{\mathcal{F}}\mathrm{V} = \overline{\mathcal{F}}0 = 0$. Of course, if $\mathrm{U} = f$ is a function this only shows that f is zero in the distribution sense, that is to say almost everywhere zero in the function sense.

Note. If $f \in \mathrm{L}^1$ and if it is known (for example from an explicit expression) that $\mathrm{C}(\lambda)$ is not only bounded but summable then $\overline{\mathcal{F}}\mathrm{C}(\lambda)$ is the function $\int_{-\infty}^{+\infty} \mathrm{C}(\lambda)e^{2i\pi\lambda x}\,d\lambda$. Theorem 6 then shows that the functions $f(x)$ and $\int_{-\infty}^{+\infty} \mathrm{C}(\lambda)e^{2i\pi\lambda x}\,d\lambda$ are equal as distributions and thus almost everywhere equal. As the integral is a continuous function this shows that the original function $f(x)$ is equal almost everywhere to a continuous function. If $f(x)$ is continuous then the two sides of (V, 1; 5) are identical. Alternatively, if $f \in \mathrm{L}^1$ and if, in the neighbourhood of the point a, f is of bounded variation, then it may be proved that

(V, 2; 34) $$\frac{f(a+0)+f(a-0)}{2} = \lim_{\mathrm{A}\to\infty} \int_{-\mathrm{A}}^{+\mathrm{A}} \mathrm{C}(\lambda)e^{2i\pi\lambda a}\,d\lambda.$$

The inversion relation enables equations (V, 2; 14), (V, 2; 16) and (V, 2; 18) to be inverted, giving

(V, 2; 35) $$\begin{cases} \mathcal{F}1 = \delta, \\ \mathcal{F}[(-2i\pi x)^m] = \delta_\lambda^{(m)}, \\ \mathcal{F}[e^{2i\pi a x}] = \delta_{\lambda-a}. \end{cases}$$

When applied to (V, 1; 31) at $a = 0$, (V, 2; 34) gives

(V, 2; 36)
$$1 = \int_{-\infty}^{+\infty} \frac{\sin 2\pi\lambda\mathrm{A}}{\pi\lambda}\,d\lambda \quad \text{or} \quad \int_0^\infty \frac{\sin k\lambda}{\lambda}\,d\lambda = \frac{\pi}{2} \quad \text{for} \quad k > 0.$$

Exercise. By using the method of residues verify the inversion relation for (V, 1; 48) :

(V, 2; 37) $$\mathcal{F}\left[\frac{2a}{a^2+4\pi^2x^2}\right] = e^{-a|\lambda|}.$$

Heuristic proof of the inversion theorem

Starting from $\mathcal{F}1 = \delta$ we shall now give the following heuristic proof of the inversion relation. A distribution U will be written as $U(x)$ and its Fourier transform V as

$$V(\lambda) = \int_{-\infty}^{+\infty} e^{-2i\pi\lambda x}\, U(x)\, dx. \tag{V, 2; 38}$$

Then $\overline{\mathcal{F}}V$ may be expressed as

$$\begin{aligned} \overline{\mathcal{F}}V &= \int_{-\infty}^{+\infty} e^{2i\pi\lambda x} V(\lambda)\, d\lambda \\ &= \int_{-\infty}^{+\infty} e^{2i\pi\lambda x}\, d\lambda \int_{-\infty}^{+\infty} U(\xi)\, e^{-2i\pi\lambda\xi}\, d\xi. \end{aligned} \tag{V, 2; 39}$$

The order of integration will be inverted (although the double integral is never summable, since $|e^{2i\pi\lambda x}\, U(\xi)\, e^{-2i\pi\lambda\xi}| = |U(\xi)|$ which, even if U is a summable function in ξ, is never summable in ξ, λ :

$$\iint |U(\xi)|\, d\xi\, d\lambda = \int_{-\infty}^{+\infty} d\lambda \int_{-\infty}^{+\infty} |U(\xi)|\, d\xi = +\infty$$

whence

$$\overline{\mathcal{F}}V = \int_{-\infty}^{+\infty} U(\xi)\, d\xi \int_{-\infty}^{+\infty} e^{-2i\pi\lambda(\xi - x)}\, d\lambda. \tag{V, 2; 40}$$

However the equation $\mathcal{F}1 = \delta$ can be (incorrectly) written as

$$\int_{-\infty}^{+\infty} e^{-2i\pi\lambda X}\, d\lambda = \delta(X). \tag{V, 2; 41}$$

Then (V, 2; 40) is written

$$\overline{\mathcal{F}}V = \int_{-\infty}^{\infty} U(\xi)\delta(\xi - x)\, d\xi = \int_{-\infty}^{+\infty} U(\xi)\delta(x - \xi)\, d\xi = U(x) \tag{V, 2; 42}$$

[Chapter III, equation (III, 2; 34)].

It is possible to justify this formalism, but the justification would be little different from the rigorous proof which has been given. It nevertheless remains true that this formalism gives a more direct approach.

4. THE PARSEVAL-PLANCHEREL EQUATION. FOURIER TRANSFORMS IN L^2

Let $f(x)$ be a twice continuously differentiable function with a bounded support. By (V, 1; 18) (for $m = 2$), $|C(\lambda)|$ is bounded above at infinity by $1/\lambda^2$ and is thus summable. From the inversion formula,

$$\text{(V, 2; 43)} \qquad \int_{-\infty}^{+\infty} |f(x)|^2\, dx = \int_{-\infty}^{+\infty} f(x)\overline{f(x)}\, dx$$

$$= \int_{-\infty}^{+\infty} f(x)\, dx \int_{-\infty}^{+\infty} \overline{C(\lambda)e^{-2i\pi\lambda x}}\, d\lambda = \iint f(x)\overline{C(\lambda)e^{-2i\pi\lambda x}}dx\, d\lambda,$$

the double integral existing by virtue of the assumptions made about f (f is summable in x and C in λ so that their product is summable in x, λ). The order of integration may now be changed to give

$$\text{(V, 2; 44)} \qquad = \int_{-\infty}^{+\infty} \overline{C(\lambda)}\, d\lambda \int_{-\infty}^{+\infty} f(x)e^{-2i\pi\lambda x}\, dx$$

$$= \int_{-\infty}^{+\infty} C(\lambda)\overline{C(\lambda)}\, d\lambda = \int_{-\infty}^{+\infty} |C(\lambda)|^2\, d\lambda.$$

The result may also be expressed by

$$\text{(V, 2; 45)} \qquad \int_{-\infty}^{+\infty} |f(x)|^2\, dx = \int_{-\infty}^{+\infty} |C(\lambda)|^2 d\lambda,$$

$$\|f\|_{L^2} = \|C\|_{L^2}.$$

Now let f be any function belonging to L^2. It may be approximated (in L^2, that is to say in the mean square) by a sequence of twice continuously differentiable functions f_n with bounded supports. It may be deduced that the Fourier transform of f [in the distribution sense, for a function in L^2 is tempered but not necessarily summable, as the example

$$f(x) = \frac{1}{\sqrt{1 + x^2}}$$

shows, so that equation (V, 1; 7) may not have any meaning] is a function $C(\lambda)$ (defined as a distribution, and thus only specified almost everywhere) whose square is summable, so that (V, 2; 45) follows. Replacing $f(x)$ and $\overline{f(x)}$ by $\overline{g(x)}$ and $g(x)$ gives an analogous result. If f and g are in L^2, so that fg is summable by Schwarz' inequality, and if $C(\lambda)$ and $D(\lambda)$ are their Fourier transforms, then

$$\text{(V, 2; 46)} \qquad \int_{-\infty}^{+\infty} f(x)\, \overline{g(x)}\, dx = \int_{-\infty}^{+\infty} C(\lambda)\, \overline{D(\lambda)}\, d\lambda,$$

$$(f, g)_{L^2} = (C, D)_{L^2}.$$

This may be summed up in a theorem which corresponds to the analogous Theorem 6 of Chapter IV on the theory of Fourier series :

THEOREM 7. (*Plancherel-Parseval*) $\mathcal{F}$ *and* $\overline{\mathcal{F}}$ *define reciprocal one-to-one mappings between* L^2_x *and* L^2_λ. *If f and g are in* L^2_x *their Fourier transforms (in the distribution sense) are in* L^2_λ *and obey equations* (V, 2; 45) *and* (V, 2; 46).

5. THE POISSON SUMMATION FORMULA

Let U_x be the distribution $\sum_{n=-\infty}^{+\infty} \delta_{x-n}$ consisting of masses $+1$ at all points whose coordinates are integers. We shall attempt to find its Fourier transform. We have

$$\mathcal{F}[\delta_{x-n}] = e^{-2i\pi n\lambda}. \tag{V, 2; 47}$$

Assuming that the operations $\mathcal{F}$ and Σ can be inverted (which is easy to justify here), then

$$\mathcal{F} \sum_{n=-\infty}^{+\infty} \delta_{x-n} = \sum_{n=-\infty}^{+\infty} \mathcal{F}\delta_{x-n} = \sum_{n=-\infty}^{+\infty} e^{-2i\pi n\lambda}. \tag{V, 2; 48}$$

However, it has been seen in the theory of Fourier series (Chapter IV) that the sum of this last series is $\sum_{n=-\infty}^{+\infty} \delta_{\lambda-n}$. Hence,

$$\mathcal{F}\left[\sum_{n=-\infty}^{+\infty} \delta_{x-n}\right] = \sum_{n=-\infty}^{+\infty} \delta_{\lambda-n}. \tag{V, 2; 49}$$

Thus we have here a distribution which, like $e^{-\pi x^2}$, is equal to its Fourier transform (apart from the name of the variable, x or λ).

Applying the definition (V, 2; 11) to this pair of distributions it may be seen that if $\varphi(\lambda) \in \mathcal{S}_\lambda$ and if $\gamma(x)$ is its Fourier transform, then

$$\sum_{n=-\infty}^{+\infty} \gamma(n) = \left\langle \sum_{n=-\infty}^{+\infty} \delta_{x-n}, \gamma \right\rangle = \left\langle \sum_{n=-\infty}^{+\infty} \delta_{\lambda-n}, \varphi \right\rangle = \sum_{n=-\infty}^{+\infty} \varphi(n), \tag{V, 2; 50}$$

whence we have the *Poisson summation formula*

$$\sum_{n=-\infty}^{+\infty} \gamma(n) = \sum_{n=-\infty}^{+\infty} \varphi(n). \tag{V, 2; 51}$$

Applying this formula to $\varphi(x) = e^{-tx^2}$ [equation (V, 1; 38)] gives

$$\sum_{n=-\infty}^{+\infty} e^{-tn^2} = \left[\sum_{n=-\infty}^{+\infty} e^{-\frac{\pi^2}{t}n^2}\right]\sqrt{\frac{\pi}{t}}. \tag{V, 2; 52}$$

This remarkable result (a functional equation for the θ function) plays an important part in the theory of elliptic functions and the heat conduction equation.

6. THE FOURIER TRANSFORM: MULTIPLICATION AND CONVOLUTION

Let S and T be two distributions with bounded supports. Their Fourier transforms are the two functions $C(\lambda)$, $D(\lambda)$ given by (V, 2; 12). $S * T$ also has a bounded support and its Fourier transform is given by the same equation

$$\text{(V, 2; 53)} \quad \begin{aligned} \mathcal{F}[S * T](\lambda) &= \langle S * T, e^{-2i\pi\lambda x}\rangle \\ &= \langle S_\xi \otimes T_\eta, e^{-2i\pi\lambda(\xi+\eta)}\rangle = \langle S_\xi \otimes T_\eta, e^{-2i\pi\lambda\xi} e^{-2i\pi\lambda\eta}\rangle \\ &= \langle S_\xi, e^{-2i\pi\lambda\xi}\rangle \langle T_\eta, e^{-2i\pi\lambda\eta}\rangle = C(\lambda)D(\lambda). \end{aligned}$$

We shall assume without proof that this result remains valid if S is any tempered distribution and T any distribution with a bounded support; $S * T$ has a meaning and it will be assumed that it is again tempered and that

$$\text{(V, 2; 54)} \qquad \mathcal{F}(S * T) = C_\lambda D(\lambda).$$

C_λ is then a distribution and $D(\lambda)$ an infinitely differentiable function (Theorem 5) so that their product has a meaning. The relation remains valid in many other cases*.

The Fourier transform of $S * T$ is the ordinary product of the transforms of S and T. The same result is obviously valid for $\overline{\mathcal{F}}$. From the Fourier inversion theorem,

$$\text{(V, 2; 55)} \qquad \begin{aligned} \mathcal{F}[S * T] &= \mathcal{F}S\mathcal{F}T, \\ \overline{\mathcal{F}}[S * T] &= \overline{\mathcal{F}}S\overline{\mathcal{F}}T, \\ \mathcal{F}[ST] &= \mathcal{F}S * \mathcal{F}T, \\ \overline{\mathcal{F}}[ST] &= \overline{\mathcal{F}}S * \overline{\mathcal{F}}T. \end{aligned}$$

THEOREM 8. *The Fourier transform changes convolution into multiplication and multiplication into convolution, in accordance with equations* (V, 2; 55) (*provided that the requisite conditions are satisfied*).

This is the essential property of the Fourier transform and the reason for its usefulness in applications, as it was for the Fourier series (Chapter IV). There are numerous applications to integral equations, partial differential equations (Laplace's equation, heat condition equation, wave equation, quantum mechanics) and probability.

Examples

1) Again consider (V, 1; 44). Obviously

$$\text{(V, 2; 56)} \qquad \left(\frac{1}{a + 2i\pi\lambda}\right)^\alpha \left(\frac{1}{a + 2i\pi\lambda}\right)^\beta = \left(\frac{1}{a + 2i\pi\lambda}\right)^{\alpha+\beta},$$

*For example, if f and g are in L^1, then $f * g$ is also in L^1 and $\mathcal{F}(f * g) = \mathcal{F}f\mathcal{F}g$.

but from this can be deduced

$$(\text{V, 2; 57})\quad Y(x)e^{-ax}\frac{x^{\alpha-1}}{\Gamma(\alpha)} * Y(x)e^{-ax}\frac{x^{\beta-1}}{\Gamma(\beta)} = Y(x)e^{-ax}\frac{x^{\alpha+\beta-1}}{\Gamma(\alpha+\beta)},$$

which is the result (III, 2; 11). In fact, however, this result was proved for any complex a, while for $a < 0$, for example, it cannot be proved by Fourier transformation, since $Y(x)e^{|a|x}\frac{x^{\alpha-1}}{\Gamma(\alpha)}$ is not tempered and has no Fourier transform. Note that the two distributions introduced on the left-hand side of (V, 2; 57) are summable functions, but neither has a bounded support.

2) Now consider the example of (V, 1; 38),

$$(\text{V, 2; 58})\qquad \mathscr{F}\left[\frac{1}{\sigma\sqrt{2\pi}}e^{-\frac{x^2}{2\sigma^2}}\right] = e^{-2\pi^2\sigma^2\lambda^2} \qquad (\sigma > 0).$$

Obviously

$$(\text{V, 2; 59})\qquad e^{-2\pi^2\sigma^2\lambda^2}e^{-2\pi^2\tau^2\lambda^2} = e^{-2\pi^2(\sigma^2+\tau^2)\lambda^2},$$

and hence

$$(\text{V, 2; 60})\qquad \frac{1}{\sigma\sqrt{2\pi}}e^{-\frac{x^2}{2\sigma^2}} * \frac{1}{\tau\sqrt{2\pi}}e^{-\frac{x^2}{2\tau^2}} = \frac{1}{\sqrt{\sigma^2+\tau^2}\sqrt{2\pi}}e^{-\frac{x^2}{2(\sigma^2+\tau^2)}},$$

which is equation (III, 2; 15) again.

3) Equation (V, 1; 48) in conjunction with (V, 1; 27) yields

$$(\text{V, 2; 61})\qquad \mathscr{F}\left[\frac{1}{\pi}\frac{a}{x^2+a^2}\right] = e^{-a|2\pi\lambda|}, \qquad a > 0.$$

Now, obviously,

$$(\text{V, 2; 62})\qquad e^{-a|2\pi\lambda|}e^{-b|2\pi\lambda|} = e^{-(a+b)|2\pi\lambda|},$$

and hence

$$(\text{V, 2; 63})\qquad \frac{1}{\pi}\frac{a}{x^2+a^2} * \frac{1}{\pi}\frac{b}{x^2+b^2} = \frac{1}{\pi}\frac{a+b}{x^2+(a+b)^2},$$

which is equation (III, 2; 23) again.

4) The facts that δ is the unit element in convolution while 1 is the unit element in multiplication are now seen to be related to the result $\mathscr{F}\delta = 1$. In the same way, bearing in mind (V, 2; 16) and (V, 2; 35), relations (V, 2; 19) and (V, 2; 20) are essentially applications of Theorem 8.

The applications to partial differential equations with constant coefficients arise from the fact that these are examples of convolution equations.

Suppose, for example, we wish to find the elementary solution of the differential operator $d^2/dx^2 - a^2$, that is, of

$$\text{(V, 2; 64)} \qquad E'' - a^2 E = \delta.$$

Suppose that E is tempered. It then has a Fourier transform C, and the transform of (V, 2; 64) gives

$$\text{(V, 2; 65)} \qquad (-4\pi^2\lambda^2 - a^2)C_\lambda = 1$$

or

$$\text{(V, 2; 66)} \qquad C(\lambda) = \frac{-1}{a^2 + 4\pi^2\lambda^2}.$$

Hence we find the solution

$$\text{(V, 2; 67)} \qquad E_x = \overline{\mathscr{F}}\left[\frac{-1}{a^2 + 4\pi^2\lambda^2}\right] = -\frac{1}{2a} e^{-a|x|}$$

[equation (V, 1; 48)].

This may be verified directly :

$$\text{(V, 2; 68)} \qquad \begin{cases} E' = \left(\dfrac{\text{sign of } x}{2}\right) e^{-a|x|}, \\ E'' = -\dfrac{a}{2} e^{-a|x|} + \delta, \end{cases}$$

(because of the discontinuity at the origin), giving (V, 2; 64). The solution we have found is completely different from that of Chapter III (convolution) belonging to $\mathscr{D}'_+$ [equation (III, 2; 88)] with $\omega = ia$, namely $Y(x)$ $(\sinh/ax)a$. The solution found here is not in $\mathscr{D}'_+$, while that of Chapter III is not tempered. In fact we have found the *only* tempered solution, since any other solution is obtained by adding a complementary function $C_1 e^{ax} + C_2 e^{-ax}$. Such a solution is tempered only for $C_1 = C_2 = 0$, since otherwise it increases exponentially for $x \to +\infty$, or for $x \to -\infty$, or for both cases. An equation such as

$$\text{(V, 2; 69)} \qquad E'' + a^2 E = \delta$$

would be much more complicated to solve by this method and indeed has an infinite number of tempered solutions.

7. OTHER EXPRESSIONS FOR THE FOURIER TRANSFORM

The transform is often written in another form. For $f(x) \in L^1$, put

$$\text{(V, 2; 70)} \qquad D(\lambda) = \frac{1}{h}\int_{-\infty}^{+\infty} e^{-i\omega\lambda x} f(x)\, dx, \quad \omega \text{ real} \neq 0, \quad h > 0.$$

Then

$$D(\lambda) = \frac{1}{h} C\left(\frac{\omega}{2\pi}\lambda\right), \tag{V, 2; 71}$$

C being what we have up to now regarded as the Fourier transform. The inversion theorem can then be written

$$f(x) = \int_{-\infty}^{+\infty} e^{+2i\pi\mu x} C(\mu)\, d\mu = h \int_{-\infty}^{+\infty} e^{+2i\pi\mu x} D\left(\frac{2\pi}{\omega}\mu\right) d\mu. \tag{V, 2; 72}$$

Putting $2\pi\mu/\omega = \lambda$,

$$f(x) = \frac{h|\omega|}{2\pi} \int_{-\infty}^{+\infty} e^{+i\omega\lambda x} D(\lambda)\, d\lambda, \tag{V, 2; 73}$$

whence the result

$$\left.\begin{aligned} D(\lambda) &= \frac{1}{h} \int_{-\infty}^{+\infty} e^{-i\omega\lambda x} f(x)\, dx \\ f(x) &= \frac{1}{k} \int_{-\infty}^{+\infty} e^{+i\omega\lambda x} D(\lambda)\, d\lambda \end{aligned}\right\}, \quad hk = \frac{2\pi}{|\omega|}. \tag{V, 2; 74}$$

The constants usually employed are

$$\left\{\begin{aligned} &\omega = \pm 2\pi, && h = k = 1, \\ &\omega = \pm 1, && h = 1, \quad k = 2\pi, \\ &\omega = \pm 1, && h = k = \sqrt{2\pi}. \end{aligned}\right. \tag{V, 2; 75}$$

3. Fourier transforms in several variables

Let $x = (x_1, x_2, \ldots, x_n) \in R^n$, $\lambda = (\lambda_1, \lambda_2, \ldots, \lambda_n) \in R^n$.

If $f(x_1, x_2, \ldots, x_n) \in L^1$, the Fourier transform is a function $C(\lambda_1, \lambda_2, \ldots, \lambda_n)$ defined by

$$\begin{aligned} &C(\lambda_1, \lambda_2, \ldots, \lambda_n) \\ &= \iint_{R^n} \cdots \int e^{-2i\pi(\lambda_1 x_1 + \lambda_2 x_2 + \cdots + \lambda_n x_n)} f(x_1, x_2, \ldots, x_n)\, dx_1 \cdots dx_n, \end{aligned} \tag{V, 3; 1}$$

which will be written for brevity as

$$C(\lambda) = \iint_{R^n} \cdots \int e^{-2i\pi\langle\lambda, x\rangle} f(x)\, dx. \tag{V, 3; 2}$$

Its properties are similar to those applying to the case of one variable.

For a function of the type $f(x) = f_1(x_1) f_2(x_2) \ldots f_n(x_n)$, since

(V, 3; 3) $$e^{-2i\pi(\lambda_1 x_1 + \cdots + \lambda_n x_n)} = \prod_{j=1}^{n} e^{-2i\pi_j \lambda_j x}$$

it is obvious that

(V, 3; 4) $$C(\lambda) = \mathscr{F}[f_1 f_2 \cdots f_n] = C_1(\lambda_1) \cdots C_n(\lambda_n) = \mathscr{F}[f_1]\mathscr{F}[f_2] \cdots \mathscr{F}[f_n].$$

In particular, if $r^2 = \sum_{j=1}^{n} x_j^2$, $\rho^2 = \sum_{j=1}^{n} \lambda_j^2$, then

(V, 3; 5) $$\mathscr{F}[e^{-\pi r^2}] = e^{-\pi \rho^2},$$

which is (V, 1; 37). Equation (V, 2; 14) remains true, but (V, 2; 16) and (V, 2; 35) must be replaced, considering the case $m = 1$ only, by

(V, 3; 6) $$\begin{cases} \mathscr{F}\left[\dfrac{\partial \delta}{\partial x_k}\right] = 2i\pi\lambda_k, \\ \mathscr{F}[-2i\pi x_k] = \dfrac{\partial \delta}{\partial \lambda_k}. \end{cases}$$

Equation (V, 1; 27) must be replaced by

(V, 3; 7) $$\mathscr{F}[f(kx_1, kx_2, \ldots, kx_n)] = \frac{1}{|k|^n} C\left(\frac{\lambda_1}{k}, \ldots, \frac{\lambda_n}{k}\right).$$

One of the relations analogous to (V, 2; 74) is

(V, 3; 8) $$hk = \left(\frac{2\pi}{|\omega|}\right)^n.$$

We shall consider in more detail the case where $f(x_1, \ldots, x_n)$ is a function of $r = \sqrt{\sum_{j=1}^{n} x_j^2}$ only:

(V, 3; 9) $$f(x_1, \ldots, x_n) = \Phi(r).$$

It will be shown that $C(\lambda_1, \lambda_2, \ldots, \lambda_n)$ is then also a function of one variable, $\rho = \sqrt{\sum_{j=1}^{n} \lambda_j^2}$. For this it is sufficient to prove that $C(\lambda)$ is invariant with respect to rotation about the origin. Let S be such a rotation and let $S\lambda$ denote the point μ into which a point λ is carried by the rotation S. Then

(V, 3; 10) $$C(\mu) = C(S\lambda) = \iint_{R^n} \cdots \int e^{-2i\pi\langle S\lambda, x\rangle} f(x)\, dx.$$

However, the scalar product is invariant under the operation S^{-1}, so that $\langle S\lambda, x\rangle = \langle S^{-1}S\lambda, S^{-1}x\rangle = \langle \lambda, S^{-1}x\rangle$. Then

$$C(S\lambda) = \iint_{R^n} \cdots \int e^{-2i\pi\langle \lambda, S^{-1}x\rangle} f(x)\, dx. \tag{V, 3; 11}$$

We make the change of variables $S^{-1}x = \xi$, obtaining $x = S\xi$. Then since the Jacobian of a rotation is always equal to 1 (conservation of volume),

$$C(S\lambda) = \iint_{R^n} \cdots \int e^{-2i\pi\langle \lambda, \xi\rangle} f(S\xi)\, d\xi. \tag{V, 3; 12}$$

However, as f is invariant under rotation, $f(S\xi) = f(\xi)$ giving $C(S\lambda) = C(\lambda)$ so that C is in fact invariant under rotation. We can write* $C(\lambda) = \Gamma(\rho)$.

It remains to evaluate $\Gamma(\rho)$. It suffices to put $\lambda_1 = \rho$, $\lambda_2 = \cdots = \lambda_n = 0$, from which

$$\Gamma(\rho) = \iint_{R^n} \cdots \int e^{-2i\pi x_1\rho}\Phi(\rho)\, dx_1\, dx_2 \cdots dx_n. \tag{V, 3; 13}$$

Applying a theorem of Chapter I on the calculation of multiple integrals by the use of spherical integrals, we have

$$\begin{aligned} \Gamma(\rho) &= \int_0^{+\infty} I(r)\, dr, \\ I(r) &= \int \Phi(r) e^{-2i\pi\rho x_1}\, dS, \\ \sum_{j=1}^{n} x_j^2 &= r^2. \end{aligned} \tag{V, 3; 14}$$

The surface integral over the sphere of radius r is evaluated by division into zones. If θ denotes the angle made by the position vector of the point $(x_1, \ldots, x_n)$ with Ox_1, the set of points of the sphere for which this angle lies between θ and $\theta + d\theta$ is termed a zone. Its "area" [$(n - 1)$ dimensional] is the product of $r\, d\theta$ with the "area" [$(n - 2)$ dimensional] of the intersection of the sphere (of radius $r \sin\theta$) by the plane $x_1 = r\cos\theta$, which is equal to

$$\frac{2\pi^{\frac{n-1}{2}}}{\Gamma\left(\frac{n-1}{2}\right)} (r \sin\theta)^{n-2}.$$

The area of an elementary zone is thus

$$\frac{2\pi^{\frac{2}{n-1}}}{\Gamma\left(\frac{n-1}{2}\right)} (r \sin\theta)^{n-2} r\, d\theta. \tag{V, 3; 15}$$

---

*Γ also appears further on as a symbol for Euler integrals but there should be no confusion.

Over all this zone the integrand in the spherical integral $\mathrm{I}(r)$ has the value $e^{-2i\pi\rho r \cos\theta}\,\Phi(r)$, so that the contribution of the elementary zone to the spherical integral is

$$\text{(V, 3; 16)} \qquad \frac{2\pi^{\frac{n-1}{2}}}{\Gamma\left(\frac{n-1}{2}\right)} e^{-2i\pi\rho r\cos\theta}\,(r\sin\theta)^{n-2}\Phi(r) r\,d\theta,$$

and finally the spherical integral is

$$\text{(V, 3; 17)} \quad \mathrm{I}(r) = \int_0^{\pi} \frac{2\pi^{\frac{n-1}{2}}}{\Gamma\left(\frac{n-1}{2}\right)} e^{-2i\pi\rho r\cos\theta}\,(r\sin\theta)^{n-2}\,\Phi(r) r\,d\theta$$

$$= r^{n-1}\Phi(r)\frac{2\pi^{\frac{n-1}{2}}}{\Gamma\left(\frac{n-1}{2}\right)}\int_0^{\pi} e^{-2i\pi\rho r\cos\theta}\,(\sin\theta)^{n-2}\,d\theta.$$

It can be seen that this integral involves a Bessel function by recalling the formula

$$\text{(V, 3; 18)} \qquad \mathrm{J}_\nu(x) = \frac{\left(\frac{x}{2}\right)^\nu}{\sqrt{\pi}\,\Gamma\left(\nu+\frac{1}{2}\right)}\int_0^{\pi} e^{-ix\cos\theta}\sin^{2\nu}\theta\,d\theta.$$

With $\nu = (n-2)/2$, $x = 2\pi\rho r$ it is found that

$$\text{(V, 3; 19)} \quad \mathrm{J}_{\frac{n-2}{2}}(2\pi\rho r) = \frac{\pi^{\frac{n-2}{2}}\rho^{\frac{n-2}{2}} r^{\frac{n-2}{2}}}{\sqrt{\pi}\,\Gamma\left(\frac{n-1}{2}\right)}\int_0^{\pi} e^{-2i\pi\rho r\cos\theta}\,(\sin\theta)^{n-2}\,d\theta$$

$$= \frac{\pi^{\frac{n-3}{2}}\rho^{\frac{n-2}{2}} r^{\frac{n-2}{2}}}{\Gamma\left(\frac{n-1}{2}\right)}\int_0^{\pi} e^{-2i\pi\rho r\cos\theta}\,(\sin\theta)^{n-2}\,d\theta.$$

It follows directly that

$$\text{(V, 3; 20)} \qquad \mathrm{I}(r) = \frac{2\pi}{\rho^{\frac{n-2}{2}}} r^{\frac{n}{2}}\,\mathrm{J}_{\frac{n-2}{2}}(2\pi\rho r)\Phi(r).$$

Hence the required formula for the Fourier transform $\Gamma(\rho)$ of $\Phi(r)$ is

$$\text{(V, 3; 21)} \qquad \Gamma(\rho) = \frac{2\pi}{\rho^{\frac{n-2}{2}}}\int_0^{+\infty} r^{\frac{n}{2}}\,\mathrm{J}_{\frac{n-2}{2}}(2\pi\rho r)\,\Phi(r)\,dr.$$

The transform $\overline{\mathcal{F}}$ is, of course, given by the same relation, since the coefficients of $\Phi(r)$ in the integral are real, and we have the inversion relation

(V, 3; 22) $$\Phi(r) = \frac{2\pi}{r^{\frac{n-2}{2}}} \int_0^{+\infty} \rho^{\frac{n}{2}} J_{\frac{n-2}{2}}(2\pi\rho r)\, \Gamma(\rho)\, d\rho.$$

These two reciprocal relations define, between functions Φ and Γ of the variables r and ρ respectively, the *Hankel transform* of order $(n-2)/2$. We shall consider the cases which are of most interest to us, $n = 1, 2, 3$.

$n = 1$. It is known that $J_{-\frac{1}{2}}(x) = \sqrt{\frac{2}{\pi x}} \cos x$ and hence

(V, 3; 23) $$\Gamma(\rho) = 2 \int_0^{+\infty} \cos(2\pi\rho r)\, \Phi(r)\, dr.$$

This result is an obvious one and the proof is trivial for $n = 1$. In fact we simply assume that f is an even function so that $C(\lambda)$ is also an even function [equation (V, 1; 28)] and the defining equation (V, 1; 7) gives

(V, 3; 24)

$$\begin{aligned} C(\lambda) &= \int_{-\infty}^{+\infty} e^{-2i\pi\lambda x} f(x)\, dx \\ &= \int_{-\infty}^{+\infty} \cos 2\pi\lambda x\, f(x)\, dx \quad \text{(the sine part of the integral being zero)} \\ &= 2\int_0^{\infty} \cos 2\pi\lambda x\, f(x)\, dx \quad \text{[which is (V, 3; 23)].} \end{aligned}$$

$n = 2$. Then

(V, 3; 25) $$\Gamma(\rho) = 2\pi \int_0^{+\infty} r J_0(2\pi\rho r)\, \Phi(r)\, dr.$$

This relation could have been easily obtained by a transformation to polar coordinates (r, θ) in (V, 3; 1) and what we have in fact obtained is only the generalization of the transformation to polar coordinates for arbitrary n.

$n = 3$. It is known that $J_{\frac{1}{2}}(x) = \sqrt{\frac{2}{\pi x}} \sin x$, so that

(V, 3; 26) $$\Gamma(\rho) = \frac{2}{\rho} \int_0^{+\infty} r \sin(2\pi\rho r)\, \Phi(r)\, dr.$$

In general it can be seen that trigonometric functions occur for odd n and Bessel functions of integer order for even n.

Example. For $n = 2$ we obtain the Fourier transform of the function Φ which is equal to 1 for $r < R$, 0 for $r > R$:

(V, 3; 27) $$\Gamma(\rho) = 2\pi \int_0^R r J_0(2\pi\rho r)\, dr.$$

Noting that Bessel functions satisfy the recurrence relations

(V, 3; 28) $$\frac{d}{dx}(x^\nu J_\nu(\lambda x)) = \lambda x^\nu J_{\nu-1}(\lambda x),$$

it may be deduced that

(V, 3; 29) $$\frac{d}{dr}(r J_1(2\pi\rho r)) = 2\pi\rho r J_0(2\pi\rho r),$$

so that the integral may be evaluated to yield

(V, 3; 30) $$\Gamma(\rho) = \frac{R}{\rho} J_1(2\pi R\rho).$$

This result is utilized in the theory of diffraction and in several other connexions. It could have been deduced directly from (V, 3; 1) by a transformation to polar coordinates.

4. A physical application of the Fourier transform : Solution of the heat conduction equation

Consider an infinitely long bar in which heat is transmitted solely by conduction. The linear specific heat (thermal capacity per unit length) is c; the thermal conductivity is γ, which means that if the temperature gradient is θ at a point of coordinate x, then the quantity of heat per unit time passing the point x from left to right is $-\gamma\theta$. It is supposed that the sources of heat are distributed along the bar and it is said that the heat source density is $\rho(x, t)$ (ρ being of either sign) if at a point x and a time t the sources give out in an interval of length $(x, x + dx)$ during a time $(t, t + dt)$ the quantity of heat $\rho dx\, dt$. Apart from these sources no heat exchange by radiation or convection takes place with the exterior. It is required to find how the temperature U at different points of the bar varies with time. U is an unknown function of x and t. We write the heat balance equation for the interval $(x, x + dx)$, which during the time $(t, t + dt)$ receives $\rho(x, t)dx\,dt$ from the source in contact with it and by conduction. Then

$$\text{(V, 4; 1)} \qquad \left[\gamma \frac{\partial U}{\partial x}(x + dx, t) - \gamma \frac{\partial U}{\partial x}(x, t)\right] dt = + \gamma \frac{\partial^2 U}{\partial x^2}\, dx\, dt,$$

since the temperature gradient at a point x is $\partial U/\partial x$.

Finally, the quantity of heat received is seen to be

$$\text{(V, 4; 2)} \qquad \left(\rho + \gamma \frac{\partial^2 U}{\partial x^2}\right) dx\, dt,$$

and as the temperature increase is $\frac{\partial U}{\partial t} dt$ and the heat capacity is $c\,dx$, it follows that

$$\text{(V, 4; 3)} \qquad c \frac{\partial U}{\partial t} - \gamma \frac{\partial^2 U}{\partial x^2} = \rho.$$

This result is called the *heat conduction equation.* It will be assumed that it remains valid even if U and ρ become distributions. The case where ρ is a distribution is very important. $\rho(x, t) = \delta(x)$ means that the only source is placed at the point 0 and furnishes a unit quantity of heat in unit time; $\rho(x, t) = \delta(x)\delta(t)$ means that the source is placed at 0 only at the instant $t = 0$ and that during this instant alone it furnishes a unit quantity of heat to the neighbourhood of 0.

Cauchy's problem for (V, 4; 3), in the case where functions only are involved, is to find for $t > 0$ a solution U for a given right-hand side $\rho(x, t)$ and a given initial temperature distribution $U(x, 0) = U_0(x)$. Our arguments will be intuitive rather than rigorous and no complete justification will be attempted.

The functions U and ρ extended by 0 for $t < 0$ will be written $\tilde{U}$ and $\tilde{\rho}$. The differentiation (in the distribution sense) relations for discontinuous functions give

$$\text{(V, 4; 4)} \qquad \frac{\partial^2 \tilde{U}}{\partial x^2} = \left\{\frac{\partial^2 \tilde{U}}{\partial x^2}\right\}, \qquad \frac{\partial \tilde{U}}{\partial t} = \left\{\frac{\partial \tilde{U}}{\partial t}\right\} + U_0(x)\delta(t),$$

so that $\tilde{U}$ satisfies the partial differential equation

$$\text{(V, 4; 5)} \qquad c \frac{\partial \tilde{U}}{\partial t} - \gamma \frac{\partial^2 \tilde{U}}{\partial x^2} = \tilde{\rho} + cU_0(x)\delta(t).$$

It will be assumed that for all t, the function ρ regarded as a function of x is tempered and thus has a Fourier transform for fixed t (zero for $t < 0$) :

$$\text{(V, 4; 6)} \qquad \tilde{\sigma}(\lambda, t) = \int_{-\infty}^{+\infty} \tilde{\rho}(x, t) e^{-2i\pi\lambda x}\, dx.$$

It will also be assumed that $U_0(x)$ is tempered so that it also has a Fourier transform given by

(V, 4; 7) $$V_0(\lambda) = \int_{-\infty}^{+\infty} U_0(x) e^{-2i\pi\lambda x}\, dx.$$

We shall then try to find whether there exists a solution of Cauchy's problem which is itself, for all t, a tempered function of x. It would then have for fixed t, a Fourier transform (zero for $t < 0$)

(V, 4; 8) $$\tilde{V}(\lambda, t) = \int_{-\infty}^{+\infty} \tilde{U}(x, t) e^{-2i\pi\lambda x}\, dx.$$

As the Fourier transformation is applied in x only with t fixed, partial differentiation with respect to t is unchanged [more precisely, the Fourier transform of $\partial\tilde{U}/\partial t$ is $\partial\tilde{V}/\partial t$, by differentiation of the integrand of (V, 4; 8)]. On the other hand, from equation (V, 2; 19) partial differentiation with respect to x is transformed to multiplication by $2i\pi\lambda$. Thus $\tilde{V}(\lambda, t)$ satisfies the equation

(V, 4; 9) $$c\frac{\partial\tilde{V}}{\partial t} + 4\pi^2\gamma\lambda^2\tilde{V} = \tilde{\sigma}(\lambda, t) + cV_0(\lambda)\delta(t).$$

This equation is now considered as defining, for any fixed λ, a differential equation in t (in the distribution sense since $\delta(t)$ appears there). As all the quantities involved have their supports in $t > 0$, it is a convolution equation in $(\mathcal{D}'_+)_t$ for fixed λ, which may be written

(V, 4; 10) $$\left(c\frac{\partial\delta}{\partial t} + 4\pi^2\gamma\lambda^2\delta(t)\right) \underset{(t)}{*} \tilde{V}(\lambda, t) = \tilde{\sigma}(\lambda, t) + cV_0(\lambda)\delta(t).$$

In the algebra $(\mathcal{D}'_+)_t$ the inverse of

(V, 4; 11) $$A = c\delta' + 4\pi^2\gamma\lambda^2\delta$$

is [equation (III, 2; 67) or Theorem 14 of Chapter III]

(V, 4; 12) $$A^{-1} = \frac{Y(t)}{c} e^{-4\pi^2 \frac{\gamma}{c}\lambda^2 t}.$$

Hence (V, 4; 10) is solved by

(V, 4; 13) $$\tilde{V}(\lambda, t) = \tilde{\sigma}(\lambda, t) \underset{(t)}{*} \frac{1}{c} Y(t)\, e^{-4\pi^2 \frac{\gamma}{c}\lambda^2 t} + V_0(\lambda) Y(t)\, e^{-4\pi^2 \frac{\gamma}{c}\lambda^2 t}.$$

If $\tilde{\sigma}(\lambda, t)$ is a function of t (for fixed λ) the first convolution product here may be written, for $t > 0$, as

(V, 4; 14) $$\frac{1}{c}\int_0^t \sigma(\lambda, \tau) e^{-4\pi^2 \frac{\gamma}{c}\lambda^2(t-\tau)}\, d\tau.$$

But (V, 4 ; 13) has the advantage of remaining valid even in the case of distributions.

Once $\tilde{V}(\lambda, t)$ is found, the Fourier transform $\overline{\mathcal{F}}$ for any fixed t gives $\tilde{U}(x, t)$ (always zero by definition for $t < 0$). Suppose for example that for $t < 0$ the bar is at zero temperature and that at the precise instant $t = 0$ a source placed at $x = 0$ instantaneously communicates a unit quantity of heat to the bar. Then $U_0(x) = 0$, $\tilde{\rho}(x, t) = \delta(x)\delta(t)$ so that $V_0(\lambda) = 0$ and

$$\tilde{\sigma}(\lambda, t) = \delta(t),$$

and (V, 4; 13) gives

(V, 4; 15) $$\tilde{V}(\lambda, t) = \frac{1}{c} Y(t) e^{-4\pi^2 \frac{\gamma}{c} \lambda^2 t}.$$

$U(x, t)$ may be obtained by equation (V, 1; 38), which gives

(V, 4; 16) $$U(x, t) = \frac{1}{c} Y(t) \frac{1}{2\sqrt{\frac{c}{\gamma}\pi t}} e^{-[x^2/4(\gamma/c)t]}.$$

This is what is known as the elementary solution of the heat conduction equation. Note that :

1) For $t > 0$, however small it may be, the temperature is > 0 all along the bar so that *heat is propagated at infinite speed.* However, for very large $|x|$ the temperature remains very small for a long time.

2) For all t the function U is represented (in the plane of x and U) by a Gaussian or bell-shaped curve. The maximum is at $x = 0$ and is equal to

$$U(0, t) = \frac{1}{c} Y(t) \frac{1}{2\sqrt{(\gamma/c)\pi t}}.$$

As soon as the source is applied at $x = 0$ the temperature of this point becomes $+\infty$ and then decreases as $1/\sqrt{t}$ for $t > 0$.

At each point $x \neq 0$ the temperature tends to 0 with t, the maximum of the bell-shaped curve at $x = 0$ becoming progressively sharper as t diminishes.

EXERCISES FOR CHAPTER V

Exercise V-1

(*a*) Let f and g be two functions defined on R and belonging to L^1. Show that

$$h(x) = f * g = \int_{-\infty}^{+\infty} f(x - y) g(y)\, dy$$

is also in L^1 and that

$$\|h\|_{L^1} \leqslant \|f\|_{L^1}\|g\|_{L^1}.$$

(*b*) Show directly that

$$\mathscr{F}h = \mathscr{F}f\mathscr{F}g.$$

Exercise V-2

(*a*) Show that the Fourier transform of an odd (or even) distribution is an odd (or even) distribution.

(*b*) Find the odd distributions satisfying $x\mathrm{T} = 1$.

(*c*) Hence deduce the Fourier transform of $\mathrm{pv}(1/x)$.

Exercise V-3

Verify that the Fourier transform of

$$f(x) = \begin{cases} |x| & \text{for } |x| < 1, \\ 0 & \text{otherwise} \end{cases}$$

is infinitely differentiable.

Exercise V-4

Consider the tempered distribution $|x|$.

(*a*) Evaluate $\delta'' * |x|$.

(*b*) Hence deduce that $\mathscr{F}(|x|)$ is of the form $\mathrm{A.\,Pf}\,(1/\lambda^2) + \mathrm{C}\delta$, where A and C are constants. Determine A.

(*c*) Evaluate $d/dx\,[\mathrm{pv}\ (1/2)]$. Hence deduce $\mathscr{F}\,[\mathrm{Pf}\,(1/x^2)]$, and then the value of the constant C. [Use the value of $\mathscr{F}(\mathrm{pv}\ 1/x)$ found in Exercise V-2.]

Exercise V-5

Show that if a distribution $\mathrm{T} \in \mathscr{S}'$ has a Fourier transform

$$\mathscr{F}(\mathrm{T}) = \sum_{n=-\infty}^{+\infty} c_n\delta_{(n)},$$

it is periodic and the c_n are its Fourier coefficients.

Exercise V-6 (Alternative proof of $\mathscr{F}1 = \delta$)

$\mathscr{S}'$ is endowed with the following topology: a family of distributions $\mathrm{T_A}$, depending on the parameter A, converges to T in $\mathscr{S}'$ as $\mathrm{A} \to \infty$ if, for every $\varphi \in \mathscr{S}$, $\langle \mathrm{T_A}, \varphi\rangle \to \langle \mathrm{T}, \varphi\rangle$ as $\mathrm{A} \to \infty$.

(i) Show that if $\mathrm{T_A} \to \mathrm{T}$ in $\mathscr{S}'$, then $\mathscr{F}\mathrm{T_A} \to \mathscr{F}\mathrm{T}$ in S′.

(ii) Consider

$$\mathrm{T_A} = \begin{cases} 1 & \text{for } |x| \leqslant \mathrm{A}, \\ 0 & \text{otherwise.} \end{cases}$$

Evaluate $\mathscr{F}\mathrm{T_A}$.

(iii) Show that $\mathrm{T_A} \to 1$ in $\mathscr{S}'$, when $\mathrm{A} \to \infty$, and that $\mathscr{F}\mathrm{T_A}$, evaluated in (ii), converges to δ in $\mathscr{S}'$. Hence deduce the result $\mathscr{F}1 = \delta$.

Exercise V-7 (Examination question, Paris, 1959)

(*a*) Obtain the Fourier transforms of the following functions or distributions* :

$$\delta_{(a)} \text{ (unit mass at point } a\text{)},$$
$$e^{2i\pi x},\ e^{-2i\pi x},\ \cos 2\pi x,\ \sin 2\pi x.$$

Consider the following function :

$$f(x) = -\frac{\cos 2\pi x}{\pi^2 x^2} + \frac{\sin 2\pi x}{2\pi^3 x^3}.$$

It is defined for $x \neq 0$. Show that it has a limit when $x \to 0$ and evaluate this limit. Let

$$g(x) = (-2i\pi x)^3 f(x).$$

What is the Fourier transform D_λ of $g(x)$? Deduce that the Fourier transform $C(\lambda)$ of $f(x)$ satisfies a differential equation, and write this down.

(*b*) Show that this differential equation has a solution $C_0(\lambda)$ which is a function of λ, and is zero outside the interval $(-1, +1)$. What is the general solution of this differential equation? By making use of the facts that $C_0(\lambda)$ has a bounded support and that $f(x)$ is a function, show that necessarily $C(\lambda) = C_0(\lambda)$. Write the Fourier inversion formula for $f(x)$ and $C(\lambda)$. Evaluate directly the integral $\overline{\mathcal{F}}C(\lambda)$ and hence recover $f(x)$.

(*c*) Consider the function $\Gamma(\lambda)$, equal to $C(\lambda)$ in the interval $(-1, +1)$ and continued outside the interval as a periodic function of period 2. Show that the previous calculation automatically gives its Fourier coefficients. Write its Fourier series and establish that it is convergent. By writing its sum for $\lambda = 1$ obtain the sum of the series $\sum_{n=1}^{\infty} \frac{1}{n^2}$.

Exercise V-8

It will be recalled that if the product of the integral functions $U(\lambda)$ and $V(\lambda)$ of the complex variable λ is zero, then at least one of the two functions $U(\lambda)$ and $V(\lambda)$ is zero. Using this property show that if S and T are two distributions with bounded supports on the line R (i.e. $\in \mathcal{E}'$) satisfying

$$S * T = 0$$

then at least one of the two distributions is zero. Naturally this becomes false if at least one of the two distributions does not belong to $\mathcal{E}'$.

Example :

$$S = 1 \in \mathcal{D}'; \qquad T = \delta' \in \mathcal{E}', \quad \text{then } S * T = 0.$$

---

*Use the preceding result and the Fourier inversion formula.

Exercise V-9
Consider the line R. Denote by H^1 the space of functions f which, together with their first derivatives df/dx (in the distribution sense), are in L^2 and consider the scalar product

$$(1)\quad ((f, g))_1 = (f, g)_{L^2} + \left(\frac{df}{dx}, \frac{dg}{dx}\right)_{L^2} = \int_{-\infty}^{+\infty} f(x)\overline{g(x)}\, dx + \int_{-\infty}^{+\infty} \frac{df}{dx}\overline{\frac{dg}{dx}}\, dx.$$

(*a*) Show that H^1, furnished with the scalar product (1), is a Hilbert space. Denote by $\|\cdot\|_1$ the norm defined by (1).

(*b*) Prove that a tempered distribution f is in H if and only if

$$(1 + \lambda^2)^{1/2}\hat{f}(\lambda) \in L^2,$$

denoting the Fourier transform of f by $\hat{f}(\lambda)$.

(*c*) For $f \in H^1$, let

$$|||f|||_1 = \|(1 + \lambda^2)^{1/2}\hat{f}(\lambda)\|_{L^2}.$$

Prove that $\|\cdot\|_1$ and $|||\cdot|||_1$ are two equivalent norms on H^1.

Exercise V-10
The Fourier transform of the distribution U in the variable

$$x = (x_1, x_2, \ldots, x_n)$$

is denoted by $\mathcal{F}U$, a distribution in the variable

$$y = (y_1, y_2, \ldots, y_n).$$

Also let

$$r = \sqrt{x_1^2 + \cdots + x_n^2}, \qquad \rho = \sqrt{y_1^2 + \ldots + y_n^2}.$$

If k is a real number such that $n/2 < k < n$ (n being the number of coordinates) establish the result

$$(1)\qquad \mathcal{F}\left(\frac{1}{r^k}\right) = \pi^{k-\frac{n}{2}} \frac{\Gamma\left(\frac{n-k}{2}\right)}{\Gamma\left(\frac{k}{2}\right)} \frac{1}{\rho^{n-k}}.$$

Start by proving that $\mathcal{F}(r^{-k})$ is proportional to $\rho^{-(n-k)}$. Next make use of the equation

$$\langle f, \mathcal{F}g\rangle = \langle \mathcal{F}f, g\rangle$$

taking $f = r^{-k}$ and g to be a suitably chosen function of $\mathcal{S}$. What happens for $k = n/2$?

Give a simple expression for $\mathcal{F}(r^{2p})$, $p \geqslant 0$. Suppose now that $n = 1$ and assume the relation

$$(2)\qquad \mathcal{F}(\log|x|) = -\frac{1}{2}\,\mathrm{Pf}\left(\frac{1}{|y|}\right) - (C + \log 2\pi)\delta,$$

where C is Euler's constant and $\mathrm{Pf}(1/|y|)$ is the distribution defined by

$$\left\langle \mathrm{Pf}\left(\frac{1}{|y|}\right), \varphi(y) \right\rangle = -\int_0^\infty \varphi'(y) \log y \, dy + \int_{-\infty}^0 \varphi'(y) \log |y| \, dy,$$

where $\varphi \in \mathcal{D}$. (Compare with Exercise II-14.) Deduce from (2) expressions for

$$\mathcal{F}\left(\mathrm{pv}\,\frac{1}{x}\right), \qquad \mathcal{F}\left(\frac{\sin x}{x}\right), \qquad \mathcal{F}(\mathrm{Y}(x)), \qquad \mathcal{F}(\mathrm{Y}(x)x^m),$$

$\mathrm{Y}(x)$ being Heaviside's function and m an integer $\geqslant 1$.

Exercise V-11

The *Hermite polynomials* $\mathrm{H}_m(x)$, where m is a positive integer or zero, are defined by the relations

$$(1) \qquad \frac{d^m}{dx^m}(e^{-2\pi x^2}) = (-1)^m \sqrt{m!}\, 2^{m-\frac{1}{4}} \pi^{\frac{m}{2}} \mathrm{H}_m(x) e^{-2\pi x^2}.$$

The *Hermite functions* $\mathcal{H}_m(x)$ are defined by

$$(2) \qquad \mathcal{H}_m(x) = \mathrm{H}_m(x) e^{-\pi x^2}.$$

(*a*) (i) Obtain

$$\mathrm{H}_0(x),\ \mathrm{H}_1(x),\ \mathrm{H}_2(x),\ \mathrm{H}_3(x).$$

(ii) By using the fact that

$$\frac{d}{dx}(e^{-2\pi x^2}) + 4\pi x e^{-2\pi x^2} = 0,$$

show that, for $m \geqslant 1$,

$$\frac{d^{m+1}}{dx^{m+1}}(e^{-2\pi x^2}) + 4\pi x \frac{d^m}{dx^m}(e^{-2\pi x^2}) + 4m\pi \frac{d^{m-1}}{dx^{m-1}}(e^{-2\pi x^2}) = 0.$$

Hence deduce the recurrence relations

$$(3) \qquad \frac{d}{dx}[\mathrm{H}_m(x)] = 2\sqrt{m\pi}\, \mathrm{H}_{m-1}(x)$$

and

$$(4) \qquad 2\sqrt{\pi(m+1)}\, \mathrm{H}_{m+1}(x) - 4\pi x \mathrm{H}_m(x) + 2\sqrt{m\pi}\, \mathrm{H}_{m-1}(x) = 0.$$

(iii) Show that, for odd m, $\mathrm{H}_m(0)$ is zero and that, for even m (i.e. $m = 2n$),

$$(5) \qquad \mathrm{H}_{2n}(0) = 2^{\frac{1}{4}} \frac{(-1)^n \sqrt{(2n)!}}{n!\, 2^n}.$$

(*b*) (i) Show that $\mathcal{H}_m(x)$ is an element of the space $\mathcal{S}$ of rapidly decreasing functions and that

$$\int_{-\infty}^{+\infty} \mathcal{H}_p(x)\mathcal{H}_q(x)\,dx = \begin{cases} 0 & \text{if } p \neq q, \\ 1 & \text{if } p = q. \end{cases} \tag{6}$$

(ii) Recalling that the Fourier transform $\mathcal{F}[f(x)]$ of a function $f(x)$ is defined by

$$\mathcal{F}[f(x)] = \int_{-\infty}^{+\infty} e^{-2\pi i\lambda x} f(x)\,dx,$$

show that

$$\mathcal{F}[e^{-\pi x^2}] = e^{-\pi\lambda^2},$$

and hence deduce that

$$\mathcal{F}[\mathcal{H}_m(x)] = (-i)^m\mathcal{H}_m(\lambda). \tag{7}$$

(*c*) T being a tempered distribution, the series

$$\sum_{n=0}^{\infty} a_m(\mathrm{T})\mathcal{H}_m$$

where

$$a_m(\mathrm{T}) = \langle \mathrm{T}, \ \mathcal{H}_m \rangle$$

is called the expansion of T in Hermite functions.

(i) Write down the Hermite function expansion of the Dirac distribution δ.

(ii) For any tempered distribution T consider the transformations $\mathcal{C}_+$ and $\mathcal{C}_-$ defined by

$$\mathcal{C}_+ (\mathrm{T}) = \frac{d\mathrm{T}}{dx} + 2\pi x\mathrm{T}, \tag{8}$$

$$\mathcal{C}_- (\mathrm{T}) = -\frac{d\mathrm{T}}{dx} + 2\pi x\mathrm{T}. \tag{9}$$

In particular the transformations (8) and (9) are defined for any function in the space $\mathcal{S}$. Show by using equations (3) and (4) that

$$\mathcal{C}_+ (\mathcal{H}_m) = 2\sqrt{\pi m}\,\mathcal{H}_{m-1} \quad \text{for} \quad m \geqslant 1,$$
$$\mathcal{C}_- (\mathcal{H}_m) = 2\sqrt{\pi(m+1)}\,\mathcal{H}_{m+1} \quad \text{for} \quad m \geqslant 0.$$

Show that the Hermite function expansion of δ found above converges in $\mathcal{S}'$ to a distribution S. Evaluate $\mathcal{C}_+$ (S) and $\mathcal{C}_-$ (S) and hence deduce that S has the form $\mathrm{C}\delta$. Then show by using equation (6) that $\mathrm{C} = 1$.

(iii) Hence deduce, using equation (7), the value of the integral

$$\int_{-\infty}^{+\infty} \mathcal{H}_m(\lambda)\,d\lambda.$$

Exercise V-12

In the plane (x, t) denote by (C) the cone

$$v^2t^2 - x^2 \geqslant 0, \quad t \geqslant 0,$$

where v is a given constant. Denote by $E_x(t)$ the distribution which is equal to $v/2$ in the cone and 0 elsewhere.

(*a*) Evaluate, for fixed t, the Fourier transform $\hat{E}_\lambda(t)$ of $E_x(t)$ with respect to x.

(*b*) By using a Fourier transformation with respect to the variable x only, find a tempered elementary solution of the operator

$$\frac{1}{v^2}\frac{\partial^2}{\partial t^2} - \frac{\partial^2}{\partial x^2}.$$

Compare the result with that of Exercise II-20.

The Laplace transform

1. Laplace transforms of functions

To each locally summable complex function $f(t)$ of the real variable t, which is zero for $t < 0$ and in addition obeys appropriate restrictive conditions, we assign a corresponding holomorphic function $F(p)$ of the complex variable p, defined by the integral

$$F(p) = \int_0^{+\infty} f(t)e^{-pt}\,dt. \tag{VI, 1; 1}$$

We then write

$$f(t) \sqsupset F(p), \tag{VI, 1; 2}$$

where $f(t)$ is the original function and $F(p)$ the image. (VI, 1; 1) is called the *Laplace integral*; $F(p)$ is the *Laplace transform* of $f(t)$.

The abscissa of absolute convergence

If $p = \xi + i\eta$, the modulus of $f(t)e^{-pt}$ is $|f(t)|e^{-\xi t}$. The summability (or absolute convergence) of the integral (VI, 1; 1) thus depends only on the real part ξ of p.

THEOREM 1. *If the integral of* (VI, 1; 1) *is summable for* $\xi = \xi_0$ *then it is summable for* $\xi \geqslant \xi_0$, *uniformly with respect to* p.

For, if $\xi \geqslant \xi_0$, $|f(t)|e^{-\xi t} \leqslant |f(t)|e^{-\xi_0 t}$.

COROLLARY. *There exists a positive or negative real number* a *such that the integral* (VI, 1; 1) *is summable for* $\xi > a$ *and not summable for* $\xi < a$.

Denote by a the greatest lower bound of the numbers ξ for which $|f(t)|e^{-\xi t}$ is summable. If $\xi > a$ then by the definition of a there is a number ξ_0 between a and ξ such that $f(t)e^{-\xi_0 t}$ is summable. Thus, by Theorem 1, $|f(t)|e^{-\xi t}$ is also summable. Now if $\xi < a$, let ξ_1 lie between ξ and a. Then $|f(t)|e^{-\xi t}$ cannot be summable, for if it were, then $|f(t)|e^{-\xi_1 t}$ would also be summable by Theorem 1 and this would be contrary to the definition of a. Thus a may be defined as the *cut* between the numbers ξ for which (VI, 1; 1) is and is not summable.

Note that for $\xi = a$ the integral may or may not be summable. The number a is called the *abscissa of summability or absolute convergence* of the Laplace integral. The region defined by $\xi > a$ is the *domain of summability* of the Laplace integral. If $a = -\infty$, the integral (VI, 1; 1) is summable for any value of the complex number p. If on the other hand $a = +\infty$, the integral of (VI, 1; 1) is never summable.

THEOREM 2. *If f is locally summable and satisfies for $t \geqslant t_0 \geqslant 0$ the inequality*

$$(\text{VI}, 1; 3) \qquad |f(t)| \leqslant Ae^{kt}, \qquad A > 0 \quad (k \textit{ positive or negative real}),$$

then the abscissa of summability a is $\leqslant k$.

It is necessary to show that (VI, 1; 1) is summable for $\xi > k$. For $t \geqslant t_0$

$$(\text{VI}, 1; 4) \qquad |f(t)|e^{-\xi t} \leqslant Ae^{-(\xi-k)t},$$

where the right-hand side is summable since $\xi - k > 0$. Thus $|f(t)|e^{-\xi t}$ is summable from t_0 to $+\infty$. However, since f is locally summable, $|f(t)|e^{-\xi t}$ is also summable from 0 to t_0 and thus, as we wished to prove, from 0 to $+\infty$.

In particular, if f is bounded the abscissa of summability a is $\leqslant 0$ or, equivalently (VI, 1; 1) is summable for $\xi > 0$. If f satisfies the conditions of local summability and *compact support* and is thus summable, the above inequality applies for any k (even with $A = 0$) when t is greater than some sufficiently large t_0, so that $a = -\infty$. $F(p)$ is then defined for any value of p. A more general condition for this is that f should satisfy the inequality (VI, 1; 3) if t is large enough, *for any* $k < 0$. For example, if

$$(\text{VI}, 1; 5) \qquad f(t) = Y(t)e^{-t^2},$$

the abscissa of summability is $-\infty$. Conversely, if

$$(\text{VI}, 1; 6) \qquad f(t) = Y(t)e^{t^2},$$

the integral (VI, 1; 1) is never summable so that $a = +\infty$.

THEOREM 3. *If a is the abscissa of summability and if $F(p)$, the Laplace transform of f, is holomorphic for $\xi > a$, then*

$$(\text{VI}, 1; 7) \qquad F^{(m)}(p) = \int_0^\infty f(t)\,(-t)^m e^{-pt}\,dt \qquad \text{for} \qquad \xi > a.$$

Alternatively, it may be stated that

$$(\text{VI}, 1; 8) \qquad (-t)^m f(t) \sqsupset F^{(m)}(p)$$

and the abscissa of summability of the Laplace transform of $(-t)^m f(t)$ is equal to that of the Laplace transform of $f(t)$.

The last part of the theorem will be proved first. As f is always supposed to be locally summable it is sufficient to consider the region where $t \geqslant 1$; then $t^m \geqslant 1$ so that the summability of $(-t)^m f(t)e^{-\xi t}$ implies that of $f(t)e^{-\xi t}$ and hence the abscissa of summability for $(-t)^m f(t)$ is not less than a. Also, for any $\varepsilon > 0$, $t^m \leqslant e^{\varepsilon t}$, provided t is large enough, and thus $|(-t)^m f(t)e^{-\xi t}| \leqslant |f(t)|e^{-(\xi-\varepsilon)t}$. For $\xi - \varepsilon > a$ (i.e. $\xi > a + \varepsilon$) the expression $(-t)^m f(t)e^{-\xi t}$ is thus summable and the abscissa of summability for $(-t)^m f(t)$ is $\leqslant a + \varepsilon$ and hence $\leqslant a$, since ε is an arbitrary positive number. Since this abscissa has been shown above to be not less than a, it can only be equal to a. The first part of the theorem follows easily. As $F(p)$ is given by the integral (VI, 1; 1) it is legitimate to differentiate it with respect to p under the $\int$ sign if the integral thus derived is uniformly summable. This is true for $\xi > a$ so that equation (VI, 1; 7), equivalent to (VI, 1; 8), is valid. Since $F(p)$ is differentiable with respect to the complex variable p for $\xi > a$, it is holomorphic in this domain. There are interesting cases in practice where the Laplace integral is convergent but not summable (and thus semi-convergent), but these will not be discussed here.

2. Laplace transforms of distributions

1. DEFINITION

If f is altered on a set of measure zero, its Laplace integral $F(p)$ is not changed. Thus, in fact, $F(p)$ is determined by the class of the function f, that is to say the distribution defined by the locally summable function f. More generally, let T be a distribution on the real axis of the variable t and with its support contained in the half-line $t \geqslant 0$, so that $T \in \mathcal{D}'_+$. Suppose also that there exists ξ_0 such that for $\xi > \xi_0$ it is true that $e^{-\xi t}T \in \mathcal{S}'$. Then T possesses a *Laplace transform*

$$\text{(VI, 2; 1)} \qquad \mathcal{C}(p) = \langle T, e^{-pt} \rangle \qquad \text{defined for} \qquad \xi > \xi_0.$$

Let $\alpha(t)$ be an infinitely differentiable function with a support bounded on the left and equal to 1 on a neighbourhood of the support of T. For $\xi > \xi_0$ there exists ξ_1 such that $\xi_0 < \xi_1 < \xi$. Then since $e^{-\xi_1 t}T \in \mathcal{S}'$ and $\alpha(t)e^{-(p-\xi_1)t} \in \mathcal{S}$ it follows that the expression

$$\langle e^{-\xi_1 t}T, \alpha(t)e^{-(p-\xi_1)t} \rangle$$

has a meaning. This expression is independent of ξ_1 and gives *by definition* what we denote by $\langle T, e^{-pt} \rangle$.

Note that the definition (VI, 2; 1) is a reasonable generalisation of (VI, 1; 1). It has been necessary to impose a certain number of restrictions on T because, although e^{-pt} is indeed an infinitely differentiable function of t, its support is not bounded so that the right-hand side of (VI, 2; 1) is not defined for arbitrary T.

It can be shown that $\mathcal{T}(p)$ is a holomorphic function of p for $\xi > \xi_0$ and that, as above,

$$(-t)^m \mathrm{T} \sqsupset \mathcal{T}^{(m)}(p). \tag{VI, 2; 2}$$

It may be seen that the Laplace transformation is linear: if $\mathrm{S} \sqsupset \mathcal{S}(p)$ for $\xi > b_1$, and $\mathrm{T} \sqsupset \mathcal{T}(p)$ for $\xi > b_2$, then

$$\mathrm{S} + \mathrm{T} \sqsupset \mathcal{S}(p) + \mathcal{T}(p) \quad \text{for} \quad \xi > \max(b_1, b_2)$$

and

$$\lambda \mathrm{S} \sqsupset \lambda \mathcal{S}(p) \quad \text{for} \quad \xi > b_1.$$

2. EXAMPLES OF LAPLACE TRANSFORMS

1)

$$\begin{cases} \delta \sqsupset 1, \\ \delta^{(m)} \sqsupset p^m, \qquad m \text{ an integer} \geqslant 0, \\ \delta_{t-a} \sqsupset e^{-ap}. \end{cases} \tag{VI, 2; 3}$$

These results come directly from (VI, 2; 1). They are valid for all p since the distributions concerned have bounded supports.

2) Let T be the distribution formed by the masses C_n $(n = 0, 1, 2, \ldots)$ at the integer points 0, 1, 2, ... Then the result

$$\delta_{t-n} \sqsupset e^{-np}$$

leads immediately to

$$\sum_{n=0}^{\infty} \mathrm{C}_n \delta_{t-n} \sqsupset \sum_{n=0}^{\infty} \mathrm{C}_n e^{-np}. \tag{VI, 2; 4}$$

Setting $e^{-p} = \mathrm{Z}$, we may write the right-hand side

$$\sum_{n=0}^{\infty} \mathrm{C}_n \mathrm{Z}^n. \tag{VI, 2; 5}$$

Thus a series with integer indices is, by a change of variables, equivalent to a Laplace transform. The domain of definition, previously a half-plane, becomes by this change of variables a circle of convergence.

3)

$$\mathrm{Y}(t) \sqsupset \frac{1}{p} \quad \text{for } \xi > 0,$$

$$\mathrm{Y}(t) \frac{t^{\alpha-1}}{\Gamma(\alpha)} \sqsupset \frac{1}{p^\alpha}, \quad \alpha > 0, \text{ for } \xi > 0, \tag{VI, 2; 6}$$

$$\mathrm{Y}(t) e^{\lambda t} \frac{t^{\alpha-1}}{\Gamma(\alpha)} \sqsupset \frac{1}{(p-\lambda)^\alpha}, \quad \alpha > 0, \ \xi > \mathcal{R}\lambda.$$

The branch of $(p-\lambda)^{\alpha}$ chosen, in the half-plane $\xi > \Re\lambda$, must be the one which, for real $p-\lambda$, has a real value > 0, and which is continuous for $\xi > \Re\lambda$.

The second result of (VI, 2; 3) and the third of (VI, 2; 6) for integer α enables an inverse to be found for any rational fraction in p, $P(p)/Q(p)$, after the latter has been expanded in partial fractions. It is worth noting that in Chapter III, § 2 the distribution δ' was called p and the function $Y(t)e^{\lambda t}\dfrac{t^{\alpha-1}}{\Gamma(\alpha)}$ for integer α, $\dfrac{1}{(p-\lambda)^{\alpha}}$. At this point p was simply a letter denoting δ'.

Here the idea is different. δ' has p and $Y(t)e^{\lambda t}\dfrac{t^{\alpha-1}}{\Gamma(\alpha)}$ has $\dfrac{1}{(p-\lambda)^{\alpha}}$ as Laplace transform, where $1/(p-\lambda)^{\alpha}$ is regarded as a *holomorphic function of the complex variable p*. However, the application is the same; it is the transformation of convolution into multiplication. We shall prove as an example the second result of (VI, 2; 6), where

$$F(p) = \int_0^{\infty} \frac{t^{\alpha-1}}{\Gamma(\alpha)} e^{-pt}\, dt. \tag{VI, 2; 7}$$

The integral is summable in the neighbourhood of zero since $\alpha > 0$. It is summable at infinity for $\xi > 0$ but not for $\xi \leqslant 0$. It is thus found that the half-plane where the function is holomorphic is no larger than the half-plane of summability, as is shown by the existence of the singularity $p = 0$ of $F(p)$.

If p is real then, setting $p = \xi$, we can make the change of variables $\xi t = u$, which gives

$$F(p) = \frac{1}{\Gamma(\alpha)\xi^{\alpha}} \int_0^{\infty} e^{-u}u^{\alpha-1}\, du = \frac{1}{\xi^{\alpha}}. \tag{VI, 2; 8}$$

If p is complex, then it will be noted that $F(p)$ must be a holomorphic function of p for $\xi > 0$. This function, being equal to $1/p^{\alpha} = 1/\xi^{\alpha}$ for $p = \xi$ real > 0, can only be the branch specified above of $1/p^{\alpha}$. This can also be shown directly.

The same change of variables $pt = u$ leads to an integration in the complex plane. If the argument of p is denoted by θ, then

$$F(p) = \frac{1}{\Gamma(\alpha)p^{\alpha}} \int_0^{e^{i\theta}\infty} e^{-u}u^{\alpha-1}\, du. \tag{VI, 2; 9}$$

It can be shown without difficulty that the integral

$$\int_0^{e^{i\theta}\infty} = \lim_{\substack{A\to\infty\\ \varepsilon\to 0}} \int_{e^{i\theta}\varepsilon}^{e^{i\theta}A}$$

may be replaced by

$$\lim_{\varepsilon \to 0} \int_{e^{i\theta}\varepsilon}^{\varepsilon} + \lim_{\substack{\varepsilon \to 0 \\ A \to \infty}} \int_{\varepsilon}^{A} + \lim_{A \to \infty} \int_{A}^{e^{i\theta}A} = \int_{0}^{\infty}$$

so that it remains true that

$$\text{(VI, 2; 10)} \qquad \int_{0}^{e^{i\theta}\infty} e^{-u}u^{\alpha-1}\,du = \Gamma(\alpha)$$

and

$$\text{(VI, 2; 11)} \qquad F(p) = \frac{1}{p^{\alpha}}.$$

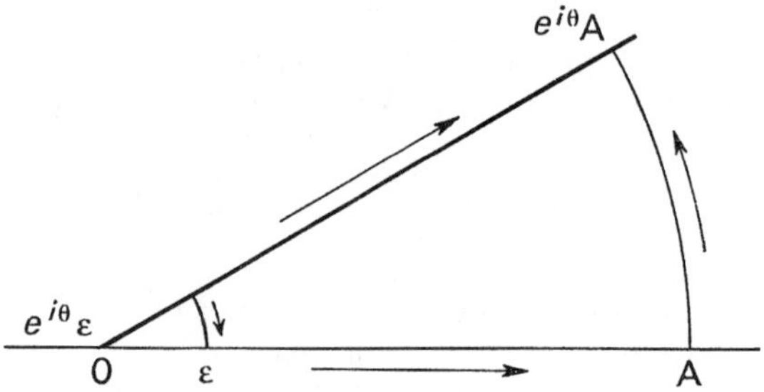

Figure (VI, 1)

The branch of p^{α} which appears here is determined by the change of variables. It is such that $p^{\alpha}t^{\alpha} = u^{\alpha}$ has argument $\alpha\theta$, so that p^{α} itself has argument $\alpha\theta$ ($p^{\alpha} = |p|^{\alpha}e^{i\theta\alpha}$), or again it may be obtained by continuity in the half-plane $\xi > 0$ from the branch which is real and > 0 for $p = \xi$ real > 0.

In particular, take $\alpha = 1$, $\lambda = \pm i\omega$:

$$\text{(VI, 2; 12)} \qquad \begin{aligned} Y(t)e^{i\omega t} &\sqsupset \frac{1}{p - i\omega} \quad \text{for } \xi > 0, \\ Y(t)e^{-i\omega t} &\sqsupset \frac{1}{p + i\omega} \quad \text{for } \xi > 0, \end{aligned}$$

whence

$$\text{(VI, 2; 13)} \qquad \begin{aligned} Y(t)\cos\omega t &\sqsupset \frac{p}{p^2 + \omega^2} \quad \text{for } \xi > 0, \\ Y(t)\sin\omega t &\sqsupset \frac{\omega}{p^2 + \omega^2} \quad \text{for } \xi > 0. \end{aligned}$$

4) We shall prove the result

$$\text{(VI, 2; 14)} \qquad Y(t)\,J_0(t) \sqsupset \frac{1}{\sqrt{p^2 + 1}} \quad \text{for } \xi > 0,$$

$J_0(t)$ being the Bessel function of order zero. The branch of

$$\sqrt{p^2+1} = \sqrt{(p+i)(p-i)}$$

is the one which, in the half-plane $\xi > 0$ where the holomorphic property applies, varies continuously from the real positive root for $p =$ real $\xi > 0$. It is necessary to evaluate

$$\text{(VI, 2; 15)} \qquad F(p) = \int_0^\infty J_0(t)e^{-pt}\,dt.$$

It is known that, for $t \geqslant 0$, $J_0(t)$ is bounded above by 1 so that the integral is certainly summable for $\xi > 0$. Because of the asymptotic relation

$$\text{(VI, 2; 16)} \qquad J_0(t) \sim \sqrt{\frac{2}{\pi t}}\cos\left(t - \frac{\pi}{4}\right),$$

it is not summable for $\xi = 0$. In addition it may be noted that the half-plane where the transform is holomorphic can be no larger than the half-plane of summability $\xi > 0$, as is shown by the existence of singularities $p = \pm i$ for $F(p)$.

Applying the series expansion

$$\text{(VI, 2; 17)} \qquad J_0(t) = \sum_{m=0}^{\infty} \frac{(-1)^m (t/2)^{2m}}{(m!)^2}$$

of $J_0(t)$ to integrate $F(p)$ term by term and assuming that this is legitimate, we obtain

$$\begin{aligned}\text{(VI, 2; 18)} \quad F(p) &= \sum_{m=0}^{\infty} \frac{(-1)^m}{(m!)^2\, 2^{2m}} \int_0^\infty t^{2m} e^{-pt}\,dt \\ &= \sum_{m=0}^{\infty} \frac{(-1)^m}{(m!)^2} \frac{1}{2^{2m}} \frac{(2m)!}{p^{2m+1}} \\ &= \sum_{m=0}^{\infty} \frac{1.3.5\ldots(2m-1)}{2.4.6\ldots 2m} \frac{(-1)^m}{p^{2m+1}} \\ &= \sum_{m=0}^{\infty} \frac{\left(-\frac{1}{2}\right)\left(-\frac{1}{2}-1\right)\cdots\left(-\frac{1}{2}-m+1\right)}{m!} \frac{1}{p^{2m+1}}.\end{aligned}$$

The last expression is a binomial series which is convergent for $|p| > 1$, giving in this case

$$\text{(VI, 2; 19)} \quad F(p) = \frac{1}{p}\left(1 + \frac{1}{p^2}\right)^{-\frac{1}{2}} = \frac{1}{p} \frac{1}{\sqrt{1 + \frac{1}{p^2}}} = \frac{1}{\sqrt{p^2+1}},$$

the branch being indicated above.

The process is legitimate if a *finite quantity* is obtained on replacing the expressions $\frac{(-1)^m}{(m!)^2}\left(\frac{t}{2}\right)^{2m} e^{-pt}$ by their moduli $\frac{1}{(m!)^2}\left(\frac{t}{2}\right)^{2m} e^{-\xi t}$. This yields

$$\text{(VI, 2; 20)} \quad \sum_{m=0}^{\infty} \frac{1}{(m!)^2} \frac{1}{2^{2m}} \frac{(2m)!}{\xi^{2m+1}} = \frac{1}{\xi} \sum_{m=0}^{\infty} \left| \frac{\left(-\frac{1}{2}\right)\left(-\frac{1}{2}-1\right)\cdots\left(-\frac{1}{2}-m+1\right)}{m!} \right| \frac{1}{\xi^{2m}},$$

which is a finite quantity for $\xi > 1$ and equal to $1/\sqrt{\xi^2+1}$. Thus the equation (VI, 2; 14) is valid for $\xi > 1$. However the Laplace integral considered is in fact summable for $\xi > 0$. Hence it must be a holomorphic function of p for $\xi > 0$ and equal to $1/\sqrt{p^2+1}$ for $\xi > 1$. Since the last expression is holomorphic for $\xi > 0$ it follows that (VI, 1; 14) is true for $\xi > 0$ although the method used to obtain it ceases to be legitimate for $0 < \xi \leqslant 1$.

5) If $f(t) \sqsupset F(p)$ then

$$\text{(VI, 2; 21)} \qquad f(\lambda t) \sqsupset \frac{1}{\lambda} F\left(\frac{p}{\lambda}\right), \qquad \lambda > 0.$$

3. THE LAPLACE TRANSFORMS AND CONVOLUTION

Let S and T be two distributions belonging to $\mathcal{D}'_+$, and possessing Laplace transforms for $\xi > b_1$ and $\xi > b_2$ respectively. Then, for

$$\text{(VI, 2; 22)} \qquad \begin{aligned} \xi > b &= \max(b_1, b_2), \\ \mathcal{S}(p) &= \langle S, e^{-pt} \rangle, \\ \mathcal{T}(p) &= \langle T, e^{-pt} \rangle. \end{aligned}$$

Hence

$$\text{(VI, 2; 23)} \quad \begin{aligned} \langle S * T, e^{-pt} \rangle &= \langle S_x \otimes T_y, e^{-p(x+y)} \rangle \\ &= \langle S_x \otimes T_y, e^{-px} e^{-py} \rangle = \langle S_x, e^{-px} \rangle \langle T_y, e^{-py} \rangle \end{aligned}$$

or

$$\text{(VI, 2; 24)} \qquad S * T \sqsupset \mathcal{S}(p)\mathcal{T}(p),$$

from which follows :

THEOREM 4. S $*$ T *has a Laplace transform for* $\xi > b = \max(b_1, b_2)$ *which is equal to the product of the Laplace transforms of* S *and* T.

This result has already been encountered for S and T with bounded supports (so that $b_1 = b_2 = -\infty$) in Chapter III, equation (III, 2; 29), page 118.

COROLLARY. *If* $T \sqsupset F(p)$, *then*

(VI, 2; 25) $$T^{(m)} \sqsupset p^m F(p).$$

For, by (III, 2; 38), $T^{(m)} = T * \delta^{(m)}$ and $\delta^{(m)} \sqsupset p^m$.
However, it must be remembered that $T * \delta^{(m)}$ is the convolution of two distributions so that even if T is a function it must be differentiated in the distribution sense.

Examples

1) $Y(t) \sqsupset 1/p$. Hence $Y' \sqsupset p.\ 1/p = 1$, so that $\delta \sqsupset 1$ (since $Y' = \delta$).

2) From the result

$$Y(t) \sin \omega t \sqsupset \frac{\omega}{p^2 + \omega^2}$$

of (VI, 2; 13) we can deduce

$$(Y(t) \sin \omega t)' = \omega Y(t) \cos \omega t \sqsupset \frac{\omega p}{p^2 + \omega^2}$$

from which it follows at once that

$$Y(t) \cos \omega t \sqsupset \frac{p}{p^2 + \omega^2}.$$

In a similar way, by differentiating the left-hand side, we obtain

$$(Y(t) \cos \omega t)' = -\omega Y(t) \sin \omega t + \delta \sqsupset \frac{p^2}{p^2 + \omega^2},$$

whence may be recovered the original result

$$Y(t) \sin \omega t \sqsupset \frac{1}{\omega}\left[1 - \frac{p^2}{p^2 + \omega^2}\right] = \frac{\omega}{p^2 + \omega^2}.$$

Also note that

(VI, 2; 26) $$(\delta'' + \omega^2 \delta) * Y(t) \frac{\sin \omega t}{\omega} = \delta,$$

and that, correspondingly,

(VI, 2; 27) $$(p^2 + \omega^2) \frac{1}{p^2 + \omega^2} = 1.$$

3) The fact that δ is the unit element in convolution corresponds to the fact that 1 is the unit element in multiplication.

4) From (III, 2; 11) it is known that

(VI, 2; 28) $$Y(t) e^{\lambda t} \frac{t^{\alpha-1}}{\Gamma(\alpha)} * Y(t) e^{\lambda t} \frac{t^{\beta-1}}{\Gamma(\beta)} = Y(t) e^{\lambda t} \frac{t^{\alpha+\beta-1}}{\Gamma(\alpha + \beta)}.$$

Using (VI, 2; 6) to obtain the Laplace transform of both sides, (VI, 2; 28) gives

(VI, 2; 29) $$\left(\frac{1}{p-\lambda}\right)^{\alpha}\left(\frac{1}{p-\lambda}\right)^{\beta}=\left(\frac{1}{p-\lambda}\right)^{\alpha+\beta},$$

which is indeed obvious.

It will be seen below that, conversely, it is possible to deduce the equation (VI, 2; 28) from the simple result (VI, 2; 29).

4. FOURIER AND LAPLACE TRANSFORMS. INVERSION OF THE LAPLACE TRANSFORM

Let us return to the case of a function $f(t)$, when

(VI, 2; 30) $$F(\xi+i\eta)=\int_0^{\infty}(f(t)e^{-\xi t})e^{-i\eta t}\,dt.$$

It can thus be seen that, for fixed ξ, $F(\xi+i\eta)$, regarded as a function of the variable η, is the *Fourier transform of* $f(t)e^{-\xi t}$. Thus a Laplace transform is equivalent to a family of Fourier transforms, namely the Fourier transforms of the functions $f(t)e^{-\xi t}$ for $\xi > a$. This fact gives rise to a large number of results.

1) It permits the deduction of Fourier transforms from Laplace transforms; the latter are easier to evaluate because the functions involved are holomorphic. For example, it follows from equation (VI, 2; 14) that the Fourier transform of

(VI, 2; 31) $$Y(t)J_0(t)e^{-\xi t}, \qquad \xi > 0$$

is

(VI, 2; 32) $$\frac{1}{\sqrt{\xi^2-\eta^2+2i\xi\eta+1}}.$$

It is seen that this result is valid for $\xi > 0$, although, as was shown during the proof of equation (VI, 2; 14), a direct calculation gives it only for $\xi > 1$. It is by using the properties of holomorphic functions that we can pass to $\xi > 0$. It is possible to go even further and to deduce, by the passage to the limit (in the distribution space) as $\xi \to 0$, the Fourier transform of the tempered distribution $Y(t)J_0(t)$.

2) *If the Laplace transform of $f(t)$ is $\equiv 0$ for $\xi > a$, then $f(t)$ is almost everywhere zero (or zero in the distribution sense).*

For, if $f(t)e^{-\xi t}$ has a zero Fourier transform, then it is almost everywhere zero and so, in consequence, is $f(t)$. Thus a holomorphic function of p can only have one inverse.

We shall accept that the following result remains true for a distribution : *if the Laplace transform* $\mathcal{T}(p)$ *of* T *is* $= 0$ *for* $\xi > b$, *then* T *is zero.*

3) The inversion formula for the Laplace transform is deduced from the Fourier inversion formula :

$$f(t)e^{-\xi t} = \frac{1}{2\pi}\int_{-\infty}^{+\infty} F(\xi + i\eta)e^{i\eta t}\,d\eta. \tag{VI, 2; 33}$$

This relation applies for all values of $\xi > a$, but for each such value must give the same function $f(t)$. Also

$$f(t) = \frac{1}{2\pi}\int_{-\infty}^{+\infty} F(\xi + i\eta)e^{(\xi+i\eta)t}\,d\eta \tag{VI, 2; 34}$$

or

$$f(t) = \frac{1}{2i\pi}\int_{\xi-i\infty}^{\xi+i\infty} F(p)e^{pt}\,dp. \tag{VI, 2; 35}$$

This is a fundamental formula and we shall proceed to investigate its conditions of validity.

THEOREM 5. *A holomorphic function* $\mathcal{T}(p)$ *is the Laplace transform of a distribution* $T \in \mathcal{D}'_+$ *if and only if there exists a half-plane* $\xi > c$ *in which the modulus of the function is bounded above by a polynomial in* $|p|$.

1) *The condition is necessary.* A general proof will not be given. We shall confine ourselves to noting that if T is a function f and if a is the abscissa of summability of its Laplace integral, then $f(t)e^{-\xi t}$ is summable for $\xi = c > a$, and then, for $\xi \geqslant c$,

$$|F(p)| \leqslant \int_0^{\infty} |f(t)|e^{-ct}\,dt = C. \tag{VI, 2; 36}$$

$|F(p)|$ is thus bounded by a constant C. If now T is a derivative of order m (in the distribution sense) of such a function f, then

$$\mathcal{T}(p) = p^m F(p) \tag{VI, 2; 37}$$

so that

$$|\mathcal{T}(p)| \leqslant C|p^m| \qquad \text{for} \qquad \xi \geqslant c.$$

Note also that $\delta^{(m)} \sqsupset p^m$, which obeys an inequality of the same type; this illustrates the discussion above since $\delta^{(m)}$ is the $(m + 1)$ st derivative of Y.

Again, the proof for the general case follows exactly the same lines, for it is shown that if T has a Laplace transform it is a derivative $f^{(m)}$ of some function f having an abscissa of summability.

2) *The condition is sufficient.* Let us first consider a particular case : Suppose that $|\mathcal{C}(p)|$ is *bounded* for $\xi > c > 0$ and that it tends to 0 with $1/|p|^2$ when $|p| \to \infty$. Then $|p|^2|\mathcal{C}(p)|$ is bounded for $\xi > c$ and there exists a constant C such that

$$|\mathcal{C}(p)| \leqslant \frac{C}{|p|^2} \quad \text{for} \quad \xi > c, \tag{VI, 2; 38}$$

giving

$$|\mathcal{C}(\xi + i\eta)| \leqslant \frac{C}{\xi^2 + \eta^2} \quad \text{for} \quad \xi > c. \tag{VI, 2; 39}$$

It is then possible to obtain the integral (VI, 2; 35),

$$f(t) = \frac{1}{2i\pi} \int_{\xi - i\infty}^{\xi + i\infty} \mathcal{C}(p) e^{pt}\, dp, \qquad \xi > c. \tag{VI, 2; 40}$$

This integral exists for fixed ξ and t, since $|e^{pt}| = |e^{(\xi + i\eta)t}| = e^{\xi t}$ which is independent of η, and $\mathcal{C}(\xi + i\eta)$ is summable in η from $\eta = -\infty$ to $\eta = +\infty$ by virtue of (VI, 2; 39). Also, this integral does not depend on the choice of ξ. For if two values ξ_1 and ξ_2, with $\xi_1 > \xi_2$, are taken, then

$$\begin{aligned} \int_{\xi_1 - i\infty}^{\xi_1 + i\infty} - \int_{\xi_2 - i\infty}^{\xi_2 + i\infty} &= \lim_{A \to \infty} \left(\int_{\xi_1 - iA}^{\xi_1 + iA} - \int_{\xi_2 - iA}^{\xi_2 + iA} \right) \\ &= \lim_{A \to \infty} \left(\int_{\xi_2 + iA}^{\xi_1 + iA} - \int_{\xi_2 - iA}^{\xi_1 - iA} \right) \end{aligned} \tag{VI, 2; 41}$$

[by Cauchy's theorem, see Figure (VI, 2)].

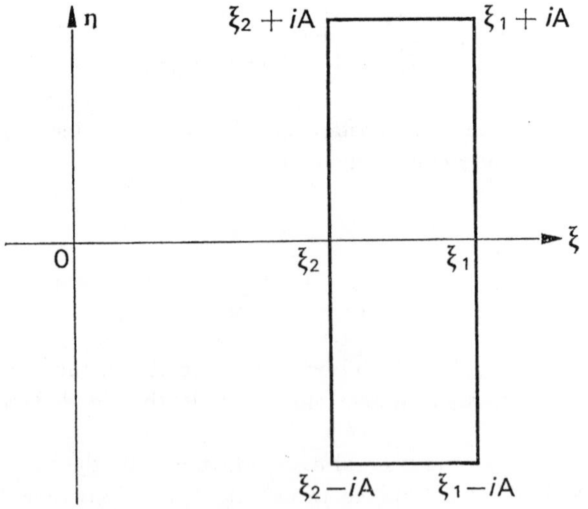

Figure (VI, 2)

Now the bracketed expression tends to 0 as $A \to \infty$, since each of its two terms is bounded above by

$$\text{(VI, 2; 42)} \qquad \int_{\xi_2+iA}^{\xi_1+iA} \frac{C|e^{pt}|d\xi}{|p|^2} \leqslant \frac{(\xi_1 - \xi_2)Ce^{\xi_1 t}}{c^2 + A^2} \to 0 \quad \text{as} \quad A \to \infty.$$

Write (VI, 2; 40) in the form

$$\text{(VI, 2; 43)} \qquad e^{-\xi t} f(t) = \frac{1}{2\pi} \int_{-\infty}^{+\infty} \mathcal{C}(\xi + i\eta) e^{i\eta t}\, d\eta.$$

The right-hand side is the Fourier integral $\overline{\mathcal{F}}$ (in η) of a summable function $\mathcal{C}(\xi + i\eta)$. The function on the left-hand side is thus continuous and hence $f(t)$ must also be continuous. It can be seen that $\mathcal{C}(\xi + i\eta)$ is the Fourier transform $\mathcal{F}$ of the function $e^{-\xi t} f(t)$, the result being understood in the sense of Fourier transforms of tempered distributions. However, if it is known that $f(t)e^{-\xi t}$ is summable then

$$\text{(VI, 2; 44)} \qquad \mathcal{C}(\xi + i\eta) = \int_{-\infty}^{+\infty} f(t) e^{-\xi t} e^{-i\eta t}\, dt$$

or

$$\text{(VI, 2; 45)} \qquad \mathcal{C}(p) = \int_{-\infty}^{+\infty} f(t) e^{-pt}\, dt.$$

It will be shown that $f(t)$ is zero for $t < 0$ and that $f(t)e^{-\xi t}$ is in fact summable for $\xi > c$. From (VI, 2; 43) it follows that

$$\text{(VI, 2; 46)} \qquad |f(t)| \leqslant \frac{e^{\xi t}}{2\pi} \int_{-\infty}^{+\infty} \frac{C}{c^2 + \eta^2}\, d\eta \leqslant \frac{Ce^{\xi t}}{2c}.$$

This result is valid for all $\xi > c$. For $t < 0$, when $\xi \to \infty$, $e^{\xi t} = e^{-\xi|t|}$ tends to zero, so that $f(t)$ is indeed zero. For $t > 0$, the minimum of $e^{\xi t}$ for $\xi > c$ is e^{ct}. Thus

$$\text{(VI, 2; 47)} \qquad |f(t)| < \frac{1}{2c} Ce^{ct}.$$

Hence $|f(t)e^{-\xi t}| < (1/2c)\, Ce^{-(\xi - c)t}$ is indeed summable for $\xi > c$. The result (VI, 2; 45) is thus valid, and as $f(t) = 0$ for $t < 0$, it gives

$$\text{(VI, 2; 48)} \qquad \mathcal{C}(p) = \int_0^{\infty} f(t) e^{-pt}\, dt \quad \text{for} \quad \xi > c$$

so that $\mathcal{T}(p)$ is proved to be a Laplace transform for $\xi > c$. We also have the inequality (VI, 2; 47) and the fact that f is continuous.

Now let us return to the general case. Let $\mathcal{T}(p)$ be a function which is holomorphic for $\xi > c > 0$ and has a modulus bounded above by a polynomial in $|p|$ of degree m. Then

$$\frac{\mathcal{T}(p)}{p^{m+2}} \tag{VI, 2; 49}$$

satisfies an inequality of type (VI, 2; 38). Thus there exists a continuous function $f(t)$ with an abscissa of summability $\leqslant c$, having the function (VI, 2; 49) as its Laplace transform. Then

$$\mathrm{T} = \left(\frac{d}{dt}\right)^{m+2} f \tag{VI, 2; 50}$$

(the derivatives being taken in the distribution sense) has, by the corollary of Theorem 4, the Laplace transform $\mathcal{T}(p)$ for $\xi > c$. This completes the required proof.

Example. Find by the above procedure the inverse of $\mathcal{T}(p) = 1$. This is obviously a holomorphic function for all complex p and is bounded above by a polynomial of degree 0 in $|p|$.

Choosing $c > 0$, the function $1/p^2$ satisfies the inequality (VI, 2; 38) for $\xi > c$, with $\mathrm{C} = 1$. It is thus the Laplace transform of a continuous function which is zero for $t < 0$. This function is obtained from

$$\begin{aligned} f(t) &= \frac{1}{2i\pi} \int_{\xi - i\infty}^{\xi + i\infty} \frac{1}{p^2} e^{pt}\, dp, \qquad \text{for} \qquad \xi > c \\ &= \frac{1}{2i\pi} \lim_{\mathrm{A} \to \infty} \int_{\xi - i\mathrm{A}}^{\xi + i\mathrm{A}} . \end{aligned} \tag{VI, 2; 51}$$

This integral will be replaced by a contour integral. For the time being, since $f(t) = 0$ for $t < 0$, we shall restrict ourselves to the case $t \geqslant 0$. Add to the segment $(\xi - i\mathrm{A}, \xi + i\mathrm{A})$ the left-hand semicircle with centre at ξ and radius A. The length of its semicircular arc is $\pi\mathrm{A}$ and on it $|e^{pt}/p^2|$ is bounded above by $e^{\xi t}/(\mathrm{A} - \xi)^2$. Thus the integral over the semicircle is bounded above by $\pi\mathrm{A}e^{\xi t}/(\mathrm{A} - \xi)^2$ and tends to 0 as $\mathrm{A} \to \infty$ (ξ being fixed beforehand).

Hence the limit of the integral $\int_{\xi - i\mathrm{A}}^{\xi + i\mathrm{A}}$ is equal to the limit of the integral over the contour Γ composed of the line segment and the semicircle so that

$$f(t) = \frac{1}{2i\pi} \lim_{\mathrm{A} \to \infty} \int_\Gamma \frac{e^{pt}}{p^2}\, dp. \tag{VI, 2; 52}$$

This integral $\int_\Gamma$ is equal to $2\pi i$ times the residue at the pole $p = 0$ inside the contour. In the neighbourhood of $p = 0$,

(VI, 2; 53) $$\frac{e^{pt}}{p^2} = \frac{1}{p^2} + \frac{t}{p} + \frac{t^2}{2!} + \cdots,$$

and thus $f(t) = t$ for $t \geqslant 0$.

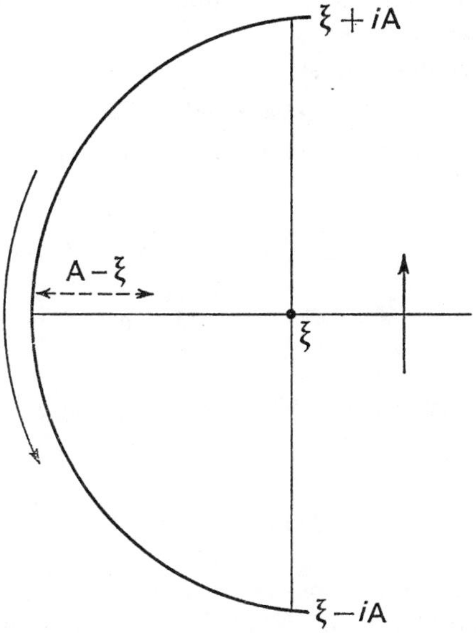

Figure (VI, 3)

For $t \leqslant 0$, it is necessary to use the right-hand semicircle with centre at ξ and radius A since it is on this semicircle that the integral tends to zero. However as there is now no singularity inside Γ it follows that $f(t) = \frac{1}{2i\pi} \lim_{A \to \infty} \int_\Gamma = 0$. Thus the final result is

(VI, 2; 54) $$tY(t) \sqsupset \frac{1}{p^2} \quad \text{for} \quad \xi > 0,$$

which is already known.
The equation (VI, 2; 50) then gives as the inverse of $\mathcal{C}(p) = 1$ the derivative

(VI, 2; 55) $$\left(\frac{d}{dt}\right)^2 (Y(t)t) = \delta,$$

whence $\delta \sqsupset 1$, which is again a known result. Of course, this could not be obtained directly by an integral of type (VI, 2; 35) since this is not a function $f(t)$, defined for any value of t, but a distribution.

3. Applications of the Laplace transform. Operational calculus

There exist many tables of Laplace transforms and their inverses. With these tables we can solve many problems by transforming convolution into multiplication. The whole complex of operations of this type is known as *operational calculus*.

We shall give several examples :

1) To evaluate $f(t) = Y(t)J_0(t) * Y(t)J_0(t)$.

Now, $Y(t)J_0(t) \sqsupset 1/\sqrt{p^2+1}$ [equation (VI, 2; 14)]. Thus the transform of f is

$$\frac{1}{\sqrt{p^2+1}}\,\frac{1}{\sqrt{p^2+1}} = \frac{1}{p^2+1}$$

so that f is the inverse of $1/(p^2+1)$. Hence $f(t) = Y(t)\sin t$ [equation (VI, 2; 13)] and

(VI, 3; 1) $$Y(t)J_0(t) * Y(t)J_0(t) = Y(t)\sin t.$$

Hence, for $t > 0$,

(VI, 3; 2) $$\int_0^t J_0(\tau)J_0(t-\tau)\,d\tau = \sin t.$$

For $t < 0$, since all the functions considered are then zero, no interesting information can be obtained. Nevertheles, since J_0 is an even function and $\sin t$ is an odd function, a change of variable $\tau = -\theta$ shows that equation (VI, 3; 2) remains valid for $t \leqslant 0$.

2) To solve a convolution equation

(VI, 3; 3) $$A * X = B, \qquad A \in \mathcal{D}'_+, \qquad B \in \mathcal{D}'_+,$$

where A and B are given and X is the required unknown. It is known that differential equations with constant coefficients and many integral and integro-differential equations are of this type, and that electric circuit theory gives rise to numerous convolution equations.

Suppose that A and B have Laplace tranforms $\mathcal{A}(p)$ and $\mathcal{B}(p)$. If there is a solution in $\mathcal{D}'_+$ and if it possesses a Laplace transfom $\mathcal{H}(p)$, then

(VI, 3; 4) $$\mathcal{A}(p)\mathcal{H}(p) = \mathcal{B}(p),$$

giving the solution

(VI, 3; 5) $$\mathcal{H}(p) = \frac{\mathcal{B}(p)}{\mathcal{A}(p)}.$$

To obtain X it only remains to find the inverse of $\mathcal{B}(p)/\mathcal{A}(p)$.

If one of the two distributions A, B has no Laplace transform (abscissa of definition $= + \infty$) this method is not applicable. If both have Laplace transforms but $\mathcal{B}(p)/\mathcal{A}(p)$ is not a Laplace transform (i.e., it does not satisfy the conditions of Theorem 5) then this method again gives no result. It proves only that there is no solution possessing a Laplace transform, although there may be one which does not possess a Laplace transform (abscissa of definition $= + \infty$).

If in fact A and B have Laplace transforms and $\mathcal{B}(p)/\mathcal{A}(p)$ has a Laplace inverse X, then X is a unique solution since it is known that a convolution equation in $\mathcal{D}'_+$ can never have more than *one* solution (Chapter III, § 2, no. 3). In particular, the elementary solution or inverse A^{-1} of the distribution A in the convolution algebra $\mathcal{D}'_+$ is given if it possesses a Laplace transform, the Laplace inverse of $1/\mathcal{A}(p)$.

Suppose, for example, that it is required to find A^{-1} for $A = Y(t)\,J_0(t)$. Here, by equation (VI, 2; 14), $\mathcal{A}(p) = 1/\sqrt{p^2+1}$. Then

$$\frac{1}{\mathcal{A}(p)} = \sqrt{p^2+1}. \tag{VI, 3; 6}$$

This function is holomorphic for $\xi > 0$ and bounded above by a polynomial in $|p|$ of degree 1 so that it is in fact a Laplace transform. Its Laplace inverse can be found immediately, for

$$\sqrt{p^2+1} = \frac{p^2+1}{\sqrt{p^2+1}},$$

whence

$$E = \left(\frac{d^2}{dt^2} + 1\right) Y(t)\,J_0(t) = (\delta'' + \delta) * Y(t)\,J_0(t). \tag{VI, 3; 7}$$

It can also be verified that the equation $E * Y(t) J_0(t) = \delta$ may be written

$$(\delta'' + \delta) * Y(t)\,J_0(t) * Y(t)\,J_0(t) = \delta,$$

which by making use of (VI, 3; 1) becomes the well-known result

$$(\delta'' + \delta) * Y(t) \sin t = \delta.$$

Bearing in mind the differential equation for J_0, namely

$$J_0'' + \frac{1}{t} J_0' + J_0 = 0, \tag{VI, 3; 8}$$

obtained by differentiation in the usual sense, we have

$$J_0''(t) + J_0(t) = \frac{-1}{t} J_0'(t) = \frac{J_1(t)}{t}, \tag{VI, 3; 9}$$

an expression which is regular at the origin, because J_1 contains t as a factor.

However, in the neighbourhood of $t = 0$,

$$J_0(t) = 1 - \frac{\left(\frac{t}{2}\right)^2}{(1\,!)^2} + \cdots$$

so that

$$(Y(t)\,J_0(t))'' = Y(t)\,J_0''(t) + \delta',$$

whence

$$\text{(VI, 3; 10)} \quad E = \left(\frac{d^2}{dt^2} + 1\right)Y(t)\,J_0(t) = Y(t)\left(\frac{J_1(t)}{t}\right) + \delta'.$$

Naturally, instead of making the calculation one should be able to find the result in the tables. Unfortunately most tables contain the inverses which are functions $f(t)$ and not distributions. It may be deduced that

$$\text{(VI, 3; 11)} \qquad Y(t)J_0(t) * \left[Y(t)\left(\frac{J_1(t)}{t}\right) + \delta'\right] = \delta.$$

This equation can be written

$$\text{(VI, 3; 12)} \quad Y(t)J_0(t) * Y(t)\left(\frac{J_1(t)}{t}\right) + Y(t)J_0'(t) + \delta = \delta$$

or, bearing in mind that $J_0' = -J_1$,

$$\text{(VI, 3; 13)} \qquad Y(t)\,J_0(t) * Y(t)\left(\frac{J_1(t)}{t}\right) = Y(t)\,J_1(t).$$

This equation can easily be verified with the help of the tables since

$$\text{(VI, 3; 14)} \qquad \begin{aligned} Y(t)\,J_0(t) &\sqsupset \frac{1}{\sqrt{p^2+1}}, \\ Y(t)\frac{J_1(t)}{t} &\sqsupset \sqrt{p^2+1} - p, \\ Y(t)\,J_1(t) &\sqsupset \frac{\sqrt{p^2+1}-p}{\sqrt{p^2+1}}. \end{aligned}$$

It gives, for all $t \geqslant 0$ (and also, by the change of variables $\tau = -\theta$, for $t \leqslant 0$),

$$\text{(VI, 3; 15)} \qquad \int_0^t \frac{J_1(\tau)}{\tau} J_0(t-\tau)\,d\tau = J_1(t).$$

From equation (VI, 3; 10) the integral equation

$$\text{(VI, 3; 16)} \qquad \int_0^t f(\tau)J_0(t-\tau)\,d\tau = g(t), \qquad t \geqslant 0$$

(where the function g on the right-hand side is given and f is the required unknown) has the solution

$$f(t) = \int_0^t g(\tau)\frac{J_1(t-\tau)}{t-\tau}\,d\tau + g'(t), \tag{VI, 3; 17}$$

on condition that g' is a function. If g' is a distribution the solution is a distribution and not a function. It can be seen that to forget the term in δ' in (VI, 3; 10) would lead to serious error. These results show how the Laplace transform may be used for any convolution problem in $\mathcal{D}'_+$.

Important note. In tables of Laplace transforms it is implicitly understood that the functions $f(t)$ are zero for $t < 0$ and they are given for $t > 0$ only, without writing down the factor $Y(t)$. If care is not taken this can cause grave errors. For example, we find in tables of Laplace transforms

$$1 \sqsupset \frac{1}{p}, \tag{VI, 3; 18}$$

which by differentiation gives the absurd result

$$0 \sqsupset 1. \tag{VI, 3; 19}$$

In fact (VI, 3; 18) means

$$Y(t) \sqsupset \frac{1}{p}, \tag{VI, 3; 20}$$

which by differentiation gives the correct result

$$\delta \sqsupset 1. \tag{VI, 3; 21}$$

Exercise. Evaluate the Laplace inverse $H_\nu(t)$ o

$$\left(\frac{1}{p^2+1}\right)^{\nu+\frac{1}{2}}, \quad \mathcal{R}p > 0, \nu > -\tfrac{1}{2},$$

the branch being chosen to be continuous with the one which is real and positive for p real and positive. (Examination question, June 1955.) It is known that

$$Y(t)e^{it}\frac{t^{\nu-\frac{1}{2}}}{\Gamma(\nu+\frac{1}{2})} \sqsupset \left(\frac{1}{p-i}\right)^{\nu+\frac{1}{2}}, \tag{VI, 3; 22}$$

$$Y(t)e^{-it}\frac{t^{\nu-\frac{1}{2}}}{\Gamma(\nu+\frac{1}{2})} \sqsupset \left(\frac{1}{p+i}\right)^{\nu+\frac{1}{2}}. \tag{VI, 3; 23}$$

The required inverse is thus the convolution product

$$\text{(VI, 3; 24)}\quad H_\nu(t) = Y(t)e^{it}\frac{t^{\nu-\frac{1}{2}}}{\Gamma(\nu+\frac{1}{2})} * Y(t)e^{-it}\frac{t^{\nu-\frac{1}{2}}}{\Gamma(\nu+\frac{1}{2})}$$

or

$$\text{(VI, 3; 25)}\quad H_\nu(t) = \frac{Y(t)}{(\Gamma(\nu+\frac{1}{2}))^2}\int_0^t e^{i(t-\tau)}e^{-i\tau}(t-\tau)^{\nu-\frac{1}{2}}\tau^{\nu-\frac{1}{2}}\,d\tau.$$

Putting $\tau = tu$,

$$\text{(VI, 3; 26)}\quad H_\nu(t) = \frac{Y(t)e^{it}t^{2\nu}}{(\Gamma(\nu+\frac{1}{2}))^2}\int_0^1 e^{-2itu}((1-u)u)^{\nu-\frac{1}{2}}\,du.$$

This integral recalls the one which yields the Bessel functions,

$$\text{(VI, 3; 27)}\quad J_\nu(t) = \frac{\left(\frac{t}{2}\right)^\nu}{\sqrt{\pi}\,\Gamma(\nu+\frac{1}{2})}\int_{-1}^{+1} e^{-itv}(1-v^2)^{\nu-\frac{1}{2}}\,dv.$$

To go from $u(1-u)$ to $1-v^2$ it is sufficient to put $u = (v+1)/2$, whence

$$\text{(VI, 3; 28)}\quad H_\nu(t) = \frac{Y(t)e^{it}t^{2\nu}}{(\Gamma(\nu+\frac{1}{2}))^2}\int_{-1}^{+1} e^{-it(1+v)}\left(\frac{1-v^2}{4}\right)^{\nu-\frac{1}{2}}\frac{dv}{2},$$

$$\text{(VI, 3; 29)}\quad H_\nu(t) = Y(t)\frac{\sqrt{\pi}}{\Gamma(\nu+\frac{1}{2})}\left(\frac{t}{2}\right)^\nu J_\nu(t) \sqsupset \left(\frac{1}{p^2+1}\right)^{\nu+\frac{1}{2}}.$$

Notes.

1) For $\nu = 0$ we again find

$$\text{(VI, 3; 30)}\quad Y(t)J_0(t) \sqsupset \frac{1}{\sqrt{p^2+1}}.$$

2) When $\nu = \frac{1}{2}$ it is known that the inverse of $1/(p^2+1)$ is $Y(t)\sin t$ (equation (VI, 3; 13)). This is in accordance with the equation

$$J_{\frac{1}{2}}(t) = \sqrt{\frac{2}{\pi t}}\sin t.$$

3) When ν tends to $-\frac{1}{2}$, $1/(p^2+1)^{\nu+\frac{1}{2}}$ tends to 1, remaining uniformly bounded above for $\mathcal{R}p \geqslant \varepsilon > 0$. It would be easy to deduce rigorously that $H_\nu(t)$ must tend to δ. We shall show this directly [from (VI, 3; 29)]. It is known that $J_\nu(t)$ remains bounded for $0 < \eta \leqslant t \leqslant A < +\infty$ and since $\Gamma(\nu+\frac{1}{2}) \to \infty$ it follows that $H_\nu(t)$ tends uniformly to 0 as $\nu \to -\frac{1}{2}$ for $0 < \eta \leqslant t \leqslant A < +\infty$.

Again, for $0 \leqslant t \leqslant \eta$, if η is small enough, $J_\nu(t)$ remains $\geqslant 0$. It is thus sufficient to show that

$$\text{(VI, 3; 31)} \qquad \int_0^\eta H_\nu(t)\,dt \to 1 \qquad \text{for} \qquad \nu \to -\tfrac{1}{2}.$$

Now, from its expansion in series, $J_\nu(t)$ may be written in the interval $(0, \eta)$ as

$$\text{(VI, 3; 32)} \qquad J_\nu(t) = \left(\frac{t}{2}\right)^\nu \frac{1}{\Gamma(\nu + 1)} + \varepsilon(t, \nu),$$

where $|\varepsilon(t, \nu)| \leqslant kt^{\nu+1}$. However, the integral $\int_0^\eta t^\nu \varepsilon(t, \nu)\,dt$ remains bounded for $\nu \to -\frac{1}{2}$, and as $\Gamma(\nu + \frac{1}{2}) \to +\infty$, the corresponding term of $H_\nu(t)$ tends to zero. Then, from the remaining part, we find

$$\text{(VI, 3; 33)} \quad \lim_{\nu \to -\frac{1}{2}} \int_0^\eta H_\nu(t)\,dt = \lim_{\nu \to -\frac{1}{2}} \frac{\sqrt{\pi}}{\Gamma\left(\nu + \frac{1}{2}\right)\Gamma(\nu + 1)} \int_0^\eta \left(\frac{t}{2}\right)^{2\nu} dt.$$

But,

$$\Gamma(\nu + 1) \to \Gamma\left(\tfrac{1}{2}\right) = \sqrt{\pi};$$

$$\int_0^\eta \left(\frac{t}{2}\right)^{2\nu} dt = \frac{\eta^{2\nu+1}}{2^{2\nu}(2\nu + 1)} \sim \frac{1}{\nu + \frac{1}{2}};$$

and

$$\Gamma\left(\nu + \tfrac{1}{2}\right) \sim \frac{1}{\nu + \frac{1}{2}};$$

so that finally

$$\text{(VI, 3; 34)} \qquad \lim_{\nu \to -\frac{1}{2}} \int_0^\eta H_\nu(t)\,dt = 1, \qquad \text{Q.E.D.}$$

EXERCISES FOR CHAPTER VI

Exercise VI-1

Let a and b be two positive numbers and let $F(p)$ be the Laplace transform of a function $f(t)$. Obtain the Laplace transform of

$$g(t) = \begin{cases} f(at - b) & \text{for} \quad t > b/a, \\ 0 & \text{for} \quad t < b/a. \end{cases}$$

Exercise VI-2

Let $f(t)$ be a function having the Laplace transform $F(p)$. Obtain the Laplace transforms of the functions

$$e^t f(t); \quad \int_0^\infty \frac{\cos(2\sqrt{tx})}{(\sqrt{\pi t})} f(x)\,dx; \quad \int_0^\infty \frac{\sin(2\sqrt{tx})}{(\sqrt{\pi x})} f(x)\,dx;$$

$$\int_0^\infty \frac{\text{ch}(2\sqrt{tx})}{(\sqrt{\pi t})} f(x)\,dx; \quad \text{and} \quad \int_0^\infty \frac{\text{sh}(2\sqrt{tx})}{(\sqrt{\pi x})} f(x)\,dx.$$

Exercise VI-3

Obtain, in terms of the Laplace transform $F(p)$ of $f(t)$, the Laplace transforms of the following functions :

$$\int_0^\infty J_0(2\sqrt{tx})\,f(x)\,dx \quad \text{and} \quad \int_0^t J_0(2\sqrt{(t-x)x})\,f(x)\,dx,$$

where J_0 is the Bessel function of index 0.

Exercise VI-4

Show that if $f(t)$ is zero for $t < 0$ and, for $t > 0$, is periodic with period T, then its Laplace transform is given by

$$F(p) = \frac{1}{1 - e^{-pT}} \int_0^T e^{-pt} f(t)\,dt.$$

Application. Obtain the Laplace transform of the function which is zero for $t < 0$ and equal to $|\sin t|$ for $t > 0$.

Exercise VI-5

Let $f(t)$ be a function which is zero for $t < 0$ and possesses the Laplace transform $F(p)$.

(*a*) Obtain the Laplace transforms of the functions

$$f\left(\frac{t}{a}\right), \quad a > 0; \quad f(t) + 1.$$

(*b*) Consider

$$f_0(t) = \begin{cases} 0 & \text{for} \quad t < 0 \\ n & \text{for} \quad n < t < n+1, \end{cases} \qquad \text{integer } n \geqslant 0.$$

Evaluate directly the Laplace transform $F_0(p)$ of $f_0(t)$. Hence deduce the Laplace transforms of the functions $f_0(t/a)$ and $f_0(t) + 1$.

(*c*) Evaluate

$$\delta' * f_0(t); \qquad \delta' * f_0\left(\frac{t}{a}\right); \qquad \delta' * (f_0(t) + 1).$$

Obtain the Laplace transforms of both sides of the resulting equations, expressing the transforms of the original expressions in terms of the transforms

$F_0(p)$, $F_1(p)$, and $F_2(p)$ of $f_0(t)$, $f_0(t/a)$ and $(f_0(t) + 1)$ respectively. Hence recover the results of part (*b*).

Exercise VI-6

Let $a > 0$ and $f(t)$ be the function which is zero for $t < 0$ and is equal to

$$\frac{t}{a} - 4n \quad \text{for} \quad 4na < t < (4n+1)a,$$
$$-\frac{t}{a} + 4n + 2 \quad \text{for} \quad (4n+1)a < t < (4n+2)a,$$
$$0 \quad \text{for} \quad (4n+2)a < t < (4n+4)a, \qquad n = 0, 1, 2, \ldots$$

(*a*) Obtain directly the Laplace transform $F(p)$ of $f(t)$.

(*b*) Evaluate $\delta'' * f(t)$.

(*c*) Obtain the Laplace transforms of both sides of the resulting equation, expressing the transform of the original expression in terms of the $F(p)$. Recover the expression for $F(p)$ found in part (*b*).

Exercise VI-7

Evaluate the Laplace transform of the function $f(t)$ which is zero for $t < 0$ and equal for $t > 0$ to

$$2^n - 1 \quad \text{for} \quad n < t < n + 1, \quad n = 0, 1, 2, \ldots$$

Find a distribution $X \in \mathfrak{D}'_+$ which is a solution of the equation

$$X * f(t) = \sum_{n=1}^{\infty} 2^{n-1} \delta_{(n)}.$$

Exercise VI-8

Solve in $\mathfrak{D}'_+$ the convolution equations

$$Y(x)e^{-x} * X_1 + Y(x) J_0(x) * X_2 = -Y(x)(\sin x + \cos x) + \delta,$$
$$Y(x) J_0(x) * X_1 + X_2 = 2Y(x)e^{-x} + \delta' - 2\delta.$$

Exercise VI-9

We accept at the outset the convention that all functions of t occurring in this problem are zero for $t < 0$.

(*a*) A function $\psi(p)$ such that

(1) $$\psi[\psi(p)] = p \text{ for any } p$$

is called a periodic function of order 2.

(i) Give examples of such functions.

(ii) Show that if $f(p)$ is a solution of (1) and if φ is any function with an inverse function φ^{-1} then $\varphi^{-1}\{f[\varphi(p)]\}$ is also a solution.

(*b*) Consider the integral equation

$$(2) \qquad f(t) + \lambda \int_0^\infty K(t, x) f(x)\, dx = g(t),$$

where $f(t)$ is an unknown function and

$$K(t, x) = \sum_{n=0}^{\infty} \frac{(-1)^n}{n!} x^n a(t) * b^n(t),$$

$b^n(t)$ being defined by

$$b^n(t) = b^{n-1}(t) * b(t).$$

The series defining $K(x, t)$ is uniformly convergent for all values of x and t lying in the interval $[0, +\infty]$.

(i) Show that the Laplace transformation reduces the solution of (2) to that of a functional equation, with unknown $\varphi(p)$, of the form

$$(3) \qquad \varphi(p) + \lambda\rho(p)\varphi[\psi(p)] = \theta(p).$$

(ii) Solve (3) when $\psi(p)$ is a periodic function of order 2. This can be shown to be the case if $K(t, x)$ is one of the functions

$$x^{-\frac{\nu}{2}} t^{\frac{\nu}{2}} J_\nu(2\sqrt{tx}) \quad \text{for} \quad \mathcal{R}(\nu) > -1,$$
$$\frac{\cos 2\sqrt{tx}}{\sqrt{t}}, \quad \frac{\sin 2\sqrt{tx}}{\sqrt{x}}.$$

(iii) Hence deduce a solution of (2) of the form

$$f(t) = h(t) * (g(t) - \lambda \int_0^\infty K(t, x) g(x) dx,$$

where $h(t)$ satisfies

$$Y(t)\, h(t) - 2Y(t) h(t) * \int_0^\infty Y(t)\, K(t, x) a(x) dx = \delta.$$

(*c*) Solve the equation

$$f(t) + \lambda \int_0^\infty t^{\frac{\nu}{2}} x^{-\frac{\nu}{2}} J_\nu(2\sqrt{tx})\, f(x)\, dx = g(t)$$

(i) for $\lambda^2 \neq 1$.

(ii) for $\lambda = 1$.

It can be shown that the general solution of the functional equation

$$f(s) = \frac{1}{s^n} f\left(\frac{1}{s}\right)$$

is

$$f(s) = h(s) + \frac{1}{s^n} h\left(\frac{1}{s}\right),$$

where h is an arbitrary function. From this we can deduce the form which $g(t)$ must have for (2) to have a solution in this case and also the form of this solution,

(iii) for $\lambda = -1$.

(*d*) (i) Show that if, in equation (3), $\rho(p)$ satisfies

$$\rho(p)\,\rho[\psi(p)] = 1, \tag{4}$$

the kernel $K(t, x)$ is reciprocal; that is to say

$$h(t) = \int_0^\infty K(t, x) k(x) dx$$

is equivalent to

$$k(t) = \int_0^\infty K(t, x)\, h(x)\, dx.$$

(ii) Show that (2) then possesses the proper values $\lambda = \pm 1$.

(iii) If $\lambda = +1$, equation (2) may be solved only if $\theta(p)$ is the solution of an equation which may be transformed into an equation in $g(t)$. Hence deduce the solutions of (2) in this case.

Exercise VI-10 (Paris 1958).

(*a*) It is given that Γ is defined by the infinite product

$$\frac{1}{\Gamma(z)} = z e^{\gamma z} \prod_{n=1}^{\infty} \left[\left(1 + \frac{z}{n}\right) e^{-z/n}\right], \tag{1}$$

where γ is Euler's constant, equal to $\lim_{m\to\infty}\left(1 + \frac{1}{2} + \cdots + \frac{1}{m} - \log m\right)$.

We propose to recover Stirling's formula by using the Laplace transform.

Put

$$\psi(z) = \frac{\Gamma'(z)}{\Gamma(z)}.$$

Give the series expansions derived from (1) for $\psi(z)$ and $\psi'(z)$; evaluate $\psi(m)$ (for integers $m \geqslant 1$) and $\lim_{m\to\infty} (\psi(m) - \log m)$.

(i) Write down without proof the Laplace transform of $Y(t)\dfrac{t^{\alpha-1}}{\Gamma(\alpha)}$, $\alpha > 0$, $Y(t)$ being Heaviside's function. Let $f(t)$ be a function which is zero for

$t < 0$, continuous and bounded for $t > 0$ and having, in the neighbourhood of 0, the asymptotic expansion for $t > 0$,

(2) $$f(t) \sim a_0 + a_1 t + a_2 t^2 + \cdots$$

(Note carefully that by this it is understood that when $t > 0$ is near enough to 0,

$$f(t) - a_0 - a_1 t - a_2 t^2 - \cdots - a_{k-1} t^{k-1}$$

remains bounded above by a quantity of the form $A_k t^k$, A_k being a suitable number > 0, for $k \geqslant 1$.) Show that, for real $p \to \infty$, its Laplace transform $F(p)$ has an asymptotic expansion, in powers of $1/p$,

(3) $$F(p) = \frac{b_0}{p} + \frac{b_1}{p^2} + \cdots$$

which is to be determined. Consider, in particular, the function

(4) $$f(t) = Y(t)\left[\frac{1}{t(e^t - 1)} - \frac{1}{t^2} + \frac{1}{2t}\right].$$

Show that it has an asymptotic expansion of type (2). Evaluate a_0 and a_1 and hence deduce b_0 and b_1 for an expansion of type (3) of $F(p)$.

(ii) Let $g(t)$ be a function having a Laplace transform $G(p)$. What is the Laplace transform of $(e^t - 1)g(t)$? Take, in particular,

(5) $$g(t) = Y(t)\frac{t}{e^t - 1}.$$

It is found that $G(p)$ satisfies a simple functional equation. From this equation and the behaviour of $G(p)$ at infinity evaluate $G(p)$, expressing it with the help of ψ'. Next consider $f(t)$ defined by (4). By considering the Laplace transform of $t^2 f(t)$ and using the fact that $F'(p) \to 0$ when real p tends to infinity, $F(p)$ may be evaluated to within an additive constant.

(iii) Bearing (3) in mind, show that the results of the second question allow us to obtain a Stirling formula (for real $p \to \infty$) of the type

(6) $$p\,! \sim p^p e^{-p}\sqrt{p}\,C\left(1 + \frac{C_0}{p} + \frac{C_1}{p^2} + \cdots\right),$$

where C is unknown and the bracketed expression is an asymptotic expansion in $1/p$. Evaluate C_0 and C_1.

(*b*) All required formulae are known except for the constant $\sqrt{2\pi}$ of Stirling's formula. Give the Fourier transform of

$$(7) \qquad Y(x)e^{-2\pi x}\frac{x^{\frac{\alpha-1}{2}}}{\Gamma\left(\frac{\alpha+1}{2}\right)}.$$

Write down Parseval's equation and evaluate the two sides independently. For the evaluation of $\int_0^\infty \frac{du}{(1+u^2)^\beta}$ set $u=\sqrt{\frac{v}{1-v}}$.

A remarkable relation between $\Gamma\left(\frac{\alpha}{2}\right)$, $\Gamma\left(\frac{\alpha+1}{2}\right)$ and $\Gamma(\alpha)$ (Legendre-Gauss) may hence be deduced.

Show that if Stirling's formula is known apart from the constant $\sqrt{2\pi}$ the Legendre-Gauss relation enables one to find this constant.

CHAPTER VII

The wave and heat conduction equations

1. Equation of vibrating strings

1. PHYSICAL PROBLEMS ASSOCIATED WITH THE EQUATION OF VIBRATING STRINGS

Transverse vibrations of a stretched string

We shall consider the small *transverse* vibrations of a homogeneous stretched string AB of length L and constant cross-sectional area σ_0.

Let the x-axis $\overrightarrow{Ox}$ be along $\overrightarrow{AB}$ with A at the origin O. A point of the string having coordinate value x $(0 \leqslant x \leqslant L)$ can be displaced perpendicularly to the string and hence its y and z coordinates can oscillate about the value zero as functions of time.

Let $\vec{u}$ be a vector in the yOz plane having coordinates y, z. Then $\vec{u}$ will be a vector function of the time t, varying about $\vec{0}$. However, this function, which gives the displacement of the point with coordinate value x, is also a function of x. Thus we finally see that the function we are required to determine is the vector function $\vec{u}(x, t)$.

Let T be the tension in the string at a point M having coordinate value x at time t. By this statement we mean that the portion of the string MB exerts forces on the portion AM, which can be represented by a unique force of magnitude T acting at M in the direction of the half tangent oriented by the arc $\overset{\frown}{MB}$. According to the principle of action and reaction the portion of the string MA exerts on the portion BM a force exactly opposite to the preceding force. As we are supposing that all deformations of the length of the string are small, we see that T is approximately equal to a mean tension T_0 which is independent of x and t.

We shall neglect the action of gravity on the string. In addition we shall not make any assumption concerning the equilibrium position of the string with respect to the vertical. Bearing this in mind we shall take the position of equilibrium of the string in a gravitational field (or any other fixed field) to be the segment of a straight line. The vector $\vec{u}$ will then be the displacement from this position of equilibrium.

Let ρ be the *linear density* of the string at the point x at time t, that is, the mass per unit length of the string or the value of the quotient $\frac{\Delta m}{\Delta x}$ as $\Delta x \to 0$, where Δm is the mass of the portion $(x, x + \Delta x)$ of the string. Since the deformations are assumed small, ρ is approximately equal to a quantity ρ_0 which depends only upon x. When we assume that the string is homogeneous we imply that ρ_0 is independent of x. However it is easy to appreciate that in practice, cases exist where ρ_0 varies continuously with respect to x, because $\rho_0 = \sigma_0 \varpi_0$, where ϖ_0 is the specific gravity (mass per unit volume) and σ_0 is the cross-sectional area of the string at the point x (supposed small). Hence a string composed of a homogeneous solid but having a cross-sectional area σ_0 which varies with x, will have a linear density ρ_0 that varies with x.

The equation that $\vec{u}(x, t)$ must satisfy can be obtained by writing for each element (MM′), $(x, x + \Delta x)$ of the string the fundamental equation of dynamics $\vec{F} = m\vec{\gamma}$. [Strictly speaking it is essential that for each small interval (x_1, x_2) of the string we are able to write that the resultant of $\vec{F}$ is equal to the resultant of $m\vec{\gamma}$ and that the moment of $\vec{F}$ is equal to the moment $m\vec{\gamma}$. This being so we will then introduce integrals in dx instead of elementary quantities proportional to Δx.] We adopt the elementary method which is more intuitive than rigorous. By means of the hypothesis that the tension is tangential to the string we readily verify that the equation of moments is automatically satisfied.

We shall neglect all quantities of the order $(\Delta x)^2$ in the presence of Δx, as Δx is assumed to be infinitely small. The precise significance of the assumption that the motion of the string is small, is as follows: If we do not make such an assumption the partial differential equation obtained will be very complicated. In particular it will be non-linear. If we acknowledge that we must be content with an approximation to this equation, and thus only seek approximate solutions to the actual physical problem, we shall neglect terms in $\left(\frac{\partial u}{\partial x}\right)^2, \frac{\partial u}{\partial x}\frac{\partial^2 u}{\partial x^2} \ldots$, which are of the "second order" in the presence of terms of the "first order" such as $\frac{\partial^2 u}{\partial x^2}$. We shall thus obtain equation (VII, 1; 8), as we shall easily show.

The smaller the values of $\frac{\partial u}{\partial x}$ and $\frac{\partial^2 u}{\partial x^2}$, the smaller will be the relative error involved. Moreover, we are able to limit the error that is introduced, for instance by replacing the exact equation by the approximate one (VII, 1; 8). For example, in the case of a violin string, by bearing in mind the usual amplitudes of vibration of such a string, we can see whether such an error is acceptable. If it is not, then it will be necessary to calculate a correcting term. The same situation exists for the case of a vibrating pendulum. The exact differential equation for the pendulum is

$\theta'' + (g/l) \sin \theta = 0$. For small oscillations about a position of equilibrium $\theta = 0$ we can replace this equation by $\theta'' + (g/l)\theta = 0$ which gives as the value for the period of the vibrations $T_0 = 2\pi\sqrt{l/g}$. This we find is only the limit of the true period as the amplitude α approaches zero. For any given motion of amplitude α we introduce, on the evaluation of the true period T, a relative error by replacing it by T_0, but this error is very rapidly reduced as α gets smaller. If this error is not acceptable then a correcting term will have to be added, and such a correcting term is known to be $T_0\alpha^2/16$. However it is possible that this term itself may not be sufficient. The true period T is a transcendental function of α which is expressible in terms of elliptic integrals as

$$T = \sqrt{\frac{2l}{g}} \int_{-\alpha}^{+\alpha} \frac{d\theta}{\sqrt{\cos\theta - \cos\alpha}}.$$

In the problem of vibrating strings, matters are more complicated, and we shall have to restrict ourselves to the case where the vibrations are infinitely small. The quantities which will have to be small are $\partial u/\partial x$ (a magnitude having no dimension which gives the gradient of the string) and $\partial^2 u/\partial^2 x$ (a magnitude of dimension l^{-1}). Under these conditions:

1) The vector $m\vec{\gamma}$ is in the plane yOz and is given by

$$\text{(VII, 1; 1)} \qquad m\vec{\gamma} = \rho_0 \frac{\partial^2 \vec{u}}{\partial t^2} \Delta x.$$

(Actually the mass of the element considered is constant throughout the motion and takes its equilibrium value $\rho_0 \Delta x$.)

2) The force exerted by the portion M′B of the string at M′ is $T(x + \Delta x)$. Its component along Ox is $T(x + \Delta x) \cos \alpha$ and its component in the plane yOz is $T\vec{\theta}$, where $\vec{\theta}$ is the projection on yOz of the unit vector along the tangent to the string at M′ oriented in the sense AB. The force exerted by the portion of the string AM at M is similar, but now it will be relative to x instead of $x + \Delta x$ and the tangent will be oriented in the opposite sense. If we neglect terms in $(\Delta x)^2$ the sum of these two forces can be written as

(VII, 1; 2) $\dfrac{\partial}{\partial x}(T \cos \alpha)\Delta x$ as the component along Ox,

$\dfrac{\partial}{\partial x}(T\vec{\theta})\Delta x$ as the component in the plane yOz.

The equation of motion along Ox is

$$\text{(VII, 1; 3)} \qquad \frac{\partial}{\partial x}(T \cos \alpha) = 0.$$

If we neglect $(\cos\alpha - 1)$ and its derivatives with respect to x, which are infinitesimals of the second order, the equation reduces to

(VII, 1; 4) $$\frac{\partial T}{\partial x} = 0.$$

Since T is assumed to be in the neighbourhood of T_0 this equation is only possible as a result of our hypothesis that $dT_0/dx = 0$, that is, T_0 is independent of x. Thus for all time t, the tension must be the same at one end of the string as at the other, and is therefore a function $T(t)$ of t oscillating about the value T_0. It is not possible to calculate this quantity without making new mechanical hypotheses. We shall see later (VII, 1; 25) that to exclude longitudinal vibrations we must have

(VII, 1; 5) $$T = T_0.$$

The equation of motion projected on yOz will give

(VII, 1; 6) $$\rho_0 \frac{\partial^2 \vec{u}}{\partial t^2} = \frac{\partial}{\partial x}(T_0\vec{\theta}).$$

The vector $\vec{\theta}$ has components dy/ds and dz/ds along the axes Oy and Oz. As the angle made by the tangent with Ox is small, dx/ds is in the neighbourhood of 1 and dy/ds and dz/ds are equivalent to dy/dx and dz/dx. Thus $\vec{\theta}$ is equivalent to $\partial\vec{u}/\partial x$.

Taking account of equation (VII, 1; 5), equation (VII, 1; 6) becomes

(VII, 1; 7) $$\rho_0 \frac{\partial^2 \vec{u}}{\partial t^2} = T_0 \frac{\partial^2 \vec{u}}{\partial x^2}.$$

If, as we have assumed, ρ_0 is independent of x (*homogeneous string*) this equation can be written

(VII, 1; 8) $$\boxed{\frac{1}{v^2}\frac{\partial^2 \vec{u}}{\partial t^2} = \frac{\partial^2 \vec{u}}{\partial x^2}}$$

where v has the dimensions of a velocity :

(VII, 1; 9) $v = \sqrt{\dfrac{T_0}{\rho_0}}$, dimensionally $\dfrac{(mlt^{-2})^{\frac{1}{2}}}{(ml^{-1})^{\frac{1}{2}}} = lt^{-1}$.

As we shall see later, v may be called the *velocity of propagation of transverse waves in the string*. We can also find a use for the tension per unit cross-sectional area, $T_0' = T_0/\sigma_0$. As $\rho_0 = \sigma_0\varpi_0$ we have

(VII, 1; 10) $$v = \sqrt{\frac{T_0}{\sigma_0\varpi_0}} = \sqrt{\frac{T_0'}{\varpi_0}}.$$

Equation (VII, 1; 8) is called the *equation of vibrating strings.* If we resolve successively along Oy and Oz we obtain for the two functions $y(x, t)$ and $z(x, t)$ the same equation

$$\text{(VII, 1; 11)} \qquad \frac{1}{v^2}\frac{\partial^2 y}{\partial t^2} = \frac{\partial^2 y}{\partial x^2}, \qquad \frac{1}{v^2}\frac{\partial^2 z}{\partial t^2} = \frac{\partial^2 z}{\partial x^2}.$$

Note that the portion Δx of the string has length $\Delta x/\cos\alpha$. If we neglect α^2 as in the previous analysis we see that this length is approximately equal to the length of the portion when at rest. Thus the density ρ does not vary; it is a constant and is equal to ρ_0 (neglecting second-order terms). Finally, the three entirely separate equations in the unknowns u, T, ρ, are

$$\text{(VII, 1; 12)} \qquad \begin{aligned} &\frac{1}{v^2}\frac{\partial^2 \vec{u}}{\partial t^2} = \frac{\partial^2 \vec{u}}{\partial x^2}, \qquad v = \sqrt{\frac{T_0}{\rho_0}}, \\ &T = T_0, \\ &\rho = \rho_0. \end{aligned}$$

The second and third equations are completely solved; only the first equation demands a solution.

Longitudinal vibrations of a solid bar

We shall now consider the *longitudinal* vibrations of a uniform, solid metal rod AB of length L and constant cross-sectional area σ_0. Let x be the coordinate value of a point M of the rod when at rest, the axis Ox of x being along $\overrightarrow{AB}$ with the origin O at A. During its motion the point M becomes displaced longitudinally. Let this displacement be u. Then $u = \overline{MM'} = \overline{OM'} - \overline{OM}$, where M′ is the position of M at time t. Here, again u is a function of x and t which we will have to determine.

Once again let T be the tension in the rod at the point x at time t, defined as for transverse vibrations. The difference between this and the preceding case is, however, important. The flexible string was only stretched due to the existence of a sufficiently large mean tension $T_0 > 0$. Here, on the contrary, the rod is not necessarily stretched: $T_0 = 0$. (Note that tension can also be called traction.) It could as well be compressed ($T_0 < 0$). The variation in tension is a reaction by the rod against any tendency to modify its shape by lengthening or shortening. We shall accept the following law, called the *fundamental law of elasticity*: For any element of length l_0 subjected to a mean tension T_0, an extension $l - l_0 = \delta l$ requires an increase in tension $T - T_0 = \delta T$ given by

$$\text{(VII, 1; 13)} \qquad \frac{\delta T}{\sigma_0} = E_0 \frac{\delta l}{l_0},$$

where E_0 is Young's modulus or coefficient of elasticity under mean tension T_0 (E_0 varies with T_0; for $T_0 = 0$ the value of E_0 will be Young's modulus at rest) and σ_0 is the cross-sectional area. Naturally this law is true only if $\delta l/l_0$ is sufficiently small. Thus it is necessary to calculate the relative dilatation $\lambda = \delta l/l_0$ at each point of the rod. The segment $(x, x + \Delta x)$ of length l_0 at rest, occupies, at time t, the position

$$\left(x + u(x, t),\ x + \Delta x + u(x + \Delta x, t)\right);$$

its length varies from $l_0 = \Delta x$ to $l = l_0 + \delta l = \Delta x + \frac{\partial u}{\partial x}\Delta x$. Thus the dilatation is

(VII, 1; 14) $$\lambda = \frac{\delta l}{l_0} = \frac{\partial u}{\partial x}.$$

(The fact that δl has to be small compared with l_0 means that $\partial u/\partial x$ has to be small.) Thus, from the law of elasticity, the tension at x at time t satisfies

(VII, 1; 15) $$T - T_0 = E_0 \sigma_0 \frac{\partial u}{\partial x}.$$

Again, let ρ be the linear density, which is approximately equal to ρ_0, the equilibrium density under a tension T_0. Writing the fundamental equation of dynamics $\vec{F} = m\vec{\gamma}$ for the element $(x, x + \Delta x)$ and resolving along Ox (no other equation is necessary, since the motion is longitudinal), we get

(VII, 1; 16) $$\rho_0 \frac{\partial^2 u}{\partial t^2}\Delta x = \frac{\partial T}{\partial x}\Delta x$$

or, using equation (VII, 1; 15),

(VII, 1; 17) $$\rho_0 \frac{\partial^2 u}{\partial t^2} = E_0 \sigma_0 \frac{\partial^2 u}{\partial x^2},$$

which is again the (scalar and not vectorial) equation of vibrating strings (VII, 1; 8) with the velocity of propagation

(VII, 1; 18) $$v = \sqrt{\frac{E_0 \sigma_0}{\rho_0}} = \sqrt{\frac{E_0}{\varpi_0}}.$$

Young's modulus plays the same part here as does T_0', the tension per unit cross-sectional area, in the transverse vibrations. Moreover, E_0 has the dimensions of $\delta T/\sigma_0$ (VII, 1; 13), the tension per unit cross-sectional area. The fundamental difference is that in the transverse case it is T_0', the mean tension per unit cross-sectional area which is used, while in the longitudinal case it is E_0, the coefficient linking the dilatation $\lambda = \delta l/l_0$ and the difference

$(T - T_0)/\sigma_0$ of the tension per unit cross-sectional area over its mean value. It is for this reason that we can imagine transverse vibrations of a flexible string when $T_0 > 0$ and longitudinal vibrations when $T_0 \leqslant 0$.

By way of an example we know that, for steel, E_0 is of the order 20,000 kg/mm. This gives in MKS units

$$v = \sqrt{\frac{20 \times 9.81}{10^{-6} \times 7.8}} \doteqdot 5\,000 \text{ m/s}.$$

From the definition of linear density each small element satisfies

(VII, 1; 19) $$\rho l = \rho_0 l_0$$

(conservation of mass); thus

(VII, 1; 20) $$\frac{\delta l}{l_0} + \frac{\delta \rho}{\rho_0} = 0.$$

Therefore

(VII, 1; 21) $$\rho - \rho_0 = \delta\rho = -\rho_0 \frac{\delta l}{l_0} = -\rho_0 \lambda = -\rho_0 \frac{\partial u}{\partial x},$$

(VII, 1; 22) $$\rho = \rho_0\left(1 - \frac{\partial u}{\partial x}\right) \quad \text{and also} \quad \varpi = \varpi_0\left(1 - \frac{\partial u}{\partial x}\right).$$

The equations of the problem are, finally,

(VII, 1; 23) $$\begin{cases} \dfrac{1}{v^2}\dfrac{\partial^2 u}{\partial t^2} = \dfrac{\partial^2 u}{\partial x^2}, \quad v = \sqrt{\dfrac{E_0\sigma_0}{\rho_0}} = \sqrt{\dfrac{E_0}{\varpi_0}}, \\ T = T_0 + E_0\,\sigma_0 \dfrac{\partial u}{\partial x}, \\ \rho = \rho_0 - \rho_0 \dfrac{\partial u}{\partial x}. \end{cases}$$

We shall solve initially the first equation (ρ_0, σ_0, E_0 being known); then if T_0 is also given the two following equations will be solved. However, we notice that $U = \partial u/\partial x$ also satisfies the equation of vibrating strings (by differentiating both sides with respect to x), and since a constant T_0 or ρ_0 also satisfies it, T and ρ satisfy the same equation:

(VII, 1; 24) $$\frac{1}{v^2}\frac{\partial^2 T}{\partial t^2} = \frac{\partial^2 T}{\partial x^2}, \qquad \frac{1}{v^2}\frac{\partial^2 \rho}{\partial t^2} = \frac{\partial^2 \rho}{\partial x^2}.$$

Longitudinal vibrations of solid rods constitute the essence of the phenomenon of propagation of sound in the rod ; the values of E (Young's modulus) give the velocity of sound in the various solids.

If we wish to examine the case of a rod subjected simultaneously to longitudinal and transverse vibrations, we can, in conformity with the (mechanical) hypotheses made in these last two problems, treat the mixed problem by calling $\vec{u}$ the transverse displacement and w the longitudinal displacement, and thus easily obtain the equations

$$\frac{1}{v_t^2}\frac{\partial^2 \vec{u}}{\partial t^2} = \frac{\partial^2 \vec{u}}{\partial x^2}, \qquad v_t = \sqrt{\frac{T_0}{\rho_0}},$$

(VII, 1; 25)
$$\frac{1}{v_l^2}\frac{\partial^2 w}{\partial t^2} = \frac{\partial^2 w}{\partial x^2}, \qquad v_l = \sqrt{\frac{E_0\sigma_0}{\rho_0}},$$
$$T = T_0 + E_0\sigma_0\frac{\partial w}{\partial x},$$
$$\rho = \rho_0 - \rho_0\frac{\partial w}{\partial x}.$$

Equation (VII, 1; 25) reduces to (VII, 1; 23) when $\vec{u} = \vec{0}$ and to (VII, 1; 12) when $w = 0$. We see that longitudinal vibrations can be excluded only when $T = T_0$. The first two equations can be solved independently of each other and correspond to two very different velocities v_1 and v_2. The third and fourth equations can be solved once the second has been solved.

Although a seismic shock may not be treated exactly as the propagation of a shock along a rod obeying the above laws, we know that the longitudinal shock is not propagated with the same velocity as the transverse shock.

Longitudinal vibrations of a column of liquid or gas

Consider a column of liquid or gas enclosed in a tube under constant pressure, and examine the longitudinal vibrations, due to which each section of liquid lying between two neighbouring vertical planes is displaced as a unit. The liquid cannot be assumed incompressible, for if this were the case, there would be no longitudinal waves. Also, as in the case of solid rods we have to allow only small deformations.

The coefficient of compressibility χ_0 of the liquid at a mean pressure $p_0 > 0$ gives the variation of pressure corresponding to a dilatation $\lambda = \delta V/V_0$:

(VII, 1; 26)
$$\frac{\delta p}{p_0} = \frac{p - p_0}{p_0} = -\frac{1}{\chi_0}\frac{\delta V}{V_0}.$$

This corresponds to a fundamental law of the type

(VII, 1; 27)
$$pV^{\frac{1}{\chi_0}} = \text{constant};$$

χ_0 is dimensionless.

As the motion is longitudinal, $\delta V/V_0 = \delta l/l_0$. On the other hand the pressure is opposed by the tension per unit cross-sectional area

$$\delta p/p_0 = \delta T/T_0, \qquad p_0 = -T_0/\sigma_0;$$

thus (VII, 1 ; 26) is identical with equation (VII, 1; 13) where

$$\text{(VII, 1; 28)} \qquad E_0 = p_0/\chi_0.$$

In the case of the transverse vibrations of a flexible string we necessarily had $T_0 > 0$; for longitudinal vibrations we could have any T whatever; here $p_0 > 0$, thus $T < 0$.

Calling u the longitudinal displacement, we have therefore equations (VII, 1; 23) modified as follows :

$$\text{(VII, 1; 29)} \qquad \begin{cases} \dfrac{1}{v^2}\dfrac{\partial^2 u}{\partial t^2} = \dfrac{\partial^2 u}{\partial x^2}, & v = \sqrt{\dfrac{p_0}{\chi_0 \varpi_0}}, \\ p = p_0 - \dfrac{p_0}{\chi_0}\dfrac{\partial u}{\partial x}, \\ \rho = \rho_0 - \rho_0 \dfrac{\partial u}{\partial x}, \end{cases}$$

and as a result p and ρ will also satisfy the equation of vibrating strings. In the case of a column of gas the vibrations are the same phenomena as those occurring in organ pipes. The law of compressibility in this case will be the adiabatic law, because the oscillations are very rapid, many hundreds of thousands per second, and they will certainly take place without any heat exchange with the exterior. Thus

$$\text{(VII, 1; 30)} \qquad pV^\gamma = \text{constant}, \qquad \gamma = \frac{C}{c},$$

where

$$\chi_0 = \frac{1}{\gamma}, \qquad E_0 = \gamma p_0.$$

Therefore we have the equations corresponding to (VII, 1; 29) with

$$\text{(VII, 1; 31)} \qquad v = \sqrt{\frac{\gamma p_0}{\varpi_0}}.$$

Assume that the gas obeys the perfect gas law, then

$$\text{(VII, 1; 32)} \qquad p_0 = \frac{\varpi_0}{M} RT_0,$$

where

$$\text{(VII, 1; 33)} \qquad v = \sqrt{\frac{\gamma}{M} RT_0}.$$

Thus

The velocity of sound in a perfect gas is proportional to the square of the absolute temperature.

Exercise
Calculate the velocity of sound in air.

If we look for the temperature T of the gas at a point x at time t, we obtain from the law (VII, 1; 32), when differentiated once,

$$\frac{\delta p}{p_0} - \frac{\delta \varpi}{\varpi_0} - \frac{\delta T}{T_0} = 0. \tag{VII, 1; 34}$$

Knowing the equations that are satisfied by p and ϖ we obtain for T,

$$-\gamma \frac{\partial u}{\partial x} + \frac{\partial u}{\partial x} - \frac{\delta T}{T_0} = 0$$

or

$$T = T_0 - (\gamma - 1) T_0 \frac{\partial u}{\partial x}. \tag{VII, 1; 35}$$

Hence we have finally the system of equations

$$\begin{gathered} \frac{1}{v^2} \frac{\partial^2 u}{\partial t^2} = \frac{\partial^2 u}{\partial x^2}, \qquad v = \sqrt{\frac{\gamma p_0}{\varpi_0}} = \sqrt{\frac{\gamma}{M} R T_0}, \\ \varpi = \varpi_0 - \varpi_0 \frac{\partial u}{\partial x}, \qquad p = p_0 - \gamma p_0 \frac{\partial u}{\partial x}, \\ T = T_0 - (\gamma - 1) \frac{\partial u}{\partial x}, \end{gathered} \tag{VII, 1; 36}$$

and as a result, ρ, ϖ, T also satisfy the equation of vibrating strings.

2. SOLUTION OF THE EQUATION OF VIBRATING STRINGS BY THE METHOD OF TRAVELLING WAVES. CAUCHY'S PROBLEM

From now on we shall not be concerned with the particular physical problem which gives rise to our equation, we shall simply call the partial differential equation

$$\frac{1}{v^2} \frac{\partial^2 u}{\partial t^2} - \frac{\partial^2 u}{\partial x^2} = 0, \tag{VII, 1; 37}$$

the equation of vibrating strings. The string being infinitely long, lying when at rest along the x axis, u is therefore a function that is defined for all x and t.

To find a general solution, make the substitution

(VII, 1; 38) $$x + vt = \xi \quad \text{or} \quad x = \frac{\xi + \eta}{2},$$

$$x - vt = \eta \quad \text{or} \quad t = \frac{1}{v}\frac{\xi - \eta}{2}.$$

We then have

(VII, 1; 39) $$\frac{\partial}{\partial x} = \frac{\partial}{\partial \xi} + \frac{\partial}{\partial \eta}, \quad \text{hence} \quad \frac{\partial^2}{\partial x^2} = \frac{\partial^2}{\partial \xi^2} + 2\frac{\partial^2}{\partial \xi \partial \eta} + \frac{\partial^2}{\partial \eta^2},$$

$$\frac{\partial}{\partial t} = v\left(\frac{\partial}{\partial \xi} - \frac{\partial}{\partial \eta}\right), \quad \text{hence} \quad \frac{\partial^2}{\partial t^2} = v^2\left(\frac{\partial^2}{\partial \xi^2} - \frac{2\partial^2}{\partial \xi \partial \eta} + \frac{\partial^2}{\partial \eta^2}\right),$$

so that

(VII, 1; 40) $$\frac{1}{v^2}\frac{\partial^2}{\partial t^2} - \frac{\partial^2}{\partial x^2} = -4\frac{\partial^2}{\partial \xi \partial \eta},$$

and in the new system of variables the equation becomes

(VII, 1; 41) $$\frac{\partial^2 u}{\partial \xi \partial \eta} = 0.$$

This equation can be solved immediately as follows :

(VII, 1; 42) $$\begin{cases} \dfrac{\partial u}{\partial \xi} = f_1(\xi), & f_1 \text{ an arbitrary function of } \xi, \\ u = f(\xi) + g(\eta), & f \text{ the primitive of } f_1 \text{ and } g \\ & \text{an arbitrary function of } \xi, \end{cases}$$

and gives as the general solution of equation (VII, 1; 37)

(VII, 1; 43) $$u(x, t) = f(x + vt) + g(x - vt),$$

where f and g are arbitrary functions of one variable. In fact the word "arbitrary" is too loose; these functions must be twice continuously differentiable in order that the change of variable (VII, 1; 39) be permissible. In any case the second derivatives of u occur in (VII, 1; 37).

If we look for solutions of (VII, 1; 37) *in the distribution sense* we are led to accept functions f and g of one variable which are not necessarily differentiable but only locally summable, or even distributions of one variable. The solution so obtained has a noteworthy physical interpretation. A function such as $g(x - vt)$ takes at the point x at time t the value which it used to have at the point $(x - vt)$ at the instant 0, as long as the function $g(x - vt)$ of x, which represents at each instant of time the entire shape of the string, is the function $g(x)$ translated by an amount $+ vt$. Alternatively we say that $u = g(x - vt)$ represents *a wave which is travelling to the right with a velocity* v. Similarly, $u = f(x + vt)$ will be *a wave travelling to the left with a velocity* v. Thus the complete wave, that is the complete solution of equation (VII, 1; 37), is a superposition of two arbitrary wave motions, one being propagated to the left and the other to the right with velocity v. This is why v is aptly called the *velocity of wave propagation.* We know that there is naturally no transport of material with velocity v. In the case of transverse vibrations of a string, u is the transverse displacement and v the velocity of longitudinal propagation. Even in the case of longitudinal vibrations there is no connection between v and the velocity of displacement of the material $\partial u/\partial t$. For example take the solution $u = x - vt$. The wave is propagated with velocity $+ v$ whilst the algebraic velocity of the particles of the material is $\partial u/\partial t = - v$. More generally, if $u = g(x - vt)$, the velocity of the particles is

$$\frac{\partial u}{\partial t} = - vg'(x - vt),$$

which is a function varying with x and t.

Cauchy's problem

Cauchy's problem can be stated as follows : To find a solution of the equation of vibrating strings satisfying the following initial conditions,

(VII, 1; 44)
$$\begin{aligned} u(x, 0) &= u_0(x) \quad \text{(shape of the string at } t = 0\text{)},\\ \frac{\partial u}{\partial t}(x, 0) &= u_1(x) \quad \text{(distribution of velocities along the string at } t = 0\text{)}. \end{aligned}$$

As the solution has the form (VII, 1; 43) it is necessary to determine f and g such that they will satisfy the initial conditions (VII, 1; 44). This will yield the equations,

(VII, 1; 45)
$$\begin{cases} u_0(x) = f(x) + g(x), \\ u_1(x) = v(f'(x) - g'(x)). \end{cases}$$

The second equation can be integrated to give

(VII, 1; 46)
$$\int_0^x u_1(\xi)\, d\xi = v(f(x) - g(x)) + C.$$

The existence of this constant C is not surprising. In the representation (VII, 1; 43) of u we can add to f and subtract from g an arbitrary constant without altering the solution. This leaves $f + g$ unchanged and modifies $f - g$ to the extent of an arbitrary constant. Whatever the choice of C the solution for u will be the same. Thus from equation (VII, 1; 46) and the first equation of (VII, 1; 45) we have

(VII, 1; 47)
$$\begin{cases} f(x) = \dfrac{1}{2}\left(u_0(x) + \dfrac{1}{v}\displaystyle\int_0^x u_1(\xi)\, d\xi\right) - \dfrac{1}{2}\dfrac{C}{v}, \\ g(x) = \dfrac{1}{2}\left(u_0(x) - \dfrac{1}{v}\displaystyle\int_0^x u_1(\xi)\, d\xi\right) + \dfrac{1}{2}\dfrac{C}{v}, \end{cases}$$

which gives as the required solution, on substituting in (VII, 1; 43),

$$\text{(VII, 1; 48)} \quad u(x, t) = \frac{1}{2}[u_0(x + vt) + u_0(x - vt)] + \frac{1}{2v}\int_{x-vt}^{x+vt} u_1(\xi)\, d\xi.$$

We see that Cauchy's problem has one solution and one solution only. However, if we require that this solution be twice continuously differentiable it is necessary to suppose that u_0 is twice differentiable and that u_1 can be differentiated once, and this is quite usually the case.

Notes on the physical aspect of the solution

1) The value of u at the point x at time t depends only on the values of u_0 and u_1 of u and $\partial u/\partial t$ at points in the interval $(x - vt, x + vt)$ at the instant $t = 0$. This agrees with what we have already seen with respect to velocity of propagation. In fact, u_1 takes on all values in the complete interval $(x - vt, x + vt)$ though u_0 takes only the values at the extremities of the range.

2) A geometric figure gives a better interpretation of this phenomenon. Take axes with abscissa x and ordinate t and suppose that $t \geqslant 0$. If through the point (x, t) we take two half-lines of slope $\pm 1/v$ directed downwards, then these lines will cut the x axis in two points, $x - vt$ and $x + vt$ (Fig. VII, 1).

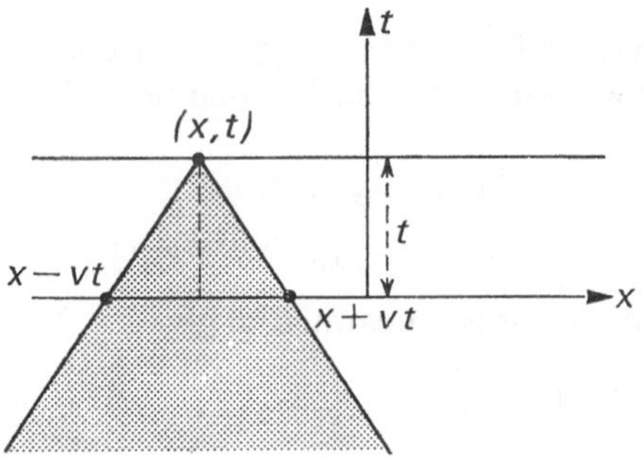

Figure (VII, 1)

The region between the two half-lines is called the "backward wave cone" of the point (x, t). If instead of solving the Cauchy problem for an initial instant $t = 0$ we solve for an initial instant τ, $-\infty < \tau < t$, the solution will be as follows:

$$\begin{aligned} u(x, t) &= \frac{1}{2}[u_0(x + v(t - \tau)) + u_0(x - v(t - \tau))] \\ \text{(VII, 1; 49)} \qquad &+ \frac{1}{2v}\int_{x-v(t-\tau)}^{x+v(t-\tau)} u_1(\xi)\, d\xi, \end{aligned}$$

where

$$u_0(x) = u(x, \tau), \qquad u_1(x) = \frac{\partial u}{\partial t}(x, \tau).$$

This time the interval of the x-axis used is $(x - v(t - \tau), x + v(t - \tau))$, which is the intersection of the line having ordinate τ with the same backward wave cone at the point (x, t).

3) We can solve Cauchy problems for forward waves as well as Cauchy problems for backward waves, which we have already solved. We say that the initial instant τ, for which we are given $u_0 = u$ and $u_1 = \partial u/\partial t$ is later than $t : \tau \geqslant t$. The solution is effected in an analogous manner and the equation to be used is again (VII, 1; 49). But allowing for the fact that $\tau \geqslant t$, it is much better to write

(VII, 1; 50)
$$u(x, t) = \frac{1}{2}[u_0(x + v(\tau - t)) + u_0(x - v(\tau - t))] - \frac{1}{2v}\int_{x-v(\tau-t)}^{x+v(\tau-t)} u_1(\xi)\, d\xi.$$

This time the interval used will be the intersection of the line $t = \tau$ with the *forward wave cone* at (x, t). That is, the region between the two straight lines having a slope of $\pm 1/v$ but directed upwards from an origin (x, t). This is natural since $t \leqslant \tau$ and what occurs at time t will influence what will take place at the future instant τ. The forward wave cone at the point (x, t) will enclose all those points (ξ, τ) whose motion will be influenced by what occurred at the point x at the time t. These considerations of the past and future are purely subjective from the point of view of the mathematical equation because the latter is invariant under the change t to $-t$.

4) The wave front is the extreme point reached by the wave. More exactly, the plane wave front at time t is the upper limit of the support of u and $\partial u/\partial t$ considered as functions of x, at time t. The coordinate value X of this plane wave front is a function $X(t)$ of t. This function satisfies a linear inequality, because the wave front cannot be propagated with a velocity greater than v. Thus for $t \geqslant \tau$,

(VII, 1; 51)
$$X(t) \leqslant X(\tau) + v(t - \tau).$$

Actually, we call u_0 and u the values of u and $\partial u/\partial t$ at the time τ. The formula (VII, 1; 49) shows that since u_0 and u_1 are zero for $x \geqslant X(\tau)$, u and $\partial u/\partial t$ are also zero for $x \geqslant X(\tau) + v(t - \tau)$. Hence equation (VII, 1; 51). But we notice that if the time scale is inverted [or by using

(VII, 1; 50) instead of (VII, 1; 49) and interchanging t and τ], we shall obtain the inequality (for $\tau \leqslant t$ only)

(VII, 1; 52) $$X(\tau) \leqslant X(t) + v(t - \tau),$$

which also gives

(VII, 1; 53) $$X(t) \geqslant X(\tau) - v(t - \tau).$$

Or, to put it in another way, the plane wave is incapable of being propagated to the left with a velocity in excess of v. Thus it will not be necessary to think of the plane wave front as being propagated to the right with a single velocity v, for, if this were so then (VII, 1; 51) would always be an equality and so also would (VII, 1 ; 52), and these two equalities contradict each other.

However, take a simple example where for $t = 0$ the initial conditions are given by

(VII, 1; 54) $$u_0(x) > 0 \text{ for } |x| < x_0, \qquad u_0(x) = 0 \text{ for } |x| \geqslant x_0; \\ u_1(x) \equiv 0.$$

The equations (VII, 1; 49) and (VII, 1; 50) for $\tau = 0$ show that the right-hand wave front is given by

(VII, 1; 55) $$X(t) = x_0 + v|t|,$$

that it travels with a velocity v for $t \geqslant 0$ and with a velocity $-v$ for $t \leqslant 0$ and that it is at time $t = 0$ that it reaches the leftmost point of its travel.

What we have said about the right-hand wave front is also true for the left-hand wave front, which satisfies the same inequalities. In the example (VII, 1; 54) the left-hand wave front is given by

(VII, 1; 56) $$Y(t) = -x_0 - v|t|,$$

and it is at $t = 0$ that it reaches the rightmost point of its travel. Therefore, it is at time $t = 0$ that the part of the string which is not at rest is the smallest possible. Note that, in this example, when $t > x^2/v$ the moving part of the string is separated into two distinct intervals

$$(-x_0 - vt, \ x_0 - vt) \text{ and } (-x_0 + vt, \ x_0 + vt).$$

5) We say that there is a *diffusion of waves*, because $u(x, t)$ depends (through its derivative u_1) on the initial conditions in the whole interval

$(x - vt, x + vt)$ and not only upon their values at the extremities of the range.

6) A particularly important case is that given by the initial conditions

$$\text{(VII, 1; 57)} \quad u_0 = 0, \qquad u_1 = v^2\delta(x), \qquad \delta = \text{Dirac distribution.}$$

If we assume without any restriction the validity of equation (VII, 1; 48), we find that for $t > 0$,

$$\text{(VII, 1; 58)} \quad u(x, t) = \begin{cases} \dfrac{v}{2} & \text{for } |x| < vt \\ 0 & \text{for } |x| > vt \end{cases} \quad \text{or} \quad u(x, t) = \frac{v}{2} \mathrm{Y}(vt - |x|),$$

where Y is Heaviside's unit function. Alternatively, we say that u is equal to $v/2$ in the forward wave cone at the origin, and is zero outside for $t \geqslant 0$. We will meet this particular solution again, later on.

String having one end fixed. Reflection of waves

We shall now assume that the string, instead of being infinitely long, has one extremity at the origin. Alternatively, we say that it is at rest along the semi-axis for $x \geqslant 0$, and that u is a function defined for $x \geqslant 0$, $-\infty < t < \infty$. Equation (VII, 1; 43) is again easily seen to be valid. The functions f and g of one variable are well defined, to within constants of opposite sign [see equation (VII, 1; 47)] *for all values of the variable,* because for x fixed and greater than zero, as t varies from $-\infty$ to $+\infty$, $x + vt$ varies from $-\infty$ to $+\infty$. Consequently equation (VII, 1; 43) gives a solution u, which is defined for *all* values of x and t. Thus, *every solution u of the equation in the region $x \geqslant 0$, $-\infty < t < \infty$, can be extended in a unique manner to a solution $\bar{u}$ which is defined over the whole (x, t) plane.*

We shall look for those motions for which the extremity $x = 0$ of the string remains at rest. The extension $\bar{u}$ is a motion that is defined for all x and t for which $\bar{u}(0, t) \equiv 0$. For this to be so, it is necessary and sufficient that

$$\text{(VII, 1; 59)} \qquad f(vt) + g(-vt) \equiv 0.$$

Alternatively we say that the functions f and g of one variable s are no longer both arbitrary, for if one is still arbitrary the other is known, since

$$\text{(VII, 1; 60)} \quad f(s) + g(-s) = 0, \qquad \text{or} \qquad g(s) = -f(-s).$$

Then, we see from equation (VII, 1; 43), that the general solution is given by

$$\text{(VII, 1; 61)} \qquad \bar{u}(x, t) = f(x + vt) - f(-x + vt),$$

where f is an arbitary function of one variable. We see that, for all t, $\bar{u}$ *is an odd function of* x, although this leads to a cancellation at $x = 0$. Thus every solution u of the equation of vibrating strings, defined for $x \geqslant 0$, $-\infty < t < +\infty$, and identically zero for $x = 0$, can be extended in a unique manner to a solution $\bar{u}$, which is defined everywhere and which for all t is an odd function of x.

$$\text{(VII, 1; 62)} \qquad \bar{u}(-x, t) = -\bar{u}(x, t).$$

Notice from equation (VII, 1; 61) that $\bar{u}$ is a superposition of two waves $\bar{u}_1$ and $\bar{u}_2$ travelling in opposite directions with velocity v and satisfying for all t

$$\text{(VII, 1; 63)} \qquad u_1(-x, t) = -\bar{u}_2(x, t).$$

We can interpret this statement by saying that there is *a reflection of the waves at the point* $x = 0$ *accompanied by a change in sign.*

Let us assume that for $t \leqslant 0$ the motion of the string (as applied to the region $x \geqslant 0$) is a wave such that u travels to the left with a velocity v. We must first of all extend this solution, for $t \leqslant 0$, to the entire x-axis in such a way that, for the wave so obtained, $\bar{u}$ will be an odd function of x. This amounts to placing on an imaginary part of the string, for which $x \leqslant 0$, a wave opposite in sign to the first and travelling to the

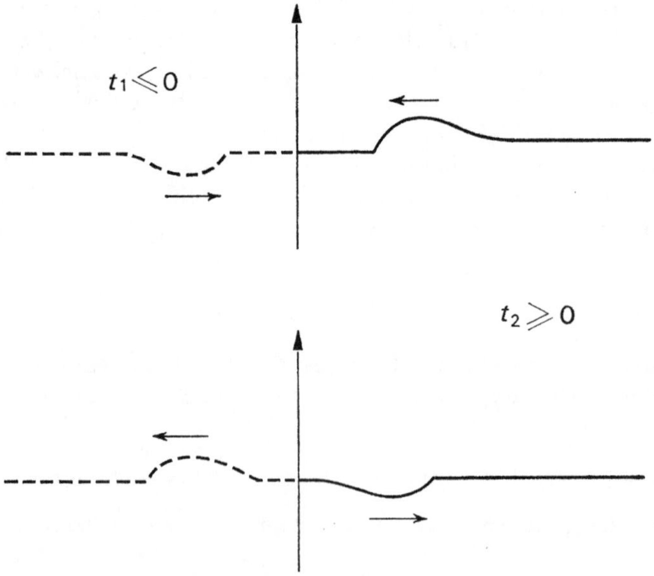

Figure (VII, 2)

right with a velocity v. Thus the solution defined everywhere for $t \leqslant 0$ can then be extended, in a unique manner for $t \geqslant 0$, by following indefinitely the propagation of the preceding waves. Consequently, while the initial wave on the actual string, for which $x \geqslant 0$, travels towards the imaginary part of the string, for which $x \leqslant 0$, and thus progressively disappears, the wave of opposite sign, situated initially on the imaginary part of the string ($x \leqslant 0$) enters progressively the region $x \geqslant 0$ occupied by the actual string. Hence, on the actual string it is as though the initial wave has been reflected at the point $x = 0$ of the string so that it returns along the string in the opposite direction together with a change in sign [Figure (VII, 2) in the (x, u) plane]. Naturally for certain values of t (and even for all values of t if the wave motion is over the whole string) there is a superposition of the incident and reflected waves, which will always give $u(0, t) \equiv 0$.

Application. A Cauchy problem is given by the initial values

$$u_0(x) = u(x, \tau),\ u_1(x) = \frac{\partial u}{\partial t}(x, \tau),$$

and the boundary condition $u(0, t) = 0$.

The solution for $t \geqslant \tau$ is obtained from equation (VII, 1; 49) after having extended u_0 and u_1 to $\bar{u}_0$ and $\bar{u}_1$ in such a way that they may be taken as odd functions of x. The equation (VII, 1; 49) introduces only the nonextended functions u_0 and u_1 if

$$x - v(t - \tau) \geqslant 0.$$

If $x - v(t - \tau) < 0$, taking account of the fact that $\bar{u}_0$ and $\bar{u}_1$ are odd functions, we modify the equation so as to use only u_0 and u_1; hence

(VII, 1; 64) $$u(x, t) = \frac{1}{2}\left[u_0(x + v(t - \tau)) - u_0(v(t - \tau) - x)\right] + \frac{1}{2v}\int_{v(t-\tau)-x}^{v(t-\tau)+x} u_1(\xi)\,d\xi.$$

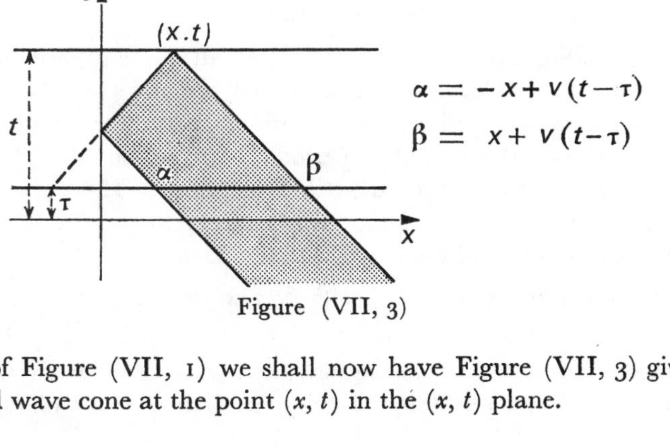

Figure (VII, 3)

Instead of Figure (VII, 1) we shall now have Figure (VII, 3) giving the backward wave cone at the point (x, t) in the (x, t) plane.

The left-hand ray of this cone undergoes a reflection at the t-axis. The values of ξ for which it is necessary to know u_0 and u_1 and so be able to determine $u(x, t)$ are in the interval (α, β), the intersection of this cone with the line $t = \tau$.

String with two ends fixed

The cases treated so far have more mathematical interest than physical interest. In practical cases the string always has a finite length. We shall now study the motion of a string of length L, represented by that part of the x-axis given by $0 \leqslant x \leqslant L$, and having both ends fixed. We shall give here a direct, synthetic solution of the problem. The ultimate solution, by Fourier analysis, has much greater practical importance. The motion of the string is again given by equation (VII, 1; 43) with the functions f and g defined for all values of the variable. However, when we write that $u(0, t) \equiv u(L, t) \equiv 0$, we get

(VII, 1; 65)
$$\begin{aligned} f(vt) + g(-vt) &= 0, \\ f(L + vt) + g(L - vt) &= 0. \end{aligned}$$

The first equation gives g in terms of f, and we thus recover the expression (VII, 1; 61) for u but with the function f satisfying

(VII, 1; 66)
$$f(L + s) - f(-L + s) = 0.$$

Alternatively, we say that the function is periodic with period 2L. Thus every solution of the equation of vibrating strings defined for $0 \leqslant x \leqslant L$, $-\infty < t < +\infty$, and satisfying $u(0, t) \equiv u(L, t) \equiv 0$ can be extended in a unique manner to a solution u defined for all values of x and t, and satisfying the relations

(VII, 1; 67)
$$\begin{aligned} &\bar{u}(-x, t) = -\bar{u}(x, t), \\ &\bar{u}(x + 2L, t) = \bar{u}(x, t) \quad \text{and} \quad \bar{u}\left(x, t + \frac{2L}{v}\right) = \bar{u}(x, t). \end{aligned}$$

Conversely these relations imply

$$\bar{u}(0, t) \equiv \bar{u}(L, t) \equiv 0.$$

The second relation expresses the fact that for all t, $\bar{u}$ is a periodic function of x of period 2L. This is not unexpected because the solution u must be odd, that is to say, antisymmetric with respect to $x = 0$ and also with respect to $x = L$ and hence it is periodic with period 2L. This is only of theoretical interest as the string has length L. However the second relation also expresses the fact that with respect to time, $\bar{u}$ is periodic of period $2L/v$, and this is physically essential. Hence :

All motions of a string of length L, *with fixed extremities, are periodic with period* $2L/v$.

This is easily understood by noting that each wave is reflected successively at the ends $x = 0$ and $x = L$ of the string, each time with a change in sign.

After a time $2L/v$ every wave, which travelled initially to the right or left, has been reflected at each extremity and has returned to its starting point with its original sign.

Application. A Cauchy problem is given by

$$u_0(x) = u(x, \tau), \qquad u_1(x) = \frac{\partial u}{\partial t}(x, \tau),$$

for $0 \leqslant x \leqslant L$ with the conditions $u(0, t) \equiv u(L, t) \equiv 0$.

To solve for $t \geqslant \tau$, we shall extend u_0 and u_1 to $\bar{u}_0$ and $\bar{u}_1$, in such a way that they will be odd and antisymmetric with respect to $x = L$ and are thus periodic with period $2L$. We shall use equation (VII, 1; 49) which, taking into account the asymmetry and periodicity, has the form

$$\text{(VII, 1; 68)} \quad u(x, t) = \frac{1}{2}[\pm u_0(\alpha) \pm u_0(\beta)] + \frac{1}{2v}\int_\alpha^\beta u_1(\xi)\, d\xi,$$

where α and β are the intersections of the line $t = \tau$ and the extreme rays of the backward wave cone at (x, t) : these rays are the half-lines from the point (x, t), directed downwards with slopes $\pm 1/v$ and being reflected at the verticals $x = 0$ and $x = L$. The sign to be taken in (VII, 1; 68) is $(-1)^\nu$, where ν is the number of reflections in question. [See Figure (VII, 4)].

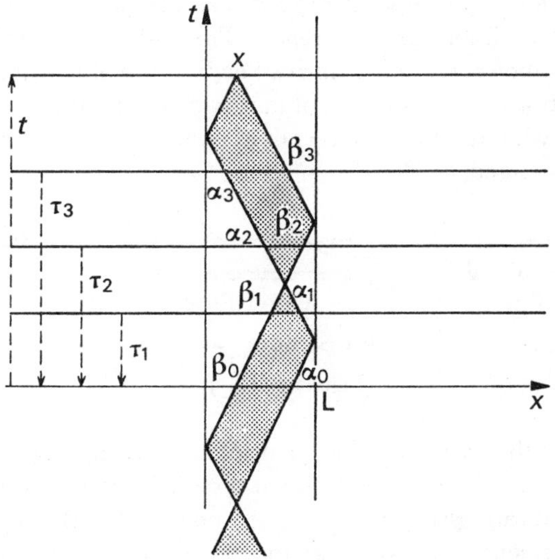

Figure (VII, 4)

The sign is -1 for α_3,	$+1$ for β_3,	if	$\tau = \tau_3$;
-1 for α_2,	-1 for β_2,	if	$\tau = \tau_2$;
-1 for α_1,	-1 for β_1,	if	$\tau = \tau_1$;
$+1$ for α_0,	-1 for β_0,	if	$\tau = \tau_0 = 0$.

Since $\alpha_1 > \beta_1$ and $\alpha_0 > \beta_0$, any integral such as $\int_\alpha^\beta u_1(\xi)\, d\xi$ can be replaced, when $\alpha > \beta$, by $-\int_\beta^\alpha u_1(\xi)\, d\xi$. The sign of $\beta - \alpha$ is $(-1)^\mu$, where μ is the number of times that the extreme rays of the cone cross between the ordinates τ and t. If we take as the value of τ the value that corresponds to the first crossing of the rays $(\tau_1 < \tau < \tau_2)$, we have $\tau = t - L/v$. Thus we have the important equation

$$u(x, t) = -u_0(L - x, \tau),$$

which corresponds to the general formula

$$\bar{u}(x, t) \equiv -\bar{u}(L - x, \quad t + L/v),$$

which follows from (VII, 1; 61), (VII, 1; 66) and (VII, 1; 67). If we add to the time half a period, L/v, the new position of the wave is antisymmetric to the old position with respect to $x = L/2$, the middle of the string. The second crossing corresponds to $\tau = t - 2L/v$ (here $\tau < 0$), and we find again that $u(x, t) = u_0(x, \tau)$, which is periodic in time with period $2L/v$.

Organ pipes

The equation for organ pipes is the same as that for vibrating strings, but the boundary conditions are different. The ends of an organ pipe can be open or closed; however, one of the two ends of the pipe will always be open so that the vibrations in the air column can be maintained by a bellows. According to whether the other end is open or closed we say that the pipe is open or closed.

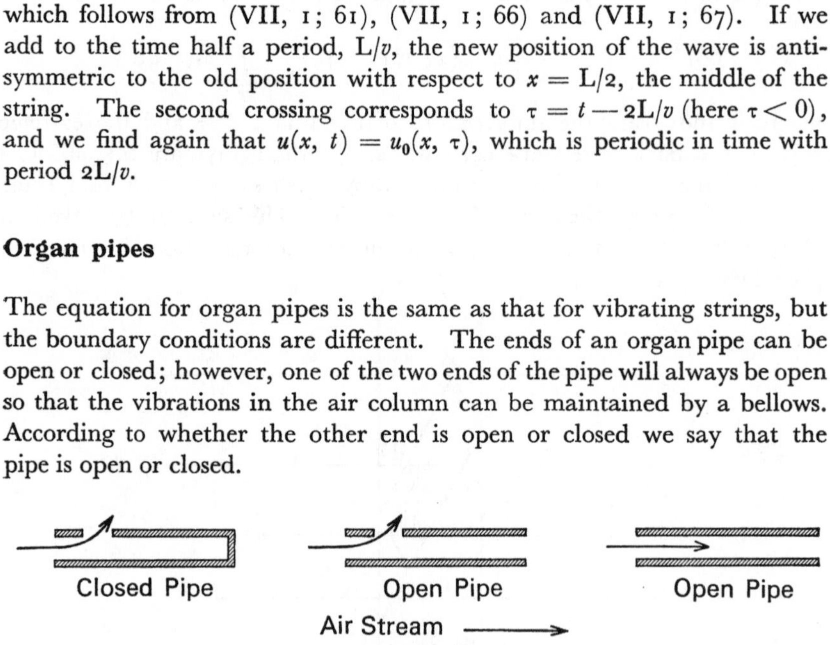

Figure (VII, 5).

As we have called u the displacement of an element of gas parallel to the axis of the pipe, the limiting condition at a closed end is always $u = 0$, as in the case of a vibrating string with a fixed point. On the other hand, now consider an open end. We shall assume that at such an end the gas pressure, in contact with the atmosphere, remains constant (equal to atmospheric pressure or slightly greater). Taking account of the third equation in (VII, 1; 36) we obtain for the limiting condition at the open end $\partial u/\partial x = 0$ (and not $u = 0$). In order that we may examine what happens at such an end we are going to consider, as in the case of a string, an endless pipe occupying the part $x \geqslant 0$ of the x axis, the end of the pipe at $x = 0$ being open.

Let u be a solution of equation (VII, 1; 37) for $x \geqslant 0$ and satisfying $\frac{\partial u}{\partial x}(0, t) \equiv 0$. From what we have seen on page 257, u is always expressible in terms of (VII, 1; 43), and can be extended to a solution $\bar{u}$ defined everywhere. Using the boundary condition at $x = 0$ we have

(VII, 1; 69) $$f'(vt) + g'(-vt) = 0.$$

This time we see that the functions f and g of one variable s satisfy

(VII, 1; 70) $$f'(s) + g'(-s) = 0.$$

By integration we deduce

(VII, 1; 71) $$f(s) - g(-s) = \text{constant}.$$

The value of this constant has no importance (see page 253) and we shall take it to be zero. The solution finally has the general form

(VII, 1; 72) $$u(x, t) = f(x + vt) + f(-x + vt).$$

We see that for all t, u is an even function of x. This is not unexpected, because $\partial\bar{u}/\partial x$ is also a solution of the same equation (VII, 1; 37) but satisfies the condition $\partial\bar{u}/\partial x\ (0, t) = 0$, thus $\partial\bar{u}/\partial x$ has to be an odd function of x. A study similar to that on page 258 shows that:

For an open pipe there is a reflection of waves with no change of sign (and naturally superposition of the incident and reflected waves occurs).

Application. The Cauchy problem

$$u_0(x) = u(x, \tau), \quad u_1(x) = \frac{\partial u}{\partial t}(x, \tau), \quad \frac{\partial u}{\partial x}(0, t) \equiv 0$$

will be solved by extending the functions u_0 and u_1 to even functions $\bar{u}_0$ and $\bar{u}_1$.

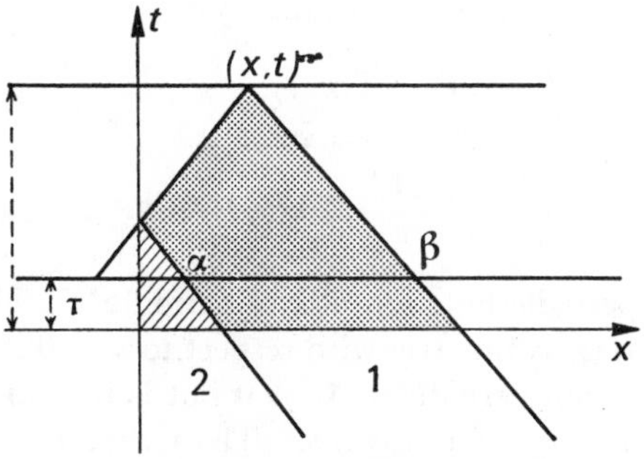

Figure (VII, 6)

In place of equation (VII, 1; 64) we shall have, if $t \geqslant \tau$, $x - v(t - \tau) < 0$:

$$\text{(VII, 1; 73)} \quad u(x, t) = \frac{1}{2}[u_0(x + v(t - \tau)) + u_0(v(t - \tau) - x)] + \frac{1}{2v}\left[2\int_0^{v(t-\tau)-x} u_1(\xi)d\xi + \int_{v(t-\tau)-x}^{x+v(t-\tau)} u_1(\xi)d\xi\right].$$

The backward wave cone at (x, t) consists of two parts, one labelled 1 and the other 2, so as to obtain the integrals of equation (VII, 1; 73). Thus we can, without difficulty, treat the problem of the open pipe and the closed pipe.

Open pipes

Here again the Fourier solution (page 276) will be the more important. The pipe occupies the part $0 \leqslant x \leqslant L$ of the x-axis, and both its ends are assumed to be open. The solution u is given by equation (VII, 1; 43), with the functions f and g defined for all values of the variable and satisfying

$$\text{(VII, 1; 74)} \quad \begin{aligned} f'(vt) + g'(-vt) &= 0, \\ f'(L + vt) + g'(L - vt) &= 0. \end{aligned}$$

The equations can be integrated as in (VII, 1; 70) but with constants of integration which are not necessarily equal. The first can be equated to zero, the second will then be some constant A_1. This is equivalent to saying that u can be extended to a solution $\bar{u}$ defined everywhere and having the form

$$\text{(VII, 1; 75)} \quad \bar{u}(x, t) = f(x + vt) + f(-x + vt),$$

with

$$\text{(VII, 1; 76)} \quad f(L + s) - f(-L + s) = A_1.$$

If $A_1 \neq 0$, f is not periodic but is the sum of $A_1 s/2L$ and a periodic function. This gives for the solution $\bar{u}$ the relations

$$\text{(VII, 1; 77)} \quad \begin{aligned} \bar{u}(-x, t) &= \bar{u}(x, t), \\ \bar{u}(x + 2L, t) &= \bar{u}(x, t), \\ \bar{u}\left(x, t + \frac{2L}{v}\right) &\equiv \bar{u}(x, t) + 2A_1. \end{aligned}$$

For fixed t, $\bar{u}$ is a periodic function of x of period $2L$. This is as expected since it is even, that is, symmetric with respect to $x = 0$ and $x = L$. With respect to time, $\bar{u}$ is not periodic if $A_1 \neq 0$ but is the sum of $At = A_1 vt/L$ and a periodic function of period $2L/v$. The term $At = A_1 vt/L$ represents a uniform flow of gas along the pipe which is quite normal as the pipe is open at both ends (Figure VII, 5). For this nearly uniform flow the solution is periodic with period $2L/v$.

Application. The solution of the Cauchy problem

$$u_0(x) = u(x, \tau), \qquad u_1(x) = \frac{\partial u}{\partial t}(x, \tau), \qquad \frac{\partial u}{\partial x}(0, t) \equiv \frac{\partial u}{\partial x}(L, t) \equiv 0$$

is very complicated. We extend u_0 and u_1 to $\bar{u}_0$ and $\bar{u}_1$ in such a way that they are even functions of x and periodic with period 2L. Equation (VII, 1; 49) then gives

(VII, 1; 78) $$u(x, t) = \frac{1}{2}[u_0(\alpha) + u_0(\beta)] + \frac{1}{2v}\int_0^L n(\xi)u_1(\xi)\,d\xi,$$

where $n(\xi)$ is the coefficient of ξ, a quantity which varies along (0, L). This coefficient can be calculated by studying the backward wave cone of (x, t), which divides the strip between $x = 0$ and $x = L$ into regions with equal coefficient (Figure VII, 7). For example, for the value τ considered in the figure we have $\alpha > \beta$ and $n(\xi) = 2$ for $0 < \xi < \beta$, $n(\xi) = 3$ for $\beta < \xi < \alpha$ and $n(\xi) = 4$ for $\alpha < \xi < L$. It is easy to calculate the value of A_1, the nonperiodic part of u, because A has the value [$n(\xi)$ being equal to 4 for $t = \tau + 2L/v$]

(VII, 1; 79) $$A = \frac{v}{L}\,\frac{1}{2}\left[u\left(x, \tau + \frac{2L}{v}\right) - u(x, \tau)\right]$$
$$= \frac{1}{2}\cdot\frac{4}{2L}\int_0^L u_1(\xi)\,d\xi = \frac{1}{L}\int_0^L u_1(\xi)\,d\xi.$$

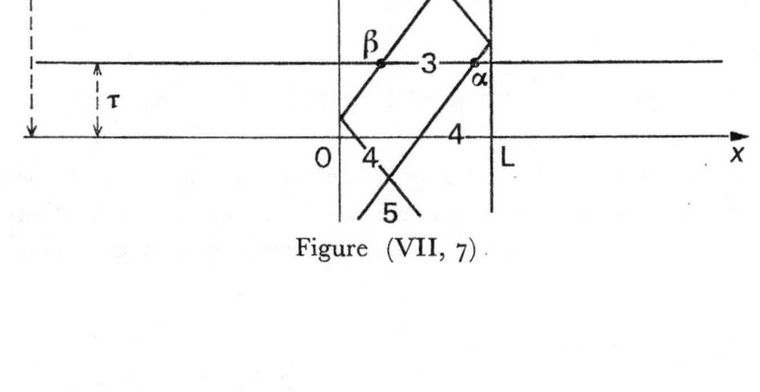

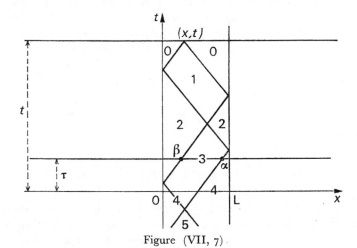

Figure (VII, 7)

Closed pipes

See the Fourier series solution (page 278). We shall assume that the pipe is open at the point $x = 0$ and closed at the point $x = L$. The solution u is given by equation (VII, 1; 43) where f and g satisfy

$$\text{(VII, 1; 80)} \qquad \begin{aligned} f(vt) + g(-vt) &= 0, \\ f'(L + vt) + g'(L - vt) &= 0. \end{aligned}$$

The second equation can always be integrated to give

$$\text{(VII, 1; 81)} \qquad f(L + vt) - g(L - vt) = 0,$$

by taking the constant of integration to be zero. Then u can be extended to give a solution $\bar{u}$ defined everywhere and having the form

$$\text{(VII, 1; 82)} \qquad \bar{u}(x, t) = f(x + vt) - f(-x + vt),$$
$$\text{(VII, 1; 83)} \qquad f(L + s) + f(-L + s) = 0.$$

This result may be expressed by saying that f is "antiperiodic" of "antiperiod" 2L, and is therefore periodic of period 4L.

Hence for $\bar{u}$ we have the two relations

$$\text{(VII, 1; 84)} \qquad \begin{aligned} \bar{u}(-x, t) &= -\bar{u}(x, t), \\ \bar{u}(x + 2L, t) &= -\bar{u}(x, t), \\ \bar{u}\left(x, t + \frac{2L}{v}\right) &= -\bar{u}(x, t). \end{aligned}$$

For fixed t, $\bar{u}$ is antiperiodic with antiperiod 2L. This is to be expected since $\bar{u}$ is antisymmetric with respect to $x = 0$ and symmetric with respect to $x = L$. With respect to the time, u is antiperiodic with antiperiod $2L/v$. This again is as expected because after a time $2L/v$ each wave has been reflected at each extremity of the pipe, has experienced a change of sign and has returned to its initial position with its sign changed.

Application. The solution of the Cauchy problem

$$u_0(x) = u(x, \tau), \qquad u_1(x) = \frac{\partial u}{\partial t}(x, \tau), \qquad u(0, t) \equiv \frac{\partial u}{\partial x}(L, t) \equiv 0$$

has the following form:

$$\text{(VII, 1; 85)} \quad u(x, t) = \frac{1}{2}[\pm u_0(\alpha) \pm u_0(\beta)] + \frac{1}{2v}\int_0^L n(\xi)u_1(\xi)\, d\xi.$$

The $\pm$ sign before $u_0(\alpha)$ or $u_0(\beta)$ is determined by $(-1)^\nu$, where ν is the number of reflections of the corresponding ray at the vertical $x = 0$. The coefficient $n(\xi)$ varies according to which region of the backward wave cone at (x, t) is being considered. For example, for the value of τ considered in Figure (VII, 8) it is necessary to take the negative sign in front of $u_0(\alpha)$ and the positive sign in front of $u_0(\beta)$ when $\alpha > \beta$. The coefficient $n(\xi)$ has the value zero except in the range $\beta < \xi < \alpha$, when it has the value 1. The integral then reduces to $\int_\beta^\alpha$.

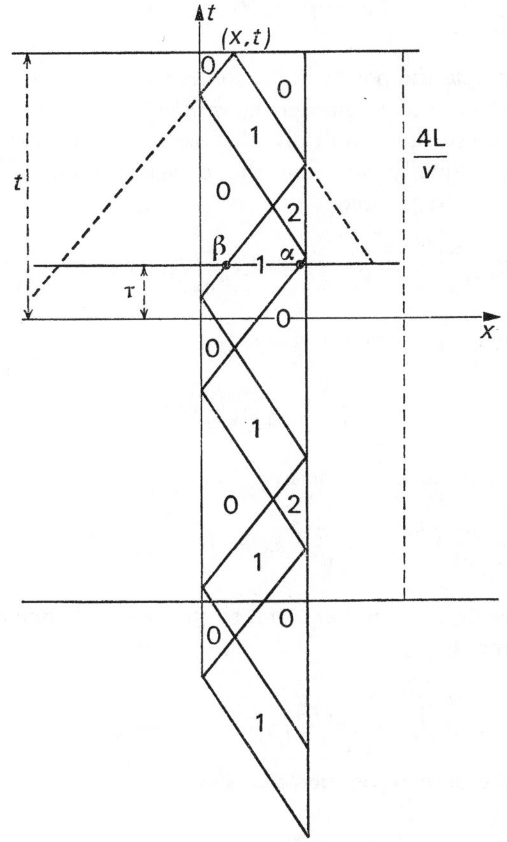

Figure (VII, 8)

Energy integrals

We shall consider the particular example given by the longitudinal vibrations of a solid rod.

The equation of energy expresses the fact that the time derivative of the total kinetic energy is equal to the rate at which work is done by the system of forces (the *power*). This equation is a consequence of the fundamental equation of dynamics: from $\vec{F} = m\vec{\gamma}$ we can deduce

$$P = \vec{F} \cdot \vec{v} = m\vec{\gamma} \cdot \vec{v} = \frac{d}{dt}\left(\frac{1}{2} mv^2\right).$$

The kinetic energy of the rod is

$$E_c = \int_0^L \frac{1}{2} \rho_0 \left(\frac{\partial u}{\partial t}\right)^2 dx. \qquad \text{(VII, 1; 86)}$$

In order to calculate the power of the forces we must examine an element $(x, x + \Delta x)$ of the rod and consider the *resultant* of the forces acting on this element. We must then multiply this resultant by the *mean* velocity of the element. Finally we sum the powers obtained from all such elements*. The resultant force is

$$E_0\sigma_0 \frac{\partial u}{\partial x}(x + \Delta x, t) - E_0\sigma_0 \frac{\partial u}{\partial x}(x, t) = E_0\sigma_0 \frac{\partial^2 u}{\partial x^2} \Delta x, \qquad \text{(VII, 1; 87)}$$

hence the power for the whole rod is

$$\int_0^L E_0\sigma_0 \frac{\partial^2 u}{\partial x^2} \frac{\partial u}{\partial t} dx. \qquad \text{(VII, 1; 88)}$$

Thus we have the equation of energy

$$\frac{d}{dt}\int_0^L \frac{1}{2} \rho_0 \left(\frac{\partial u}{\partial t}\right)^2 dx = \int_0^L E_0\sigma_0 \frac{\partial^2 u}{\partial x^2} \frac{\partial u}{\partial t} dx \qquad \text{(VII, 1; 89)}$$

which can be written in another form, so that we have one equation valid for both problems, as

$$\frac{d}{dt}\int_0^L \frac{1}{2} \frac{1}{v^2} \left(\frac{\partial u}{\partial t}\right)^2 dx = \int_0^L \frac{\partial^2 u}{\partial x^2} \frac{\partial u}{\partial t} dx, \qquad \text{(VII, 1; 90)}$$

having chosen the units to be such that $E_0\sigma_0 = 1$.

---

*We could perhaps argue that we ought to consider for each element the forces which act at its extremities and multiply each of these by the velocity of its point of application, and then sum the resultants obtained for the various elements. We would thus replace

$$\left(\frac{\partial}{\partial x}\frac{\partial u}{\partial x}\right)\frac{\partial u}{\partial t} \quad \text{by} \quad \frac{\partial}{\partial x}\left(\frac{\partial u}{\partial x}\frac{\partial u}{\partial t}\right).$$

However, this method is incorrect, and is not a consequence of the equation $\vec{F} = m\vec{\gamma}$ applied to the various elements. All the intermediate terms will cancel and only the first and last will remain. We would thus obtain a false equation, which is equation (VII, 1; 93) without the second term on the left-hand side.

Note that this equation is an immediate consequence of the equation of vibrating strings, because the left-hand side is simply

$$\int_0^L \frac{1}{v^2}\frac{\partial^2 u}{\partial t^2}\frac{\partial u}{\partial t}\,dx. \tag{VII, 1; 91}$$

The equality (VII, 1; 90) is an equality between quantities which depend only on t and not on x. Therefore we have a non-linear equation, which is in fact *quadratic*, between quantities of the second order with respect to the magnitude of u, and this is why we can neglect only quantities of the third order. Integrating the left-hand side of (VII, 1; 90) by parts we get

(VII, 1; 92)

$$\left[\frac{\partial u}{\partial x}\frac{\partial u}{\partial t}\right]_{x=0}^{x=L} - \int_0^L \frac{\partial u}{\partial x}\frac{\partial^2 u}{\partial x\,\partial t}\,dx = \left[\frac{\partial u}{\partial x}\frac{\partial u}{\partial t}\right]_{x=0}^{x=L} - \frac{d}{dt}\int_0^L \frac{1}{2}\left(\frac{\partial u}{\partial x}\right)^2 dx.$$

Hence we obtain the new equation

$$\frac{d}{dt}\int_0^L \frac{1}{2}\left(\frac{1}{v^2}\left(\frac{\partial u}{\partial t}\right)^2 + \left(\frac{\partial u}{\partial x}\right)^2\right) dx = \left(\frac{\partial u}{\partial x}\frac{\partial u}{\partial t}\right)_{x=0}^{x=L}. \tag{VII, 1; 93}$$

The right-hand side is simply the power of the external forces, the tension and pressure acting on the two ends of the rod. Thus we see that the first law of thermodynamics is satisfied and that the system possesses at each instant of its motion an energy which can be written

$$E = \frac{1}{2}\int_0^L \left(\frac{1}{v^2}\left(\frac{\partial u}{\partial t}\right)^2 + \left(\frac{\partial u}{\partial x}\right)^2\right) dx. \tag{VII, 1; 94}$$

This energy depends on the state (the position and velocity) of the system and is the sum of a kinetic energy E_k and an internal energy E_i. The equality (VII, 1; 93) when integrated with respect to the time t from $t = t_1$ to $t = t_2$ will express the fact that *the work done by the exterior forces is equal to the variation in the energy* $E_2 - E_1$.

If we always have $\partial u/\partial x = 0$ at the ends of the rod, let $u = 0$, then we see by differentiating with respect to time that $\partial u/\partial t = 0$, and hence the work done by the external forces is zero and *the energy* E *remains constant for all time.*

Cauchy's problem continued. Uniqueness of solution

The general Cauchy problem can be stated as follows : To find a solution u of the equation of vibrating strings which satisfies :

1. *Initial conditions, where the values of* u *and* $\partial u/\partial t$ *for* $t = 0$ *are known functions* u_0 *and* u_1 *respectively of* x.

2. *Boundary conditions, where we are given for all t the values of u and $\partial u/\partial x$ at $x = 0$ and $x = \mathrm{L}$.*

We shall now show that this problem has *at most* one solution. The difference U between any two solutions is also a solution of the problem, because the equation of vibrating strings is linear. Consequently, U will satisfy the initial conditions $\mathrm{U}_0 = \mathrm{U}_1 = 0$, and also the boundary conditions

$$\mathrm{U}(x, t) = 0 \qquad \text{or} \qquad \frac{\partial \mathrm{U}}{\partial x}(x, t) = 0$$

for $x = 0$ and $x = \mathrm{L}$. Under these conditions,

$$\left[\frac{\partial \mathrm{U}}{\partial x}\frac{\partial \mathrm{U}}{\partial t}\right]_{x=0}^{x=\mathrm{L}}$$

is always zero for all time. The energy equation (VII, 1; 93) then expresses the fact that the integral (VII, 1; 94) is constant for all time. Now, at time $t = 0$, the energy has the value

$$\text{(VII, 1; 95)} \qquad \frac{1}{2}\int_0^{\mathrm{L}}\left(\frac{1}{v^2}\mathrm{U}_1^2(x) + \mathrm{U}_0'^2(x)\right)dx = 0.$$

Thus the energy is zero for all time t. As this is the integral of a function greater than or equal to zero, the latter is zero almost everywhere. But it is continuous (a solution of the equation having been assumed twice continuously differentiable); thus

$$\frac{\partial \mathrm{U}}{\partial t} \qquad \text{and} \qquad \frac{\partial \mathrm{U}}{\partial x}$$

are identically zero. Thus U is a constant independent of x and t. Since it is zero for $t = 0$, it is identically zero, and as it is the difference of two solutions, these two solutions must be identical.

In the previous pages we obtained solutions to various Cauchy problems from the general solution (VII, 1; 43). Now we shall obtain solutions by another method due to Fourier.

3. SOLUTION OF CAUCHY'S PROBLEM BY FOURIER ANALYSIS

Cauchy's problem for a vibrating string when both ends are fixed

In this case we are going to show that a solution exists and then we shall find that solution. The boundary conditions here are

$$u(0, t) \equiv u(\mathrm{L}, t) \equiv 0,$$

and the initial conditions are as usual u_0 and u_1, where obviously

$$u_0(0) = u_0(L) = u_1(0) = u_1(L) = 0.$$

We say that the string has a *stationary wave motion* if, for that wave motion, $u(x, t)$ can be expressed as the product of a function of x with a function of t:

(VII, 1; 96) $$u(x, t) = U(x)V(t).$$

We shall look for the stationary waves on a string that has both ends fixed. For such a motion the equation of vibrating strings can be written

(VII, 1; 97) $$\frac{1}{v^2} UV'' = U''V$$

or

(VII, 1; 98) $$\frac{1}{v^2} \frac{V''}{V} = \frac{U''}{U} = -\lambda,$$

λ being a constant since it must depend only on x and at the same time depend only on t. Conversely, if λ is an arbitrary constant the solutions $U(x)$ and $V(t)$ of the differential equations

(VII, 1; 99) $$V'' + \lambda v^2 V = 0, \qquad U'' + \lambda U = 0$$

define a stationary wave motion. We shall assume that $V(t)$ is not identically zero. If this were not so then the string would remain at rest for all time. Thus $u(0, t)$ and $u(L, t)$ can only be zero, whatever may be the value of t, if $U(0) = U(L) = 0$. This will restrict the values that λ can take. Actually the second equation (VII, 1; 99) can be solved to give

(VII, 1; 100) $$U(x) = A \cos\sqrt{\lambda}x + B \sin\sqrt{\lambda}x \quad \text{for} \quad \lambda \neq 0$$

(if $\lambda < 0$, $\sqrt{\lambda}$ is imaginary and the cos and sin will be replaced by cosh and sinh). The condition $U(0) = 0$ gives $A = 0$. Thus taking account of the fact that $B \neq 0$ [if it were zero we would again have $u(x, t) \equiv 0$], the condition $U(L) = 0$ gives

(VII, 1; 101) $$\sin\sqrt{\lambda}L = 0.$$

This proves that $\lambda > 0$, because a hyperbolic sine cannot be zero unless its argument is zero. Thus we have

(VII, 1; 102) $$\sqrt{\lambda}L = k\pi,$$

where k is an integer, or

$$\sqrt{\lambda} = \frac{k\pi}{L}.$$

The solution so obtained is, to within a constant factor,

$$\text{(VII, 1; 103)} \qquad \sin k\pi \frac{x}{L}.$$

The constant factor can be set equal to 1 without changing the result because every constant factor of U can be included in V. For the same reason we need only take $k > 0$, because replacing k by $-k$ is effectively multiplying U by -1.

The case when $\lambda = 0$ requires a separate study, because for $\lambda = 0$ the solution of the second equation of (VII, 1; 99) no longer has the same form. It is in fact

$$\text{(VII, 1; 104)} \qquad U(x) = Ax + B.$$

No linear function can be zero for $x = 0$ and $x = L$ unless it is identically zero. Thus $\lambda = 0$ does not yield stationary waves [other than $u(x, t) \equiv 0$]. It now remains to solve the first equation of (VII, 1; 99) irrespective of which of the solutions gives an acceptable stationary wave motion. The general solution is

$$\text{(VII, 1; 105)} \qquad \begin{aligned} V(t) &= A \cos v\sqrt{\lambda}t + B \sin v\sqrt{\lambda}t \\ &= A \cos k\pi \frac{vt}{L} + B \sin k\pi \frac{vt}{L}. \end{aligned}$$

Thus the stationary waves on a string having both ends fixed depend on some arbitrary integer $k > 0$ and for a fixed value of k, upon two real parameters A and B. The most general form for such a stationary wave is

$$\text{(VII, 1; 106)} \quad u(x, t) = \sin k\pi \frac{x}{L} \left(A \cos k\pi \frac{vt}{L} + B \sin k\pi \frac{vt}{L} \right).$$

The stationary waves corresponding to an integer k have a period (in time) given by

$$\text{(VII, 1; 107)} \qquad T = \frac{2\pi}{k\pi \frac{v}{L}} = \frac{2L}{kv}$$

and the frequency is

$$\text{(VII, 1; 108)} \qquad N = \frac{1}{T} = k \frac{v}{2L}.$$

These frequencies are all multiples of a "fundamental frequency" $N_0 = v/2L$. For a periodic vibration which travels with a velocity v we also introduce the *wavelength* $\Lambda = vT$, where

$$\text{(VII, 1; 109)} \qquad \Lambda = \frac{2L}{k} \qquad \text{for} \qquad L = k\frac{\Lambda}{2}.$$

This length of the waves is the period in x of the solution $\bar{u}$ which is the extension of u (see page 260).

Thus for a string having both ends fixed the wavelength of the stationary waves are such that L is an integral multiple of half wavelengths. This becomes intuitively obvious if we extend U beyond $x = 0$ and $x = L$ by antisymmetry. Then due to the periodicity of period 2L we have the following figure (VII, 9). This means of extending U beyond the physical limits of the string (which, naturally amounts to extending the function $\sin k\pi x/L$) has been justified previously by quite general arguments.

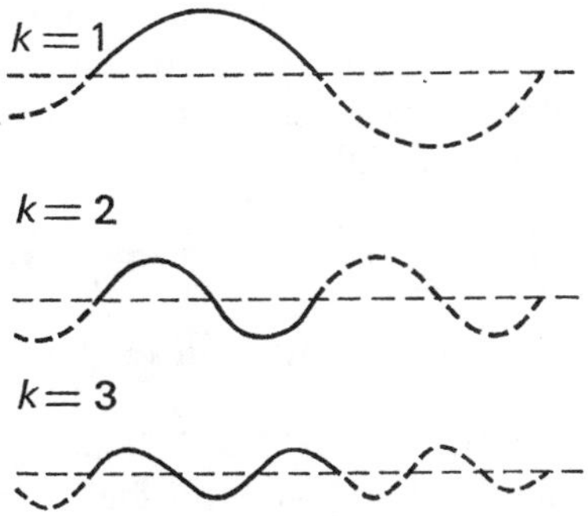

Figure (VII, 9)

We will now show that we can solve the Cauchy problem given above by assuming that the solution $u(x, t)$ is an *infinite linear combination of stationary waves on a string fixed at both ends.* If we find such a solution it will be the only one, as a result of the uniqueness theorem already proved and as a consequence of our discussion on page 260. Moreover, as every solution of the equation is determined in terms of its initially known values u_0 and u_1, we have also to show that every solution of the equation of vibrating strings, for a string with both ends fixed, is a combination of the stationary solutions. We recognize from this the fact that every solution has a period $2L/v$ which is a common multiple of the fundamental period of all stationary waves. Thus we look for solutions of the form

$$\text{(VII, 1; 110)} \quad u(x, t) = \sum_{k=1}^{\infty} \sin k\pi \frac{x}{L} \left(A_k \cos k\pi \frac{vt}{L} + B_k \sin k\pi \frac{vt}{L} \right).$$

Applying the initial conditions that u must satisfy, we have

$$\text{(VII, 1; 111)} \quad u_0(x) = u(x, 0) = \sum_{k=1}^{\infty} A_k \sin k\pi \frac{x}{L},$$

$$\text{(VII, 1; 112)} \quad u_1(x) = \frac{\partial u}{\partial t}(x, 0) = \sum_{k=1}^{\infty} k\pi \cdot \frac{v}{L} B_k \sin k\pi \frac{x}{L}.$$

Now every function of x, defined over the interval (0, L) can be (formally at least) expanded in a unique Fourier series in terms of the $\sin k\pi x/L$. In fact, we extend such a function over $(-L, 0)$ in such a way that it is an odd function. Then for all the real axis it is a function having period $2L$. [In such a case the function is also antisymmetric with respect to $x = L$, because $f(-x) = -f(x)$ and $f(x + 2L) = f(x)$ imply that $f(2L - x) = -f(x)$. Conversely, an antisymmetric function with respect to $x = 0$ and $x = L$ has a period $2L$ because $f(-x) = -f(x)$ and $f(2L - x) = -f(x)$ imply that $f(x + 2L) = f(x)$.] Thus it is possible to expand the function in terms of $\cos k\omega x$ and $\sin k\omega x$, where $\omega = 2\pi/2L = \pi/L$. As the function is an odd function it can be expanded in terms of $\sin k\pi x/L$ alone. The equations (VII, 1; 111) and (VII, 1; 112) then give the coefficients A_k and B_k as

(VII, 1; 113)
$$A_k = \frac{2}{L}\int_0^L u_0(\xi) \sin k\pi \frac{\xi}{L}\, d\xi,$$
$$B_k = \frac{2}{k\pi v}\int_0^L u_1(\xi) \sin k\pi \frac{\xi}{L}\, d\xi.$$

Notice from equation (VII, 1; 110) that A_k and B_k have the dimensions of u, that is the dimension l. This is also true for u_0 whilst u_1 has the dimensions of $ut^{-1} = lt^{-1}$. Hence the equations (VII, 1; 113) are homogeneous, sin being dimensionless and $d\xi$ having the dimension l. Finally, then, the explicit solution to the Cauchy problem can be written as

(VII, 1; 114)
$$u(x, t) = \sum_{k=1}^{\infty} \sin k\pi \frac{x}{L}\left[\left(\cos k\pi \frac{vt}{L}\right)\frac{2}{L}\int_0^L u_0(\xi) \sin k\pi \frac{\xi}{L}\, d\xi + \left(\sin k\pi \frac{vt}{L}\right)\frac{2}{k\pi v}\int_0^L u_1(\xi) \sin k\pi \frac{\xi}{L}\, d\xi\right].$$

If we assume that the known initial quantities u_0 and u_1 are continuously differentiable once, then the Fourier series (VII, 1; 111) and (VII, 1; 112) are uniformly convergent. But this does not settle the question of the convergence of the series (VII, 1; 114) : and above all neither does the differentiablity of u with respect to x and y up to the second order, on its own, allow us to assert that the series of stationary waves defining u is truly a solution of the equation of vibrating strings. These are difficulties which we shall not pursue. The consideration of convergence and derivatives in the distribution sense is completely ignored; so also is the hypothesis that the given initial quantities u_0 and u_1 are functions sufficiently differentiable with respect to x.

Calculation of energy. Since energy is independent of t it can be calculated for $t = 0$ and has the value

(VII, 1; 115)
$$E = \frac{1}{2}\int_0^L \left(\frac{1}{v^2} u_1^2 + u_0'^2\right) dx.$$

Using Parseval's formula we can write this as

(VII, 1; 116)
$$E = \frac{L}{4}\sum_{k=1}^{\infty}\left(B_k^2\frac{k^2\pi^2}{L^2} + A_k^2\frac{k^2\pi^2}{L^2}\right) = \sum_{k=1}^{\infty}\frac{\pi^2}{4L}k^2(A_k^2 + B_k^2).$$

If we call u_k the kth *stationary mode*, as it appears in $uu = \sum_{k=1}^{\infty} u_k$, then the energy E_k due to u_k is exactly the quantity which occurs in the sum on the right-hand side. Alternatively we say

(VII, 1; 117)
$$u(x, t) = \sum_{k=1}^{\infty} u_k(x, t),$$
$$E = \sum_{k=1}^{\infty} E_k.$$

Thus the energy of a mode is equal to the sum of the energies of the stationary modes of which it is composed.

This is not an obvious fact because the energy is a quadratic expression and hence nonlinear. It follows from the orthogonality of the functions $\sin k\pi x/L$ and $\cos k\pi x/L$, or, again, from Parseval's formula. Notice that the energy of a stationary mode E_k is proportional to $A_k^2 + B_k^2$, that is, to the square of the amplitude, and to k^2, the square of the frequency. Note also that we can, once again, demonstrate the constancy of the energy with respect to the time. Because if we calculate E at time t, we obtain an integral which can be evaluated by using, once again, Parseval's formula; but it is necessary here to use the coefficients of the expansion of $\partial u/\partial x$ and $\partial u/\partial t$ (in terms of $\sin k\pi x/L$ and $\cos k\pi x/L$) at time t, which is equivalent to replacing the numbers A_k and B_k by

(VII, 1; 118)
$$A_k' = A_k \cos k\pi\frac{vt}{L} + B_k \sin k\pi\frac{vt}{L},$$
$$B_k' = -A_k \sin k\pi\frac{vt}{L} + B_k \cos k\pi\frac{vt}{L}.$$

Now,

(VII, 1; 119)
$$A_k'^2 + B_k'^2 = A_k^2 + B_k^2.$$

Thus the energy is constant.

We often say that the stationary components u_k, which occur in the expression for u have a "physical meaning". We mean by this that each motion u of simple physical models (resonators of fixed frequency) can be decomposed into its components u_k. This fact is of very great importance in practice.

Cauchy's problem for an open organ pipe

We postulate that at each open end the pressure is constant. Hence we see from (VII, 1; 36) that this signifies that

$$\frac{\partial u}{\partial x}(0, t) \equiv \frac{\partial u}{\partial x}(L, t) \equiv 0,$$

which is a new class of boundary conditions (see page 264). If instead of u we take the pressure p to be the unknown we know that it also satisfies the equation of vibrating strings. In this case we shall have the boundary conditions

$$p(0, t) = p_1, \quad p(L, t) = p_2,$$

which, if we put

$$\pi = p - p_1 - \frac{x}{L}(p_2 - p_1),$$

gives for π a vibrating-string equation with the boundary conditions $\pi(0, t) \equiv \pi(L, t) \equiv 0$; these are the same as for the previous problem solved alove.

The stationary waves satisfy the same equation (VII, 1; 99). The boundary conditions $U'(0) = U'(L) = 0$ yield once more, for $\lambda \neq 0$, equation (VII, 1; 102), and to within a factor

(VII, 1; 120) $$U(x) = \cos k\pi \frac{x}{L}.$$

For reasons already given we are able to limit ourselves to values of λ such that $\lambda > 0$. In this case $\lambda = 0$ does give a stationary wave because the solution (VII, 1; 104) readily gives $U'(0) = U'(L) = 0$, if U is an arbitrary constant, which amounts to the same as using (VII, 1; 120) with $k \neq 0$. Once k has been chosen, the general solution of the first equation of (VII, 1; 99) is always (VII, 1; 105), for $k \neq 0$. When $k = 0$, the solution is

(VII, 1; 121) $$V(t) = At + B.$$

Finally we find that for $k \neq 0$ the stationary waves are

(VII, 1; 122) $$u_k(x, t) = \cos k\pi \frac{x}{L}\left(A \cos k\pi \frac{vt}{L} + B \sin k\pi \frac{vt}{L}\right).$$

Their frequencies and wavelengths are the same as for the vibrating string with both ends fixed (which is not surprising since the modified pressure is then also stationary and satisfies the boundary conditions of vibrating strings). But the function $U(x)$ extended fully to the right (it is a cosine series) has a completely different form.

For $k = 0$ the stationary wave is

(VII, 1; 123) $$u_0(x, t) = At + B,$$

which is not periodic (if $A \neq 0$).

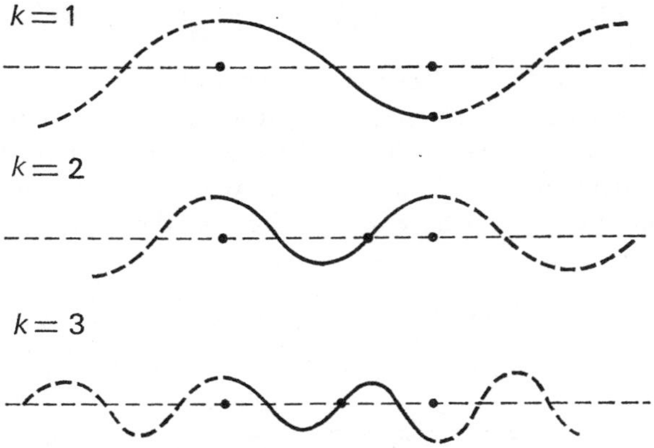

Figure (VII, 10)

Thus we shall solve the Cauchy problem by looking for a solution u, consisting of stationary waves, which will require that u_0 and u_1 are expanded in a series in terms of $\cos k\pi x/L$. This is possible, because if we extend the functions symmetrically with respect to 0 as periodic functions of period $2L$, their Fourier expansion in terms of $\cos k\pi x/L$ and $\sin k\pi x/L$ is also of period $2L$. Further, because the functions are even they can be expressed in terms of cosines only. Thus, finally, we have

(VII, 1; 124) $$u(x,t) = \sum_{k=1}^{\infty} \cos k\pi \frac{x}{L}\left[\left(\cos k\pi \frac{vt}{L}\right)\frac{2}{L}\int_0^L u_0(\xi)\cos k\pi\frac{\xi}{L}\,d\xi + \left(\sin k\pi\frac{vt}{L}\right)\frac{2}{k\pi v}\int_0^L u_1(\xi)\cos k\pi\frac{\xi}{L}\,d\xi\right] + \frac{1}{L}\int_0^L u_0(\xi)\,d\xi + \frac{t}{L}\int_0^L u_1(\xi)\,d\xi.$$

The non-periodic term in t corresponds to a translation of the entire mass of gas, which is blown into one end and out of the other. The mean velocity of circulation of the gas along the pipe is then

(VII, 1; 125) $$V_0 = \frac{1}{L}\int_0^L u_1(\xi)\,d\xi,$$

and we thus recover the result obtained on page 265, equation (VII, 1; 79).

This velocity, naturally, bears no relation to the velocity of wave propagation v. We know that the propagation of a wave does not involve the transport of material. The energy E is again the sum of the energies E_k and includes E_0.

Cauchy's problem for a closed pipe

We shall call a pipe that is closed at one end and open at the other (Figure VII, 11) a closed pipe. At the open end air is blown in but leaves via a neighbouring orifice. The column of air enclosed in the pipe is excited into forced oscillations. We shall assume that the closed end is at $x = 0$, and that the open end is at $x = L$. The boundary conditions this time are

$$(u, t) \equiv 0, \qquad \frac{\partial u}{\partial x}(L, t) = 0.$$

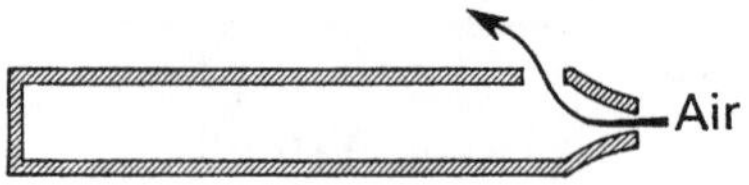

Figure (VII, 11)

The stationary waves are always given by (VII, 1; 99) but U must satisfy $U(0) = 0$, $U'(L) = 0$. Thus λ is real and > 0 because if $\lambda < 0$ the cosine and sine become cosh and sinh. The condition $U(0) = 0$ will yield a sinh solution, but then U' is a cosh expression which never vanishes and can never be zero, because then $U(x) = Ax + B$, and hence $U(0) = 0$ implies $B = 0$ and $U'(L) = 0$ implies $A = 0$. Thus U is a sine, U' is a cosine, and the condition $U'(L) = 0$ can be written

(VII, 1; 126) $$\cos \sqrt{\lambda} L = 0,$$

(VII, 1; 127) $$\sqrt{\lambda} L = k\pi + \pi/2 = (2k + 1)(\pi/2).$$

The function $U(x)$ is, to within a factor,

(VII, 1; 128) $$U(x) = \sin(2k + 1)\frac{\pi}{2}\frac{x}{L}.$$

We can restrict k to integral values $\geqslant 0$ because k and $-(k + 1)$ yield opposite results. The stationary modes corresponding to a given k are

(VII, 1; 129)

$$u_k(x, t) = \sin(2k + 1)\frac{\pi}{2}\frac{x}{L}\left[A \cos(2k + 1)\frac{\pi}{2}\frac{vt}{L} + B \sin(2k + 1)\frac{\pi}{2}\frac{vt}{L}\right].$$

The period T, frequency N and wavelength Λ satisfy

$$(\text{VII, 1; 130})\qquad T = \frac{4L}{(2k+1)v}, \qquad N = (2k+1)\frac{v}{4L},$$
$$\Lambda = \frac{4L}{2k+1}, \qquad L = (2k+1)\frac{\Lambda}{4}.$$

When k varies the frequencies are seen to be *odd multiples* of the fundamental frequency $N_0 = v/4L$ which is half the fundamental frequency of the open pipe of the same length. Thus *the wavelength of the pipe is an odd number of quarter wavelengths.* The function $U(x)$, extended along the whole axis (it is a sine solution) will have the form shown in Figure (VII, 12).

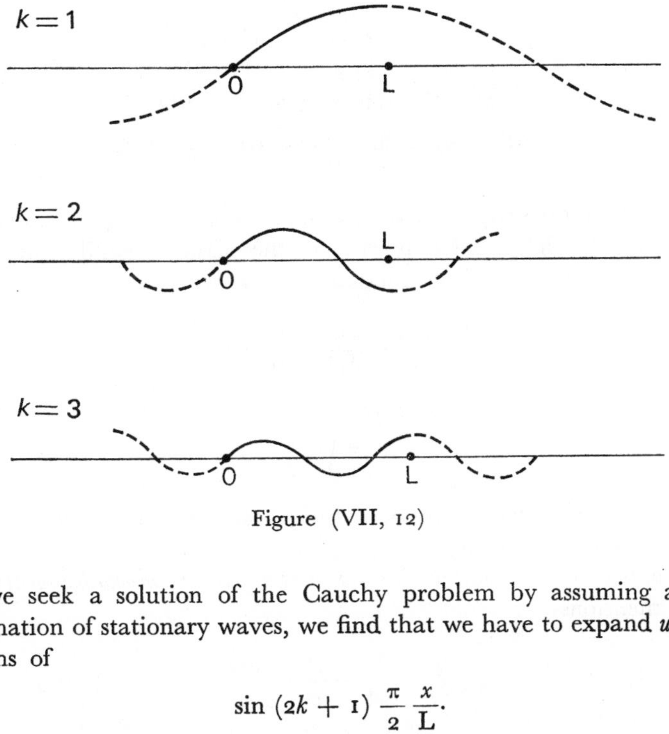

Figure (VII, 12)

If we seek a solution of the Cauchy problem by assuming a linear combination of stationary waves, we find that we have to expand u_0 and u_1 in terms of

$$\sin(2k+1)\frac{\pi}{2}\frac{x}{L}.$$

To do this we extend u_0 and u_1 into $(-L, 0)$ in such a way that they are antisymmetric (odd), and then into $(L, 3L)$ in such a way that they are symmetric about $x = L$. The functions so formed are expandable in terms of $\cos m\pi x/2L$ and $\sin m\pi x/2L$ and have a period $4L$. Since the functions are odd, only the sine terms are non-zero. Further, as the functions are symmetric about $x = L$ only those values of $m = 2k + 1$ will yield non-zero coefficients.

Thus we have the solution

$$\text{(VII, 1; 131)} \quad u(x, t) = \sum_{k=0}^{\infty} \sin (2k+1) \frac{\pi}{2} \frac{x}{L}$$
$$\times \left[\left(\cos (2k+1) \frac{\pi}{2} \frac{vt}{L} \right) \frac{2}{L} \int_0^L u_0(\xi) \sin (2k+1) \frac{\pi}{2} \frac{\xi}{L} d\xi \right.$$
$$\left. + \left(\sin (2k+1) \frac{\pi}{2} \frac{vt}{L} \right) \frac{2}{(2k+1)\frac{\pi}{2} v} \int_0^L u_1(\xi) \sin (2k+1) \frac{\pi}{2} \frac{\xi}{L} d\xi \right],$$

where again

$$E = \sum_{k=0}^{\infty} E_k.$$

2. Vibrating membranes and waves in three dimensions

In the study of the small transverse vibrations of a homogeneous stretched membrane, the deformation normal to the plane of equilibrium (which will be the (x, y) plane) is a function $u(x, y, t)$ which satisfies the equation

$$\text{(VII, 2; 1)} \qquad \frac{1}{v^2} \frac{\partial^2 u}{\partial t^2} - \frac{\partial^2 u}{\partial x^2} - \frac{\partial^2 u}{\partial y^2} = 0,$$

in which

$$\text{(VII, 2; 2)} \qquad v = \sqrt{\frac{F}{\rho}}$$

where F is the surface tension, ρ the surface specific gravity, and v has the dimensions of a velocity.

We shall call the equation (VII, 2; 1) *the vibrating membrane equation.* In three dimensions

$$\text{(VII, 2; 3)} \qquad \frac{1}{v^2} \frac{\partial^2 u}{\partial t^2} - \frac{\partial^2 u}{\partial x^2} - \frac{\partial^2 u}{\partial y^2} - \frac{\partial^2 u}{\partial z^2} = 0$$

is the *wave equation.* In this section we propose to study equations (VII, 2; 1) and (VII, 2; 3). For simplicity we put

$$\text{(VII, 2; 4)} \qquad \Box_2 = \frac{1}{v^2} \frac{\partial^2}{\partial t^2} - \frac{\partial^2}{\partial x^2} - \frac{\partial^2}{\partial y^2}$$

and

$$\text{(VII, 2; 5)} \qquad \Box_3 = \frac{1}{v^2} \frac{\partial^2}{\partial t^2} - \frac{\partial^2}{\partial x^2} - \frac{\partial^2}{\partial y^2} - \frac{\partial^2}{\partial z^2}.$$

1. THE SOLUTION OF THE VIBRATING MEMBRANE EQUATION AND THE WAVE EQUATION IN THREE DIMENSIONS BY THE METHOD OF TRAVELLING WAVES. CAUCHY PROBLEMS

Let us recall equation (VII, 1; 37). In the (x, t) plane we denote by (C) the forward wave-cone

(VII, 2; 6) $$v^2t^2 - x^2 \geqslant 0, \qquad t \geqslant 0.$$

Let $E(x, t)$ be the function

(VII, 2; 7) $$E(x, t) = \begin{cases} \dfrac{v}{2} \text{ in (C)}, \\ 0 \text{ elsewhere.} \end{cases}$$

A very simple calculation shows that, in the distribution sense, we have

(VII, 2; 8) $$\frac{1}{v^2}\frac{\partial^2 E}{\partial t^2} - \frac{\partial^2 E}{\partial x^2} = \delta.$$

Alternatively we can say that, $E(x, t)$, as defined in (VII, 2; 7), is an elementary solution of the operator

(VII, 2; 9) $$\Box_1 = \frac{1}{v^2}\frac{\partial^2}{\partial t^2} - \frac{\partial^2}{\partial x^2}.$$

Let $f(\xi)$ be a sufficiently regular function of the variable ξ, and let

(VII, 2; 10) $$I(x, t; f) = \frac{1}{v^2}\int_{-\infty}^{+\infty} E(x-\xi, t) f(\xi)\, d\xi = \frac{1}{2v}\int_{x-vt}^{x+vt} f(\xi)\, d\xi.$$

Note. The above integral is the convolution integral in ξ for fixed t, of $E(\xi, t)$ and $f(\xi)$.

It is very easy to verify that, whatever the function f (provided that it is sufficiently regular), $I(x, t; f)$ is a solution of equation (VII, 1; 37) and that the solution (VII, 1; 48) of the Cauchy problem (VII, 1; 44) is simply

(VII, 2; 11) $$u(x, t) = I(x, t; u_1) + \frac{\partial}{\partial t} I(x, t; u_0).$$

We shall use an identical method to study equation (VII, 2; 3).

Elementary solution of $\Box_3$

In R^4 denote by (Γ) the forward wave-cone

(VII, 2; 12) $$v^2t^2 - (x^2 + y^2 + z^2) \geqslant 0, \qquad t \geqslant 0,$$

and call the bounding surface Σ.

Let $E(x, y, z, t)$ be the distribution over Σ defined by

$$\text{(VII, 2; 13)} \quad \langle E, \varphi\rangle = \int_0^\infty \frac{dt}{4\pi t} \iint_{x^2+y^2+z^2=v^2t^2} \varphi(x, y, z, t)\, dS.$$

The second integral is taken over the sphere having the origin as the centre and radius vt in the hyperplane $t =$ constant; dS is the element of surface area. Let

$$\text{(VII, 2; 14)} \quad \bar{\varphi}(s, t) = \frac{1}{4\pi s^2} \iint_{x^2+y^2+z^2=s^2} \varphi(x, y, z, t)\, dS$$

[$\bar{\varphi}(s, t)$ is defined for fixed t, and is the mean value of the function

$$(x, y, z) \rightarrow \varphi(x, y, z, t)$$

over the sphere centre O and radius s, in the hyperplane $t =$ constant.] Then we have

$$\text{(VII, 2; 15)} \qquad \overline{\frac{\partial^2\varphi}{\partial t^2}} = \frac{\partial^2\bar{\varphi}}{\partial t^2}$$

and

$$\text{(VII, 2; 16)} \quad \overline{\Delta\varphi} = \Delta\bar{\varphi} = \frac{\partial^2\bar{\varphi}}{\partial s^2} + \frac{2}{s}\frac{\partial\bar{\varphi}}{\partial s} \qquad \text{[cf. equation (II, 2; 59)].}$$

The equation (VII, 2; 13) can be written

$$\text{(VII, 2; 17)} \qquad \langle E, \varphi\rangle = \int_0^\infty v^2 t\bar{\varphi}(vt, t)\, dt.$$

Taking account of (VII, 2; 15) and (VII, 2; 16) we have

$$\begin{aligned}\text{(VII, 2; 18)} \quad \langle \Box_3 E, \varphi\rangle = \langle E, \Box_3\varphi\rangle &= \frac{1}{v^2}\left\langle E, \frac{\partial^2\varphi}{\partial t^2}\right\rangle - \langle E, \Delta\varphi\rangle \\ &= \int_0^\infty \left[\frac{1}{v^2} v^2 t \frac{\partial^2\bar{\varphi}}{\partial t^2} - v^2 t\left(\frac{\partial^2\bar{\varphi}}{\partial s^2} + \frac{2}{s}\frac{\partial\bar{\varphi}}{\partial s}\right)\right]_{s=vt} dt \\ &= \int_0^\infty \left[\frac{1}{v^2}\frac{\partial^2}{\partial t^2}(sv\bar{\varphi}) - \frac{\partial^2}{\partial s^2}(sv\bar{\varphi})\right]_{s=vt} dt.\end{aligned}$$

But

$$\text{(VII, 2; 19)} \quad \left(\frac{1}{v^2}\frac{\partial^2 U}{\partial t^2} - \frac{\partial^2 U}{\partial s^2}\right)_{s=vt} = \frac{1}{v}\frac{d}{dt}\left[\left(\frac{1}{v}\frac{\partial U}{\partial t} - \frac{\partial U}{\partial s}\right)_{s=vt}\right].$$

It follows that

$$\begin{aligned}\text{(VII, 2; 20)} \quad \langle E, \Box_3\varphi\rangle &= \int_0^\infty \frac{1}{v}\frac{d}{dt}\left[\left\{\frac{1}{v}\frac{\partial}{\partial t}(sv\bar{\varphi}) - \frac{\partial}{\partial s}(sv\bar{\varphi})\right\}_{s=vt}\right] dt \\ &= \frac{1}{v}\left[\left\{\frac{1}{v}\frac{\partial}{\partial t}(sv\bar{\varphi}) - \frac{\partial}{\partial s}(sv\bar{\varphi})\right\}_{s=vt}\right]_{t=0}^{t=\infty}\end{aligned}$$

and since $\varphi(x, y, z, t)$ has a bounded support, this last expression is equal to

$$\text{(VII, 2; 21)} \quad -\left[\frac{1}{v^2} vs \frac{\partial\bar{\varphi}}{\partial t} - \frac{1}{v}\left(v\bar{\varphi} + sv\frac{\partial\bar{\varphi}}{\partial s}\right)\right]_{s=t=0}$$
$$= \bar{\varphi}(0, 0) = \varphi(0, 0, 0, 0) = \langle\delta, \varphi\rangle.$$

Consequently the distribution $E(x, y, z, t)$ defined by equation (VII, 2; 13) is an elementary solution of $\Box_3$.

Solution of the Cauchy problem for the wave equation in three dimensions

In this case Cauchy's problem is stated as follows : Find a solution of equation (VII, 2; 3) that satisfies the given initial conditions

$$\text{(VII, 2; 22)} \qquad u(x, y, z, 0) = u_0(x, y, z),$$
$$\frac{\partial u}{\partial t}(x, y, z, 0) = u_1(x, y, z).$$

We shall give an explicit solution to this problem and shall assume that the solution obtained is the only one.

First we shall show that, for all functions $f(\xi, \eta, \zeta)$ which are sufficiently regular,

(VII, 2; 23)

$$I(x,y,z,t;f) = \frac{1}{v^2}\iiint_{R^3} E(x-\xi, y-\eta, z-\zeta, t)\, f(\xi, \eta, \zeta)\, d\xi\, d\eta\, d\zeta$$
$$= \frac{1}{4\pi v^2 t}\iint_{\gamma(x,y,z;\,vt)} f(\xi, \eta, \zeta)\, dS,$$

where $\gamma(x, y, z; vt)$ is the sphere with centre (x, y, z) and radius vt, in the hyperplane $t =$ constant, is a solution of equation (VII, 2; 3). Now equation (VII, 2; 23) can be written

$$\text{(VII, 2; 24)} \quad I(x, y, z, t; f) = \frac{1}{4\pi v^2 t}\iint_{\gamma(0,0,0;\,vt)} f(x-\xi, y-\eta, z-\zeta)\, dS$$

from which it follows that

$$\text{(VII, 2; 25)} \qquad \Delta I(x,y,z,t;f) = I(x,y,z,t;\Delta f).$$

But we also have

$$\text{(VII, 2; 26)} \qquad I(x, y, z, t; f) = t\bar{f}(vt),$$

where $\bar{f}(r)$ is the mean value of the function $f(r)$ over the sphere with centre (x, y, z) and radius r. We will treat $\bar{f}(r)$ as a function of r alone. Then, if we replace r by the distance in R^3 from the origin, it becomes a function on R^3 such that its Laplacian can be put in the classical form (II, 2; 50). Thus we have successively :

$$\text{(VII, 2; 27)} \qquad \frac{\partial I}{\partial t}(x, y, z, t; f) = \bar{f}(vt) + vt\bar{f}'(vt)$$

and

$$\text{(VII, 2; 28)} \quad \frac{\partial^2 I}{\partial t^2}(x, y, z, t; f) = 2v\bar{f}'(vt) + v^2 t\bar{f}''(vt)$$
$$= v^2 t\left(\frac{2}{vt}\bar{f}'(vt) + \bar{f}''(vt)\right) = v^2 t\Delta\bar{f}(vt)$$
$$= v^2 t\overline{\Delta f}(vt) = v^2 I(x, y, z, t; \Delta f).$$

The equations (VII, 2; 25) and (VII, 2; 28) show that $I(x, y, z, t; f)$ is indeed a solution of equation (VII, 2; 3).

As $t \to 0$ we see that

$$\text{(VII, 2; 29)} \quad I(x, y, z, t; f) \to 0 \qquad \text{from (VII, 2; 26),}$$

$$\text{(VII, 2; 30)} \quad \frac{\partial}{\partial t} I(x, y, z, t; f) \to \bar{f}(0) = f(x, y, z) \qquad \text{from (VII, 2; 27),}$$

$$\text{(VII, 2; 31)} \quad \frac{\partial^2}{\partial t^2} I(x, y, z, t; f) \to 0 \qquad \text{from (VII, 2; 28).}$$

It is then immediately verified that

$$\text{(VII, 2; 32)} \quad u(x, y, z, t) = I(x, y, z, t; u_1) + \frac{\partial}{\partial t} I(x, y, z, t; u_0)$$

is a solution of the Cauchy problem (VII, 2; 22). In fact as $t \to 0$,

$$\text{(VII, 2; 33)} \quad I(x, y, z, t; u_1) \to 0 \qquad \text{from (VII, 2; 29),}$$

$$\text{(VII, 2; 34)} \quad \frac{\partial}{\partial t} I(x, y, z, t; u_0) \to u_0(x, y, z) \qquad \text{from (VII, 2; 30).}$$

Thus $u(x, y, z, t)$ given by (VII, 2; 32) converges to $u_0(x, y, z)$. Furthermore

$$\text{(VII, 2; 35)} \quad \frac{\partial}{\partial t} I(x, y, z, t; u_1) \to u_1(x, y, z) \qquad \text{from (VII, 2; 30)}$$

and

$$\text{(VII, 2; 36)} \quad \frac{\partial^2}{\partial t^2} I(x, y, z, t; u_0) \to 0 \qquad \text{from (VII, 2; 31).}$$

Thus $\partial u(x, y, z, t)/\partial t$ tends to $u_1(x, y, z)$ as $t \to 0$. Q.E.D.

In full, equation (VII, 2; 32) can be written

$$\text{(VII, 2; 37)} \quad u(x, y, z, t) = \frac{1}{4\pi v^2 t} \iint_{\Upsilon(x, y, z; vt)} u_1(\xi, \eta, \zeta)\, dS$$
$$+ \frac{\partial}{\partial t}\left(\frac{1}{4\pi v^2 t} \iint_{\Upsilon(x, y, z; vt)} u_0(\xi, \eta, \zeta)\, dS\right).$$

Solution of the Cauchy problem for vibrating membranes. Method of descent

Let $g(\xi, \eta)$ be a sufficiently regular function of the two variables ξ and η. Considered as a function of the three variables ξ, η, and ζ, $g(\xi, \eta)$ is a function which is independent of ζ, which can be written

(VII, 2; 38) $$g(\xi, \eta) = g(\xi, \eta) \otimes 1_\zeta = \tilde{g}(\xi, \eta, \zeta).$$

Now evaluate $I(x, y, z, t; \tilde{g})$ with the aid of (VII, 2; 23) :

(VII, 2; 39) $$I(x, y, z, t; \tilde{g}) = \frac{1}{4\pi v^2 t} \iint_{\gamma(x,\, y,\, z;\, vt)} (g(\xi, \eta) \otimes 1_\zeta)\, dS.$$

For the calculations which follow we refer to Figure (VII, 13).

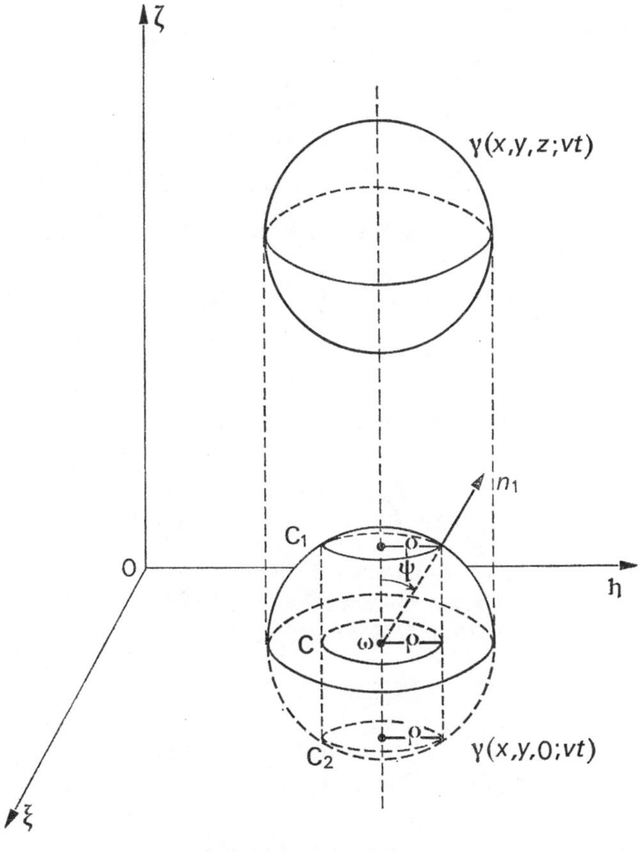

Figure (VII, 13)

The integral which appears as the right-hand side of (VII, 2; 39) is independent of z and we have

(VII, 2; 40)

$$I(x, y, z, t; \tilde{g}) = I(x, y, 0, t; \tilde{g}) = \frac{1}{4\pi v^2 t} \iint_{\gamma(x, y, 0; vt)} (g(\xi, \eta) \otimes 1_\zeta)\, dS.$$

For $0 \leqslant \rho \leqslant vt$, we denote by $\bar{g}(\rho)$ the mean value of the function $g(\xi, \eta)$ on the circle C (in the (ξ, η) plane) which has centre $\omega = (x, y)$ and radius ρ. Thus $\bar{g}(\rho)$ is also the mean value of $g(\xi, \eta, \zeta)$ on one of two parallel circles C_1 and C_2, which have radius ρ and are on the sphere $\gamma(x, y, 0; vt)$. Denote by $\overrightarrow{\omega n_1}$ the exterior normal to the sphere at a point of C_1 and set

(VII, 2; 41)
$$\psi = (\overrightarrow{O\zeta}, \overrightarrow{\omega n_1}).$$

We then have

(VII, 2; 42)
$$\rho = vt \sin \psi,$$

hence

(VII, 2; 43)
$$d\psi = \frac{d\rho}{\sqrt{v^2t^2 - \rho^2}}.$$

Since

(VII, 2; 44)
$$\iint_{\gamma(x, y, 0; vt)} (g(\xi, \eta) \otimes 1_\zeta)\, dS = 2 \int_0^{\frac{\pi}{2}} vt\, d\psi (2\pi\rho \bar{g}(\rho)),$$

we finally have

(VII, 2; 45)
$$I(x, y, z, t; \tilde{g}) = \frac{1}{v} \int_0^{vt} \frac{\bar{g}(\rho)}{\sqrt{v^2t^2 - \rho^2}}\, \rho\, d\rho.$$

This expression depends only on x, y, t, and g. Put

(VII, 2; 46)
$$J(x, y, t; g) = I(x, y, z, t; \tilde{g}) = \frac{1}{v} \int_0^{vt} \frac{\bar{g}(\rho)}{\sqrt{v^2t^2 - \rho^2}}\, \rho\, d\rho.$$

For all functions $g(\xi, \eta)$ which are sufficiently regular, $J(x, y, t; g)$ is a solution of equation (VII, 2; 1), the equation of vibrating membranes. Actually we know from equation (VII, 2; 3) that

(VII, 2; 47)
$$\square_3 I(x, y, z, t; \tilde{g}) = 0.$$

However, for a function g of two variables ξ and η only, $I(x, y, z, t; \tilde{g})$ is independent of z, and we have

(VII, 2; 48)
$$\frac{\partial^2}{\partial z^2} I(x, y, z, t; \tilde{g}) = 0$$

and it follows that

(VII, 2; 49) $\quad \Box_3 I(x, y, z, t; \tilde{g}) = \Box_2 I(x, y, z, t; \tilde{g}) = 0.$

Hence

(VII, 2; 50) $\quad \Box_2 J(x, y, t; g) = \Box_2 I(x, y, z, t; \tilde{g}) = 0.$

Cauchy's problem for the vibrating membrane can be stated as follows: To find a solution of the equation (VII, 2; 1) which satisfies the following initial conditions:

(VII, 2; 51)
$$u(x, y, 0) = u_0(x, y),$$
$$\frac{\partial}{\partial t} u(x, y; 0) = u_1(x, y).$$

One solution of the problem (VII, 2; 51) is given by

(VII, 2; 52) $\quad u(x, y, t) = J(x, y, t; u_1) + \dfrac{\partial}{\partial t} J(x, y, t; u_0).$

In fact, by putting

(VII, 2; 53)
$$\tilde{u}_0(\xi, \eta, \zeta) = u_0(\xi, \eta) \otimes 1_\zeta$$

and

(VII, 2; 54)
$$\tilde{u}_1(\xi, \eta, \zeta) = u_1(\xi, \eta) \otimes 1_\zeta,$$

equation (VII, 2; 52) can be written

(VII, 2; 55) $\quad u(x, y, t) = I(x, y, z, t; \tilde{u}_1) + \dfrac{\partial}{\partial t} I(x, y, z, t; \tilde{u}_0).$

We immediately deduce that, when $t \to 0$,

(VII, 2; 56)
$$u(x, y, t) \to \tilde{u}_0(x, y, z) = u_0(x, y),$$

using equations (VII, 2; 33) and (VII, 2; 34), and

(VII, 2; 57)
$$\frac{\partial}{\partial t} u(x, y, t) \to \tilde{u}_1(x, y, z) = u_1(x, y)$$

from equations (VII, 2; 35) and (VII, 2; 36).

For a sufficiently regular function $g(\xi, \eta)$ Consider the expression

(VII, 2; 58)
$$\frac{1}{2\pi v} \iint_{D(x, y;\, vt)} \frac{g(\xi, \eta)}{\sqrt{v^2 t^2 - (x - \xi)^2 - (y - \eta)^2}}\, d\xi\, d\eta,$$

where $D(x, y; vt)$ is a disc with centre (x, y) and radius vt. Let

(VII, 2; 59) $\quad x - \xi = \rho \cos \theta, \qquad y - \eta = \rho \sin \theta.$

Then (VII, 2; 58) is equal to

$$\frac{1}{2\pi v}\int_0^{vt}\frac{\rho\,d\rho}{\sqrt{v^2t^2-\rho^2}}\int_0^{2\pi} g(\xi,\eta)\,d\theta.$$

But $\int_0^{2\pi} g(\xi,\eta)\,d\theta$ is simply $2\pi\tilde{g}(\rho)$. Thus it follows that

$$\text{(VII, 2; 60)}\quad \frac{1}{2\pi v}\iint_{D(x,y;\,vt)}\frac{g(\xi,\eta)}{\sqrt{v^2t^2-(x-\xi)^2-(y-\eta)^2}}\,d\xi d\eta = J(x,y,t;g).$$

The solution (VII, 2; 52) of the Cauchy problem (VII, 2; 51) for the vibrating membranes can thus be written as

(VII, 2; 61)

$$u(x,y,t) = \frac{1}{2\pi v}\left\{\iint_{D(x,y;\,vt)}\frac{1}{\sqrt{v^2t^2-(x-\xi)^2-(y-\eta)^2}}\,u_1(\xi,\eta)\,d\xi d\eta + \frac{\partial}{\partial t}\iint_{D(x,y;\,vt)}\frac{1}{\sqrt{v^2t^2-(x-\xi)^2-(y-\eta)^2}}\,u_0(\xi,\eta)\,d\xi d\eta\right\}.$$

It can be proved that the Cauchy problem (VII, 2; 51) has one and only one solution, given by (VII, 2; 61).

Notes

1) In the case of waves in three dimensions the value of u [given by equation (VII, 2; 37)] at the point (x, y, z) at time t depends only on the values of u_0 and u_1 of u and $\partial u/\partial t$ on the sphere with centre (x, y, z) and radius vt at time 0. There is no diffusion of the waves.

On the other hand, in the case of waves in two dimensions, the value of u [given by equation (VII, 2 ; 61)] at the point (x, y) at time t depends upon the initial conditions u_0 and u_1 over the whole disc that has centre (x, y) and radius vt at time 0. In this case there is a diffusion of the waves.

2) *Case of waves in three dimensions.* Let (x, y, z, t) be a point in R^4. The *backward wave-cone* is the cone which has the following equation

$$v^2(t-\tau)^2-(x-\xi)^2-(y-\eta)^2-(z-\zeta)^2 \geqslant 0, \qquad t \geqslant \tau,$$

and the *forward wave-cone* the cone that has the equation

$$v^2(t-\tau)^2-(x-\xi)^2-(y-\eta)^2-(z-\zeta)^2 \geqslant 0, \qquad t \leqslant \tau.$$

If, instead of solving the Cauchy problem for an initial instant 0, we solve for an initial time t_0, that is, we seek solutions of equation (VII, 2; 3) that satisfy the conditions

$$u_1(x, y, z, t_0) = u_0(x, y, z) \qquad \text{and} \qquad \frac{\partial}{\partial t}u(x, y, z; t_0) = u_1(x, y, z),$$

we see that the equation (VII, 2; 32) giving $u(x, y, z, t)$ must be replaced by

$$u(x, y, z, t) = \mathrm{I}(x, y, z, t - t_0; u_1) + \frac{\partial}{\partial t}\mathrm{I}(x, y, z, t - t_0; u_0) \quad \text{for} \quad t > t_0$$

and by

$$u(x, y, z, t) = -\left[\mathrm{I}(x, y, z, t_0 - t; u_1) + \frac{\partial}{\partial t}\mathrm{I}(x, y, z, t_0 - t; u_0)\right] \quad \text{for} \quad t < t_0.$$

By transforming these expressions by means of (VII, 2; 23) we see that what occurs at time t at the point (x, y, z) depends only upon what took place at time t_0 on the *surface* of the sphere $\gamma(x, y, z; v(t - t_0))$ intersected by the *surface* of the backward wave-cone at the point (x, y, z, t) when $\tau = t_0$, if $t > t_0$, and upon what occurs on the *surface* of the sphere $\gamma(x, y, z; v(t_0 - t))$ intersected by the surface of the forward wave-cone when $\tau = t_0$, if $t < t_0$.

3) *Case of waves in two dimensions.* Let (x, y, z) be a point in R^3. Then the backward (resp. forward) wave-cone at this point is the cone

$$v^2(t - \tau)^2 - (x - \xi)^2 - (y - \eta)^2 \geqslant 0, \qquad t \geqslant \tau$$

$$[\text{resp.} \quad v^2(t - \tau)^2 - (x - \xi)^2 - (y - \eta)^2 \geqslant 0, \qquad t \leqslant \tau].$$

If we solve the Cauchy problem for the initial instant t_0, then $u(x, y, t)$ is given by

$$u(x, y, t) = \mathrm{J}(x, y, t - t_0; u_1) + \frac{\partial}{\partial t}\mathrm{J}(x, y, t - t_0; u_0) \quad \text{for} \quad t > t_0,$$

and by

$$u(x, y, t) = -\left[\mathrm{J}(x, y; t_0 - t; u_1) + \frac{\partial}{\partial t}\mathrm{J}(x, y, t_0 - t; u_0)\right] \quad \text{for} \quad t < t_0.$$

Then, using equation (VII, 2; 60), we see that what occurs at time t at the point (x, y) depends only upon what occurred at the time t_0 contained *in* the disc $\mathrm{D}(x, y; v(t - t_0))$ formed by the intersection of the plane $\tau = t_0$ and the backward wave-cone at the point (x, y, t) if $t > t_0$, and upon what occurred *in* the disc $\mathrm{D}(x, y; v(t_0 - t))$ formed by the intersection of the plane $\tau = t_0$ and the forward wave-cone at the point (x, y, t) if $t < t_0$.

Bounded membranes

In the Cauchy problem (VII, 2; 51) we assumed that the membrane was of infinite extent, the given functions being defined over the whole (x, y) plane.

Now suppose that the membrane is bounded, i.e. when at rest the membrane occupies a domain G, in the (x, y) plane, with boundary Γ*. A Cauchy problem will then be given by

a) the initial values

$$\left.\begin{aligned} u_0(x, y) &= u(x, y, 0) \\ u_1(x, y) &= \frac{\partial}{\partial t} u(x, y, 0) \end{aligned}\right\} \quad \text{for} \quad (x, y) \in G; \tag{VII, 2; 62}$$

b) the boundary conditions (we shall concern ourselves only with Dirichlet type boundary conditions)

$$u(x, y, t) \equiv 0 \qquad \text{on } \Gamma \text{ for all time.} \tag{VII, 2; 63}$$

Whenever possible, we shall start by extending u_0 and u_1 to $\bar{u}_0$ and $\bar{u}_1$ in such a way that for every point $(x, y) \in \Gamma$ the function $u(x, y, t)$, given by (VII, 2; 61), in which we have replaced u_0 and u_1 by $\bar{u}_0$ and $\bar{u}_1$ respectively, is zero irrespective of the value of t.

For all (x, y, t), $u(x, y, t)$ will then be given by

(VII, 2; 64)

$$u(x, y, t) = \frac{1}{2\pi v}\left\{ \iint_{D(x, y;\, vt)} \frac{1}{\sqrt{v^2t^2 - (x - \xi)^2 - (y - \eta)^2}} \bar{u}_1(\xi, \eta)\, d\xi\, d\eta \right.$$
$$\left. + \frac{\partial}{\partial t} \iint_{D(x, y;\, vt)} \frac{1}{\sqrt{v^2t^2 - (x - \xi)^2 - (y - \eta)^2}} \bar{u}_0(\xi, \eta)\, d\xi\, d\eta \right\}.$$

Example. If G is the half-plane $x > 0$ then Γ is the axis $x = 0$. Thus we will have

$$\left.\begin{aligned} \bar{u}_0(x, y) &= -u_0(-x, y) \\ \bar{u}_1(x, y) &= -u_1(-x, y) \end{aligned}\right\} \quad \text{for} \quad x < 0. \tag{VII, 2; 65}$$

This is manifestly a generalization of the case of a string having one fixed end.

In the case when G is unbounded we shall assume the uniqueness of the solution. However, when G is bounded we shall actually show that the solution is unique.

Energy integrals. Uniqueness of solutions in the Cauchy problem

We assume, as in practice, that G is a finite domain bounded by the closed curve. By multiplying both sides of (VII, 2; 1) by $\partial u/\partial t$ and integrating over G we obtain

$$\frac{d}{dt} \iint_G \frac{1}{2v^2}\left(\frac{\partial u}{\partial t}\right)^2 dx\, dy = \iint_G \Delta u \frac{\partial u}{\partial t}\, dx\, dy. \tag{VII, 2; 66}$$

*There is no possibility of confusing Γ and the future wave-cone.

By integrating the right-hand side by parts and using (VII, 2; 63) we obtain

$$\text{(VII, 2; 67)}\quad \frac{d}{dt}\iint_G \frac{1}{2}\left\{\frac{1}{v^2}\left(\frac{\partial u}{\partial t}\right)^2+\left(\frac{\partial u}{\partial x}\right)^2+\left(\frac{\partial u}{\partial y}\right)^2\right\}dx\,dy=0,$$

and in this case the energy integral is

$$\text{(VII, 2; 68)}\quad E=\frac{1}{2}\iint_G\left[\frac{1}{v^2}\left(\frac{\partial u}{\partial t}\right)^2+\left(\frac{\partial u}{\partial x}\right)^2+\left(\frac{\partial u}{\partial y}\right)^2\right]dx\,dy.$$

From equation (VII, 2; 67) we see that this integral is constant with respect to time.

Now we establish the uniqueness of the solution of the Cauchy problem given by (VII, 2; 62) and (VII, 2; 63). The difference, U, between two solutions of equation (VII, 2; 1) is also a solution. It satisfies the initial conditions $U_0(x, y) = U_1(x, y) = 0$, and the boundary conditions $U(x, y, t) = 0$ on Γ. Consequently,

$$\text{(VII, 2; 69)}\quad E=\frac{1}{2}\iint_G\left(\frac{1}{v^2}U_1^2+\left(\frac{\partial U_0}{\partial x}\right)^2+\left(\frac{\partial U_0}{\partial y}\right)^2\right)dx\,dy=0.$$

The energy, which is zero initially, therefore remains zero for all time. Thus

$$\text{(VII, 2; 70)}\quad \frac{\partial U}{\partial t}\equiv 0,\qquad \frac{\partial U}{\partial x}\equiv 0,\qquad \text{and}\qquad \frac{\partial U}{\partial y}\equiv 0.$$

Hence U is a constant. As it is zero initially it must be zero for all time. Hence U is identically zero. Q.E.D.

2. SOLUTION OF THE CAUCHY PROBLEM FOR VIBRATING MEMBRANES BY THE METHOD OF HARMONICS*

The domain G is assumed to be finite and bounded by the closed curve Γ. We say that a solution of (VII, 2; 1) is a *stationary solution* if it has the form

$$\text{(VII, 2; 71)}\quad u(x, y, t) = U(x, y)V(t).$$

For such a function the equation (VII, 2; 1) becomes

$$\text{(VII, 2; 72)}\quad \frac{\Delta U}{U}=\frac{V''}{v^2 V}=\lambda,$$

*This method is also valid for the three-dimensional wave equation. This case will only be dealt with by examples.

where λ can only be a constant. Conversely, if λ is a constant the solutions $U(x, y)$ and $V(t)$ of the equations

(VII, 2; 73) $$\Delta U - \lambda U = 0,$$

(VII, 2; 74) $$V'' - \lambda v^2 V = 0$$

define a stationary solution of (VII, 2; 1).

We propose to seek the stationary solutions of (VII, 2; 1) which satisfy the conditions given by (VII, 2; 63) on the bounding contour. We can assume that $V(t)$ is not identically zero. This is a necessary assumption if the membrane is not to remain at rest. Thus we can only apply (VII, 2; 63), whatever the value of t, if

(VII, 2; 75) $$U(x, y) = 0 \quad \text{on} \quad \Gamma.$$

As in the case of vibrating strings this restricts the possible values of λ. We shall assume that the bounding curve of G is "regular" and we shall use the following theorem, which we give without proof.

DIRICHLET'S THEOREM. *Consider the solutions of the partial differential equation*

(VII, 2; 73) $$\Delta U - \lambda U = 0,$$

satisfying the boundary condition

(VII, 2; 75) $$U(x, y) = 0 \quad \text{on} \quad \Gamma.$$

There exist non-zero solutions for only a denumerable set of values of λ, and these values of λ are real and strictly negative:

$$\lambda_1 = -\omega_1^2, \quad \lambda_2 = -\omega_2^2, \quad \ldots, \quad \lambda_n = -\omega_n^2, \ldots$$

For each value λ_n of λ the equation has a finite number $p(n)$ of solutions which are linearly independent and belong to $L^2(G)$. If we denote by

$$U_{n1}, U_{n2}, \ldots, U_{n, p(n)}$$

a set of linearly independent solutions corresponding to λ_n such that $(u_{ni}, u_{nj}) = \delta_{ij}$ in $L^2(G)$, then the $U_{k, j(k)}$ $(k = 1, 2, \ldots; j(k) = 1, 2, \ldots, p(k))$ form a Hilbert basis of $L^2(G)$.*

---

*The λ_n so found are the eigenvalues of the problem (VII, 2; 73) (VII, 2; 75): the $U_{n1}, U_{n2}, \ldots, U_{np(n)}$ are the eigenvectors. If $p(n) = 1$, λ_n is a simple eigenvalue; if $p(n) > 1$, λ_n is a multiple eigenvalue and $p(n)$ is the multiplicity.

Note. It is easy to see that λ must be real and negative. In fact, integration by parts, taking account of (VII, 2; 75), gives

$$\text{(VII, 2; 76)} \quad \iint_G \left\{ \left(\frac{\partial U}{\partial x}\right)^2 + \left(\frac{\partial U}{\partial y}\right)^2 \right\} dx\, dy = -\iint_G U \Delta U \, dx\, dy,$$

and the right-hand side, using (VII, 2; 73), equals

$$\text{(VII, 2; 77)} \quad -\lambda \iint_G U^2 \, dx\, dy,$$

whence the result. It remains to solve the equation

$$\text{(VII, 2; 78)} \quad V'' + \omega^2 v^2 V = 0,$$

which has a solution

$$\text{(VII, 2; 79)} \quad V(t) = a \cos \omega v t + b \sin \omega v t.$$

The most general form of the stationary wave solution is therefore

$$\text{(VII, 2; 80)} \quad u_n(x, y, t) = \sum_{j=1}^{p(n)} U_{nj}(x, y)\, (a_{nj} \cos \omega_n v t + b_{nj} \sin \omega_n v t).$$

Now we look for a solution of (VII, 2; 1) which has the form

$$\text{(VII, 2; 81)} \quad u(x, y, t) = \sum_{n=1}^{\infty} \sum_{j=1}^{j=p(n)} U_{nj}(x, y)\, (a_{nj} \cos \omega_n v t + b_{nj} \sin \omega_n v t).$$

We must also express that $u(x, y, t)$ satisfies the initial conditions (VII, 2; 62), hence

$$\text{(VII, 2; 82)} \quad u_0(x, y) = \sum_{n=1}^{\infty} \sum_{j=1}^{p(n)} U_{nj} a_{nj},$$

$$\text{(VII, 2; 83)} \quad u_1(x, y) = \sum_{n=1}^{\infty} \sum_{j=1}^{p(n)} U_{nj} b_{nj} \omega_n v.$$

Thus we have to expand the functions $u_0(x, y)$ and $u_1(x, y)$ in a series in terms of the functions U_{nj}. This is possible provided that u_0 and u_1 are in $L^2(G)$, since the U_{nj} form a total set in $L^2(G)$.

3. PARTICULAR CASES OF RECTANGULAR AND CIRCULAR MEMBRANES

Rectangular membranes

In this case G is the rectangle

$$\text{(VII, 2; 83)} \quad 0 \leqslant x \leqslant A, \quad 0 \leqslant y \leqslant B.$$

We look for $U(x, y)$ of the form

$$U(x, y) = L(x)M(y). \tag{VII, 2; 84}$$

The equation (VII, 2; 73) then becomes

$$\frac{L''}{L} + \frac{M''}{M} + \omega^2 = 0 \tag{VII, 2; 85}$$

and the condition (VII, 2; 75) can be expressed as

$$L(0) = L(A) = 0, \tag{VII, 2; 86}$$

$$M(0) = M(B) = 0. \tag{VII, 2; 87}$$

Equation (VII, 2; 85) will only be true if

$$L'' + \alpha L = 0, \tag{VII, 2; 88}$$

$$M'' + \beta M = 0, \tag{VII, 2; 89}$$

where α and β are constants which satisfy

$$\alpha + \beta = \omega^2. \tag{VII, 2; 90}$$

The problem of (VII, 2; 88) and (VII, 2; 86) has the following solutions:

$$\sin\frac{k\pi x}{A}, \text{ where } \sqrt{\alpha} = \frac{k\pi}{A} \quad \text{and } k \text{ is an integer} > 0 \tag{VII, 2; 91}$$

(compare with page 271).

Similarly the problem of (VII, 2; 89) and (VII, 2; 87) has for its solutions

$$\sin\frac{l\pi y}{B}, \text{ where } \sqrt{\beta} = \frac{l\pi}{B} \quad \text{and } l \text{ is an integer} > 0. \tag{VII, 2; 92}$$

Thus we obtain in this case all the eigenvalues of the problem (VII, 2; 73) and (VII, 2; 75):

$$\lambda_{kl} = -\omega_{kl}^2 = -\pi^2\left(\frac{k^2}{A^2} + \frac{l^2}{B^2}\right), \tag{VII, 2; 93}$$

where k and l are positive integers. We find in this particular case that the values of λ are negative real numbers. The eigenfunctions are

$$U_{kl} = \sin\frac{k\pi x}{A}\sin\frac{l\pi y}{B}. \tag{VII, 2; 94}$$

We obtain for the function $V(t)$

$$V_{kl}(t) = a_{kl}\cos\omega_{kl}vt + b_{kl}\sin\omega_{kl}vt \tag{VII, 2; 95}$$

where ω_{kl} is given by (VII, 2; 93).

The stationary modes corresponding to the integers k and l have a frequency (in time) of

(VII, 2; 96)
$$N = \frac{v}{2}\sqrt{\frac{k^2}{{}^2A} + \frac{l^2}{B^2}},$$

which is a generalization of equation (VII, 1; 108) which was obtained for vibrating strings. If $A \neq B$, the above frequency is not a multiple of the fundamental frequency

(VII, 2; 97)
$$N_0 = \frac{v}{2}\sqrt{\frac{1}{A^2} + \frac{1}{B^2}}.$$

However when $A = B$ (a square membrane)

(VII, 2; 98)
$$N = \sqrt{\frac{k^2 + l^2}{2}}\, N_0 \qquad \text{with} \qquad N_0 = \frac{\sqrt{2}}{2A}\, v.$$

We look for a solution of the equation (VII, 2; 1) in the form

(VII, 2; 99)
$$u(x, y, t) = \sum_{k,\, l=1}^{\infty} \sin\frac{k\pi x}{A} \sin\frac{l\pi y}{B} (a_{kl} \cos \omega_{kl}vt + b_{kl} \sin \omega_{kl}vt).$$

In order that this should satisfy the initial conditions we must have

(VII, 2; 100)
$$u_0(x, y) = \sum_{k,\, l=1}^{\infty} a_{kl} \sin\frac{k\pi x}{A} \sin\frac{l\pi y}{B},$$
$$u_1(x, y) = \sum_{k,\, l=1}^{\infty} v b_{kl}\pi \sqrt{\frac{k^2}{A^2} + \frac{l^2}{B^2}} \sin\frac{k\pi x}{A} \sin\frac{l\pi y}{B}.$$

Thus we must expand $u_0(x, y)$ and $u_1(x, y)$ in a Fourier series of two variables (see the example in Chapter IV) in terms of $\sin\frac{k\pi x}{A} \sin\frac{l\pi y}{B}$. Now, every function f of x and y defined in the rectangle G admits of a unique expansion in this form. In fact, for $0 \leqslant y \leqslant B$ we can extend $x \to f(x, y)$ to $(-A, 0)$ as an odd function of x. Then for $y \in (-B, 0)$ and $x \in (-A, +A)$ we may define

$$f(x, y) = -f(x, -y).$$

Thus we obtain a function defined over the rectangle

$$(-A \leqslant x \leqslant A, -B \leqslant y \leqslant B).$$

We then extend it to a periodic function of period $2A$ with respect to x and of period $2B$ with respect to y.

The coefficients a_{kl} and b_{kl} are given by the formulae

(VII, 2; 101) $$\frac{a_{kl}}{4} = \frac{1}{AB}\int_0^A dx \int_0^B u_0(x, y) \sin\frac{k\pi x}{A} \sin\frac{l\pi y}{B}\, dy$$

and

$$\frac{b_{kl}\omega_{kl}v}{4} = \frac{1}{AB}\int_0^A dx \int_0^B u_1(x, y) \sin\frac{k\pi x}{A} \sin\frac{l\pi y}{B}\, dy.$$

This demonstrates, in our particular case, that the system of functions U_{kl}, defined in (VII, 2; 94) is complete. Thus we obtain a solution to our problem and we know that it is the only one.

Circular membranes

In this case we take G to be the disc

(VII, 2; 102) $$x^2 + y^2 \leqslant R^2.$$

We shall now restrict ourselves to studying the solutions of (VII, 2; 1) which possess rotational symmetry, that is, solutions which have the form $u(r, t)$ where $r = \sqrt{x^2 + y^2}$, and which satisfy the initial conditions

(VII, 2; 103) $$u_0(r) = (r, 0), \qquad u_1(r) = \frac{\partial}{\partial t} u(r, 0),$$

and the boundary conditions

(VII, 2; 104) $$u(R, t) \equiv 0.$$

We shall look for $u(r, t)$ of the form

(VII, 2; 105) $$u(r, t) = U(r)V(t).$$

The equation (VII, 2; 73) then becomes [compare with equation (II, 2; 59)]

(VII, 2; 106) $$U'' + \frac{1}{r}U' + \omega^2 U = 0,$$

and the condition (VII, 2; 104) reduces to

(VII, 2; 107) $$U(R) = 0.$$

In (VII, 2; 106) change the variable according to

(VII, 2; 108) $$\rho = \omega r.$$

Then $\tilde{U}(\rho) = U(\rho/\omega)$ is a solution of the equation

(VII, 2; 109)
$$\tilde{U}'' + \frac{1}{\rho}\tilde{U}' + \tilde{U} = 0.$$

Thus, since we are concerned only with regular solutions, we have (see Chapter IX and Exercise IX, 5)

(VII, 2; 110)
$$\tilde{U}(\rho) = J_0(\rho),$$

where J_0 is the Bessel function of zero order. Hence we have as a solution of the equation (VII, 2; 106),

(VII, 2; 111)
$$U(r) = J_0(\omega r).$$

For equation (VII, 2 ; 107) to be satisfied we must have ω such that

(VII, 2; 112)
$$J_0(\omega R) = 0.$$

Now we know that all the zeros of J_0 are real (Chapter IX). Therefore the condition (VII, 2; 112) leads to a unique set of real values of ω. On the other hand, in (VII, 2; 111) ω and $-\omega$ give the same function $U(r)$. Thus we need retain only the positive values of ω. Let $z_1, z_2, \ldots$ be the set of positive zeros of J_0. Then the values of ω are given by

(VII, 2; 113)
$$\omega_n R = z_n \qquad \text{or} \qquad \omega_n = \frac{z_n}{R}.$$

This shows once more that in this particular case, the $\lambda_n = -\omega_n^2$ are real quantities < 0. The function $U_n(r)$ corresponding to a particular ω_n is

(VII, 2; 114)
$$U_n(r) = J_0\left(\frac{z_n r}{R}\right).$$

We obtain for the function $V(t)$

(VII, 2; 115)
$$V_n(t) = a_n \cos\frac{z_n}{R} vt + b_n \sin\frac{z_n}{R} vt.$$

The stationary mode corresponding to a particular integer n has a frequency

(VII, 2; 116)
$$N = \frac{z_n}{2\pi R} v.$$

Thus we shall look for $u(r, t)$ in the form

(VII, 2; 117)
$$u(r, t) = \sum_{n=1}^{\infty} J_0\left(\frac{z_n r}{R}\right)\left(a_n \cos\frac{z_n}{R} vt + b_n \sin\frac{z_n}{R} vt\right).$$

We find that the result which satisfies the initial conditions (VII, 2; 103) is

(VII, 2; 118)
$$u_0(r) = \sum_{n=1}^{\infty} J_0\left(\frac{z_n r}{R}\right) a_n,$$
$$u_1(r) = \sum_{n=1}^{\infty} J_0\left(\frac{z_n r}{R}\right) \frac{z_n}{R} b_n v.$$

Thus we must expand $u_0(r)$ and $u_1(r)$ in a series in terms of $J_0(z_n r/R)$. This is possible, as a result of Dirichlet's theorem, if u_0 and u_1 are sufficiently regular. The coefficients a_n and b_n are then completely determined; so also is $u(r,t)$. Thus we obtain a solution to our problem, and we know that it is the only solution.

4. THE WAVE EQUATION IN R^n

The equation

(VII, 2; 119)
$$\frac{1}{v^2}\frac{\partial^2 u}{\partial t^2} - \Delta u = 0,$$

where

(VII, 2; 120)
$$\Delta u = \sum_{i=1}^{n} \frac{\partial^2 u}{\partial x_i^2},$$

is called the *wave equation* in R^n. A detailed study of this equation is beyond the scope of this book. We shall restrict ourselves to dealing with certain aspects of it through examples.

3. The heat conduction equation

We have already come across the equation of heat conduction (Chapter V, Section 4). We propose to study here, by methods similar to those already used for the equations of vibrating strings and membranes, the solution of the simplified equation

(VII, 3; 1)
$$\frac{\partial u}{\partial t} - \frac{\partial^2 u}{\partial x^2} = 0.$$

1. SOLUTION BY THE METHOD OF PROPAGATION. CAUCHY'S PROBLEM

From Chapter V, equation (V, 4; 16) we know one elementary solution of the operator

(VII, 3; 2)
$$\frac{\partial}{\partial t} - \frac{\partial^2}{\partial x^2},$$

namely

$$\frac{Y(t)}{2\sqrt{\pi t}} e^{-\frac{x^2}{4t}}. \tag{VII, 3; 3}$$

On the other hand, it can be immediately verified that, no matter what function $f(\xi)$ is used, provided that it is sufficiently regular, the function

$$I(x, t; f) = \int_{-\infty}^{+\infty} \frac{e^{-\frac{(x-\xi)^2}{4t}}}{2\sqrt{\pi t}} f(\xi)\, d\xi \tag{VII, 3; 4}$$

is, for $x \in \mathrm{R}$ and $t > 0$, a solution of the equation (VII, 3; 1). In addition, we shall show that when $t \to 0$, $I(x, t; f) \to f(x)$. From (V, 1; 38) we have

$$\frac{e^{-\frac{\xi^2}{4t}}}{2\sqrt{\pi t}} = \mathscr{F}[e^{-4\pi^2 t\lambda^2}]. \tag{VII, 3; 5}$$

Thus we have

$$I(x, t; f) = \mathscr{F}[e^{-4\pi^2 t\lambda^2}] \underset{(\xi)}{*} f. \tag{VII, 3; 6}$$

But when $t \to 0$, $e^{-4\pi^2 t\lambda^2} \to 1$ in $\mathscr{S}'$, and it follows (see the examples in Chapter V) that $\mathscr{F}[e^{-4\pi^2 t\lambda^2}] \to \delta$ in $\mathscr{S}'$ and hence in $\mathscr{D}'$. Then, using the continuity of the convolution integral, we deduce that

$$I(x, t; f) \to \delta * f = f(x). \tag{VII, 3; 7}$$

Cauchy's problem

1) *The infinite conductor.* Cauchy's problem in this case is as follows: To find a solution of equation (VII, 3; 1) which satisfies the initial conditions

$$u_0(x) = u(x, 0). \tag{VII, 3; 8}$$

We obtain as our solution, using equation (VII, 3; 7),

$$u(x, t) = \int_{-\infty}^{+\infty} \frac{e^{-\frac{(x-\xi)^2}{4t}}}{2\sqrt{\pi t}} u_0(\xi)\, d\xi. \tag{VII, 3; 9}$$

2) *The semi-infinite conductor, with one end at the origin.* Now Cauchy's problem is : To find the solution of equation (VII, 3; 1) which satisfies the initial conditions (VII, 3; 8) and a boundary condition which may be

$$u(0, t) \equiv 0 \tag{VII, 3; 10}$$

or

$$\frac{\partial u}{\partial x}(0, t) \equiv 0 \tag{VII, 3; 11}$$

or may be an even more complicated condition, which will not concern us here.

When we have the boundary condition (VII, 3; 10) we extend u_0 to an odd function $\bar{u}_0$. When the boundary condition is (VII, 3; 11) we extend u_0 to an even function u_1. In the first case the solution of the Cauchy problem will be given by equation (VII, 3; 9), where u_0 will be replaced by $\bar{u}_0$, and in the second case the solution will again be given by equation (VII, 3; 9) where u_0 will now be replaced by u_1.

3) *Finite conductor with ends at* $x = 0$ *and* $x = L$. The Cauchy problem consists in finding a solution of equation (VII, 3; 1) satisfying the initial conditions (VII, 3; 8) and two boundary conditions. We will restrict ourselves to the following boundary conditions; either

$$u(0, t) = u(L, t) \equiv 0 \tag{VII, 3; 12}$$

or

$$\frac{\partial u}{\partial x}(0, t) = \frac{\partial u}{\partial x}(L, t) \equiv 0. \tag{VII, 3; 13}$$

It will also be necessary to extend u_0 in such a way that (VII, 3; 12) or (VII, 3; 13) is always satisfied. However, we do not insist on this point as we shall give an explicit solution of the problem by the method of harmonics.

Uniqueness of the Cauchy problem

Consider the case of a conductor of length L. The difference between two solutions of the Cauchy problem (VII, 3; 8), (VII, 3; 12) or (VII, 3; 8), (VII, 3; 13) is a solution of the equation (VII, 3; 1) satisfying the initial conditions

$$U_0(x) = 0 \tag{VII, 3; 14}$$

and the boundary conditions

(VII, 3; 15) $$U(0, t) = U(L, t) \equiv 0$$

or

(VII, 3; 16) $$\frac{\partial U}{\partial x}(0, t) = \frac{\partial U}{\partial x}(L, t) \equiv 0.$$

From equation (VII, 3; 1) we deduce that

(VII, 3; 17) $$\int_0^L \frac{\partial U}{\partial t} U \, dx = \int_0^L \frac{\partial^2 U}{\partial x^2} U \, dx$$

and hence that

(VII, 3; 18) $$\frac{1}{2}\frac{d}{dt}\int_0^L U^2 dx = \left[\frac{\partial U}{\partial x} U\right]_0^L - \int_0^L \left(\frac{\partial U}{\partial x}\right)^2 dx.$$

For the boundary conditions (VII, 3; 15) or (VII, 3; 16) the term $\left[\frac{\partial U}{\partial x} U\right]_0^L$ is zero. Thus, put

(VII, 3; 19) $$G(t) = \int_0^L U^2(x, t) \, dx.$$

We deduce from (VII, 3; 19) and (VII, 3; 18) that $G(t)$ is a positive decreasing function. But $G(0)$ is zero, from (VII, 3; 14), and thus it follows that $G(t)$ is identically zero, and hence U is zero. Q. E.D.

2. THE SOLUTION OF CAUCHY'S PROBLEM BY THE METHOD OF HARMONICS

A *stationary solution* of equation (VII, 3; 1) is a solution of the form

(VII, 3; 20) $$u(x, t) = U(x)V(t).$$

For such a function the equation becomes

(VII, 3; 21) $$\frac{U''}{U} = \frac{V'}{V} = -\lambda,$$

where λ is a constant. Conversely, if λ is a constant, the solutions $U(x)$ and $V(t)$ of the differential equations

(VII, 3; 22) $$U'' + \lambda U = 0,$$

(VII, 3; 23) $$V' + \lambda V = 0,$$

define a stationary solution of the equation (VII, 3; 1). We shall first look for stationary solutions of equation (VII, 3; 1) which satisfy the conditions (VII, 3; 12).

We may assume that $V(t)$ is not identically zero because, if it were, $u(x, t)$ would be identically zero. Thus the conditions (VII, 3; 12) can only be satisfied if we have

(VII, 3; 24) $$U(0) = U(L) = 0.$$

The solutions of (VII, 3; 22) satisfying (VII, 3; 24) can only be

(VII, 3; 25) $$\sin\sqrt{\lambda}x \quad \text{with} \quad \sqrt{\lambda} = \frac{k\pi}{L}, \quad k \text{ an integer} > 0.$$

Equation (VII, 3; 23) then has the solution

(VII, 3; 26) $$a_k e^{-k^2\pi^2 t/L^2}.$$

The general form of the stationary mode is then

(VII, 3; 27) $$u_k(x, t) = a_k e^{-k^2\pi^2 t/L^2} \sin\frac{k\pi x}{L}.$$

Thus we look for a solution of the form

(VII, 3; 28) $$u(x, t) = \sum_{k=1}^{\infty} a_k e^{-k^2\pi^2 t/L^2} \sin\frac{k\pi x}{L}.$$

Writing the result, which must satisfy (VII, 3; 8), we have

(VII, 3; 29) $$u_0(x) = \sum_{k=1}^{\infty} a_k \sin\frac{k\pi x}{L}.$$

It now remains to expand $u_0(x)$ in a Fourier series in terms of $\sin k\pi x/L$, which is possible (see page 273). The coefficients a_k are given by

(VII, 3; 30) $$a_k = \frac{2}{L}\int_0^L u_0(x) \sin\frac{k\pi x}{L}\, dx.$$

Thus here again we obtain an explicit solution of the Cauchy problem (VII, 3; 8), (VII, 3; 12) and we know that it is the only one. In order to solve the Cauchy problem (VII, 3; 8), (VII, 3; 13) by this method we shall have to solve the same equations (VII, 3; 22) and (VII, 3; 23), but the condition (VII, 3; 24) will be replaced by the conditions

(VII, 3; 31) $$\frac{dU}{dx}(0) = \frac{dU}{dx}(L) = 0.$$

The solutions of (VII, 3; 22) which satisfy (VII, 3; 31) are

(VII, 3; 32) $$\cos\sqrt{\lambda}x, \quad \text{where} \quad \sqrt{\lambda} = \frac{k\pi}{L}, \quad k \text{ an integer} \geqslant 0.$$

The solutions of (VII, 3; 23) are then

$$\text{(VII, 3; 33)} \qquad V_0(t) = a_0, \qquad V_k(t) = a_k e^{-k^2\pi^2 t/L^2} \quad \text{for} \quad k > 0.$$

Thus we look for a solution of the form

$$\text{(VII, 3; 34)} \qquad u(x, t) = a_0 + \sum_{k=1}^{\infty} a_k e^{-\frac{k^2\pi^2 t}{L^2}} \cos \frac{k\pi x}{L}.$$

Finally, the result which satisfies (VII, 3; 8) is

$$\text{(VII, 3; 35)} \qquad u_0(x) = \sum_{k=0}^{\infty} a_k \cos \frac{k\pi x}{L}.$$

An expansion of $u_0(x)$ as a Fourier series in $\cos k\pi x/L$ completely determines the coefficients a_k. They are given by the equations

$$\text{(VII, 3; 36)} \qquad a_0 = \frac{1}{L} \int_0^L u_0(x)\, dx,$$

$$a_k = \frac{2}{L} \int_0^L u_0(x) \cos \frac{k\pi x}{L}\, dx.$$

EXERCISES FOR CHAPTER VII

Exercise VII-1 (Plucked strings)
Consider a vibrating string, of length L, fixed at both ends $x = 0$ and $x = L$ (Figure VII, 14). Initially the string is plucked at two points A and C in such a way that the initial velocity is zero and the initial position

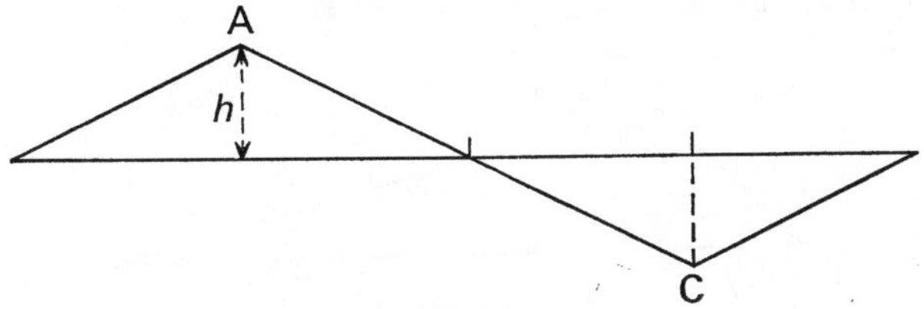

$u_0(x)$
Figure (VII, 14)

of the string is defined by the curve given by the function $u_0(x) = \frac{4h}{L} v_0(x)$, where $h > 0$ and

$$v_0(x) = \begin{cases} x, & 0 \leqslant x \leqslant L/4, \\ -(x - L/2), & L/4 \leqslant x \leqslant 3L/4, \\ x - L, & 3L/4 \leqslant x \leqslant L. \end{cases}$$

By using the method of stationary modes find the position of the string after time t. What are the frequencies of the harmonics obtained? According to what ratio do their intensities decrease?

Exercise VII-2.

This is a similar problem (Figure VII, 15) to the preceding one, but now we assume that the string is plucked at only one point and that its initial position is defined by the curve given by the function $u_0(x) = \frac{2h}{L} v_0(x)$,

$$v_0(x) = \begin{cases} x, & 0 \leqslant x \leqslant L/2, \\ L - x, & L/2 \leqslant x \leqslant L. \end{cases}$$

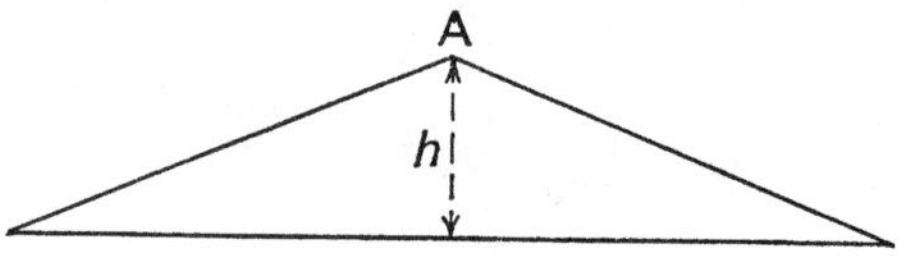

$u_0(x)$
Figure (VII, 15)

Exercise VII-3.
Another similar problem (Figure VII, 16) but now with

$$u_0(x) = \begin{cases} \frac{h}{l} x, & 0 \leqslant x \leqslant l, \\ \frac{h}{l - L}(x - L), & l \leqslant x \leqslant L, \end{cases}$$

where L and h are given by $h > 0$, $0 < l < L$.

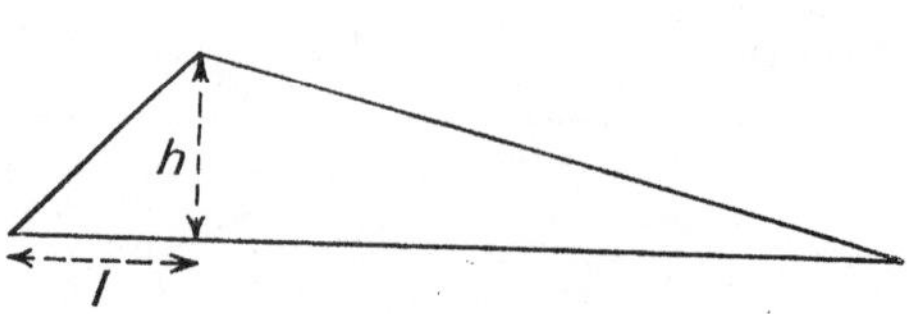

$u_0(x)$
Figure (VII, 16)

Exercise VII-4

Find by the method of harmonics the position at time t of a vibrating string of length L with both ends fixed, whose initial velocity is given as zero and whose initial displacement is defined by the curve given by the function $u_0(x) = x(L - x)$. What frequencies have the harmonics, and in what ratio do their intensities decrease?

Exercise VII-5 (Struck strings)

Find by the method of harmonics the position at time t of a vibrating string of length L with both ends fixed, whose initial position is that part of the x axis for which $0 \leqslant x \leqslant L$ and whose initial velocity is represented by a Dirac distribution at a point B having coordinate value b, where

$$0 < b < L \quad \left(\frac{\partial u}{\partial t}(x, 0) = \delta_{(b)}\right).$$

Exercise VII-6

Find the solution of the equation of vibrating strings satisfying the following Cauchy conditions:

$$u_0(x) = \begin{cases} 1 & \text{if } 2p < x < 2p + 1, \\ 0 & \text{if } 2p + 1 < x < 2p + 2, \end{cases} \qquad p \text{ an integer,}$$

$$u_1(x) = \sum_{n=-\infty}^{+\infty} \delta_{(n)}.$$

Exercise VII-7

Find the solution of

$$\frac{\partial^2 u}{\partial x^2} - \frac{1}{v^2}\frac{\partial^2 u}{\partial t^2} = \sin(\alpha x - \omega t)$$

satisfying

$$u(x, 0) = 0 \quad \text{and} \quad \frac{\partial u}{\partial t}(x, 0) = 0,$$

where $\dfrac{\omega}{\alpha} = c$.

Exercise VII-8

Examine the propagation of a plane wave which travels with a velocity v_1 when $x > 0$ and with a velocity v_2 when $x < 0$ and which satisfies the conditions

$$u(x, 0) = 0, \qquad \frac{\partial u}{\partial t}(x, 0) = \psi(x),$$

$\psi(x)$ being zero for $x \leqslant 0$.

Exercise VII-9

Let

(1) $$\Lambda_n = \frac{\partial^n}{\partial y^n} + a_1 \frac{\partial^{n-1}}{\partial y^{n-1}} \frac{\partial}{\partial x} + \cdots + a_p \frac{\partial^{n-p}}{\partial y^{n-p}} \frac{\partial^p}{\partial x^p} + \cdots + a_n \frac{\partial^n}{\partial x^n},$$

where the a_i are given complex numbers. It is required to find solutions of

(2) $$\Lambda_n(u(x, y)) = 0$$

in the form $u(x, y) = f(x + \lambda y)$, where λ is a root of the characteristic equation

(3) $$\lambda^n + a_1\lambda^{n-1} + \cdots + a_p\lambda^{n-p} + \cdots + a_n = 0.$$

In the case when the characteristic equation (3) has n real roots $\lambda_1, \ldots, \lambda_n$, equation (1) can be written

$$\Lambda_n = \left(\frac{\partial}{\partial y} - \lambda_1 \frac{\partial}{\partial x}\right)\left(\frac{\partial}{\partial y} - \lambda_2 \frac{\partial}{\partial x}\right) \cdots \left(\frac{\partial}{\partial y} - \lambda_n \frac{\partial}{\partial x}\right).$$

(*a*) Show that, if all the λ_i are distinct, every solution of (2) has the form

(4) $$u(x, y) = \sum_{k=1}^{n} F_k(x + \lambda_k y),$$

where F_k are arbitrary functions. This can be proved by induction on n.

(*b*) How must (4) be modified to cater for the case when (3) has repeated roots?

Exercise VII-10

By considering the wave equation in R^n,

$$\frac{1}{v^2}\frac{\partial^2 u}{\partial t^2} - \Delta u = 0; \qquad \Delta = \sum_{i=1}^{n} \frac{\partial^2}{\partial x_i^2},$$

for the region

$$0 \leqslant x_i \leqslant A_i, \qquad i = 1, 2, \ldots n,$$

find all stationary solutions of the form

$$u(x_1, x_2, \ldots, x_n, t) = \left[\prod_{i=1}^{n} U_i(x_i)\right] V(t)$$

for which

$$U_i(0) = U_i(A_i) = 0, \qquad i = 1, 2, \ldots n.$$

For $n = 2$ find those solutions which satisfy the initial conditions

$$\frac{\partial}{\partial t} u(x_1, x_2, 0) = 0, \qquad u(x_1, x_2, 0) = x_1 x_2 (A_1 - x_1)(A_2 - x_2).$$

Exercise VII-11
For the equation

$$\frac{\partial^4 u}{\partial x^4} + \frac{\partial^2 u}{\partial t^2} = 0,$$

find every stationary solution of the form

$$u(x, t) = U(x)V(t)$$

which satisfies in the domain $0 \leqslant x \leqslant 1$ the boundary conditions

$$u(0, t) = u(1, t) = 0 \qquad \text{and} \qquad \frac{\partial^2 u}{\partial x^2}(0, t) = \frac{\partial^2 u}{\partial x^2}(t, 1) = 0.$$

Find those solutions which satisfy the initial conditions

$$\frac{\partial u}{\partial t}(x, 0) = 0, \qquad u(x, 0) = x^4 - 2x^3 + x.$$

Exercise VII-12
By considering the heat conduction equation

$$(1) \qquad \frac{\partial u}{\partial t} - \frac{\partial^2 u}{\partial x^2} = 0$$

for a straight conductor of length L occupying the region $0 \leqslant x \leqslant L$, find all the stationary solutions of equation (1) which satisfy the boundary conditions

$$\frac{\partial}{\partial x} u(0, t) = \frac{\partial}{\partial x} u(L, t) = 0.$$

Exercise VII-13
This is a similar problem to the preceding one but with the boundary conditions

$$u(0, t) = 0, \qquad \frac{\partial u}{\partial x}(L, t) = 0.$$

Exercise VII-14
Another similar problem but with the boundary conditions

$$\frac{\partial u}{\partial x}(0, t) = 0, \qquad u(L, t) = 0.$$

Exercise VII-15 (Heat conduction equation in two dimensions)
Consider the heat conduction equation for a plane slab (P) :

$$c\frac{\partial u}{\partial t} = \gamma \Delta u, \qquad \Delta = \frac{\partial^2}{\partial x^2} + \frac{\partial^2}{\partial y^2}$$

where $u(x, y, t)$ is the temperature at a point (x, y) of the slab at time t, c the thermal capacity per unit area, and γ the thermal conductivity.

(*a*) Assume that (P) is the plane region defined by $x^2 + y^2 \leqslant R^2$. Find all stationary solutions of the form

$$u(x, y, t) = U(r)\, V(t), \qquad t = \sqrt{x^2 + y^2},$$

for which u is always zero at the edge of the slab ($r = R$).

(*b*) Assume that (P) is the plane region defined by $R_1^2 \leqslant x^2 + y^2 \leqslant R_2^2$. Find all stationary solutions of the form $U(r)V(t)$ for which $u(x, y, t)$ is always zero at the edges of the slab ($r = R_1$ and $r = R_2$).

(*c*) Assume that (P) is the plane region defined by $0 \leqslant x \leqslant a$, $0 \leqslant y \leqslant b$. Find all the stationary solutions of the form $U(x)V(y)W(t)$ for which $u(x, y, t)$ is always zero at the edges of the slab.

Exercise VII-16 (Heat conduction equation in three dimensions)
Find all the stationary solutions of the equation

$$c\frac{\partial u}{\partial t} = \gamma \Delta u, \qquad \Delta = \frac{\partial^2}{\partial x^2} + \frac{\partial^2}{\partial y^2} + \frac{\partial^2}{\partial z^2}$$

which have the form

$$u(x, y, z, t) = U\left(\sqrt{x^2 + y^2}\right) V(z)\, W(t),$$

for which $u(x, y, z, t)$ is zero at the boundaries of the domain between the two planes $z = 0$ and $z = b$ and the two cylinders

$$x^2 + y^2 \leqslant R_1^2, \qquad x^2 + y^2 \leqslant R_2^2.$$

Exercise VII-17 (Cauchy's problem for the equation of vibrating membranes)
We will be concerned with the following problem: to solve directly the equation of vibrating membranes

$$(1) \qquad \frac{1}{v^2}\frac{\partial^2 u}{\partial t^2} - \Delta u = 0, \qquad \Delta = \frac{\partial^2}{\partial x^2} + \frac{\partial^2}{\partial y^2}$$

so that the solution satisfies the initial conditions

$$(2) \qquad \begin{aligned} u(x, y, 0) &= u_0(x, y), \\ \frac{\partial u}{\partial t}(x, y, 0) &= u_1(x, y). \end{aligned}$$

(*a*) Let $f(\xi, \eta)$ be a sufficiently regular function, and

$$(3) \qquad I(x, y, t; f) = \frac{1}{2\pi v}\iint_{\gamma(x,\, y;\, vt)} \frac{1}{\sqrt{v^2t^2 - (x-\xi)^2 - (y-\eta)^2}} f(\xi, \eta)\, d\xi\, d\eta,$$

where $\gamma(x, y, vt)$ is the disc with centre (x, y) and radius vt.

(i) Show that

$$\Delta I(x, y, t; f) = I(x, y, t; \Delta f).$$

ii) By changing the variables according to

$$x - \xi = r \cos \theta,$$
$$y - \eta = r \sin \theta,$$

show that

$$(4) \qquad \mathrm{I}(x, y, t; f) = \frac{1}{v} \int_0^{vt} \frac{\bar{f}(r)}{\sqrt{v^2t^2 - r^2}}\, r\, dr,$$

where

$$\bar{f}(r) = \frac{1}{2\pi} \int_0^{2\pi} f(\xi, \eta)\, d\theta$$

is the mean value of f on the circle with centre (x, y) and radius r. Show that

$$\frac{\partial^2}{\partial t^2} \mathrm{I}(x, y, t; f) = v \int_0^{vt} \left(\frac{\bar{f}'(r)}{r} + \bar{f}''(r) \right) \frac{r\, dr}{\sqrt{v^2t^2 - r^2}}.$$

In the course of this calculation we shall first of all integrate by parts the second integral on the right-hand side of (4) and then the integral

$$\int_0^{vt} \frac{\bar{f}'(r)\, dr}{\sqrt{v^2t^2 - r^2}}.$$

(iii) Deduce from (i) and (ii) that $\mathrm{I}(x, y, t; f)$ is a solution of (1).

(*b*) (i) Show that, when $t \to 0$,

1. $\mathrm{I}(x, y, t; f) \to 0$;
2. $\frac{\partial}{\partial t} \mathrm{I}(x, y, t; f) \to \bar{f}(0) = f(x, y)$;
3. $\frac{\partial^2}{\partial t^2} \mathrm{I}(x, y, t; f) \to 0$.

(ii) Deduce that the solution of (1) which satisfies the conditions (2) is

$$u(x, y, t) = \mathrm{I}(x, y, t; u_1) + \frac{\partial}{\partial t} \mathrm{I}(x, y, t; u_0).$$

Exercise VII-18

(*a*) Consider the Schrödinger equation in three dimensions

$$\left[-\frac{\hbar^2}{2m} \Delta + \mathrm{V}(\vec{r}) \right] \psi(\vec{r}, t) = i\hbar \frac{\partial}{\partial t} \psi(\vec{r}, t)$$

and call the states represented by the wave function ψ in the form

$$\psi(\vec{r}, t) = \Phi(\vec{r}) e^{-i(\mathrm{E}t/\hbar)}$$

stationary states.

Show that the search for stationary states reduces to the problem of seeking the eigenvalues and eigenvectors of a certain operator, and give that operator. Also give the differential equation satisfied by $\Phi(\vec{r})$.

(*b*) Reduce the general problem to a problem in one dimension, and discuss the case of the harmonic oscillator where the potential is given by

$$V(x) = \frac{1}{2} Kx^2, \qquad K \text{ being a constant.}$$

(i) Give the equation satisfied by $\Phi(x)$.

(ii) By putting $K = \omega^2 m$ make a simple change of variable which will ensure that the only parameter is the quantity $\frac{2E}{\hbar\omega}$. Label this new variable z.

(iii) Make the substitution

$$\Phi(z) = e^{\alpha z^2} f(z).$$

This will make the coefficient of f'' and f independent of z. Only the coefficient of f' will depend on z.

(iv) Look for a solution in the form of the series

$$f(z) = \sum a_n z^n.$$

Give the recurrence relation between the coefficients a_n.

(v) Find the necessary and sufficient conditions for this series to be a polynomial. Deduce the eigenvalues and the stationary states.

(vi) If N is the degree of the polynomial corresponding to an eigenvalue E_N, express the coefficients of this polynomial, known as a Hermite polynomial of degree N, as functions of a_N, which is arbitrarily put equal to 2^N.

Call $H_N(z)$ the Hermite polynomial.

(vii) Give the normalized eigenvalues of the problem. Show that any two distinct eigenvalues an "orthogonal" by using the relation

$$\int_{-\infty}^{+\infty} H_\lambda(z) H_\mu(z) e^{-z^2}\, dz = 2^\lambda \lambda\,!\,(\pi)^{1/2} \delta_{\lambda\mu},$$

where λ and μ are positive integers and $\delta_{\lambda\mu}$ the Kronecker delta.

CHAPTER VIII

The gamma function

1. THE FUNCTION $\Gamma(z)$

The gamma function defined as an integral

Let

$$\text{(VIII, 1; 1)} \qquad \Gamma(z) = \int_0^{+\infty} e^{-t} t^{z-1}\, dt, \qquad z = x + iy.$$

The integral is summable for $x > 0$ (as $t \to \infty$ it is always summable, due to the factor e^{-t}; as $t \to 0$, $|e^{-t}\, t^{z-1}| \sim t^{x-1}$ and hence it is summable for $x - 1 > -1$ or $x > 0$). The summability is uniform for

$$0 < \alpha \leqslant x \leqslant A < +\infty,$$

because then

$$|e^{-t}\, t^{z-1}| \leqslant \begin{cases} t^{\alpha-1} & \text{for} \quad 0 \leqslant t \leqslant 1, \\ t^{A-1}\, e^{-t} & \text{for} \quad 1 \leqslant t < +\infty, \end{cases}$$

and consequently

$$\int_0^1 t^{\alpha-1}\, dt \quad \text{and} \quad \int_1^\infty t^{A-1}\, e^{-t}\, dt$$

are finite.

Thus $\Gamma(z)$ is a continuous function of z for $x > 0$, $e^{-t}\, t^{z-1}$ being continuous in z (and even continuous in t and z) for $t > 0$, $x > 0$ (Lebesgue's theorem: Chapter I, Theorem 45). *Formally*, we can differentiate (VIII, 1; 1) to give

$$\text{(VIII, 1; 2)} \qquad \Gamma'(z) = \int_0^{+\infty} e^{-t}\, t^{z-1} \log t\, dt.$$

This equation is valid for $x > 0$ *if the integral on the right-hand side is itself uniformly summable* for $0 < \alpha \leqslant x \leqslant A < \infty$. This follows as above because

$$\int_0^1 t^{\alpha-1} |\log t|\, dt \qquad \text{and} \qquad \int_1^\infty t^{A-1} \log t\, e^{-t}\, dt$$

are finite. We can calculate higher order derivatives in the same way. We notice in particular that

$$\text{(VIII, 1; 3)} \qquad \Gamma''(z) = \int_0^{+\infty} e^{-t}\, t^{z-1} (\log t)^2\, dt.$$

$\Gamma(z)$ is therefore a function which, for $x > 0$, possesses all derivatives (infinitely differentiable) with respect to the complex variable z (such a function is said to be holomorphic).

Functional equation

$$\Gamma(z+1) = \int_0^{+\infty} e^{-t} t^z \, dt = [-t^z e^{-t}]_{t=0}^{t\to+\infty} + z \int_0^{\infty} e^{-t} t^{z-1} \, dt.$$

(VIII, 1; 4) $$\boxed{\Gamma(z+1) = z\Gamma(z)} \qquad \text{for} \qquad x > 0.$$

Hence

$$\Gamma(z+n) = (z+n-1)(z+n-2)\ldots z\Gamma(z) \quad \text{for} \quad x > 0,$$

(VIII, 1; 5) $$\Gamma(n) = (n-1)!\,\Gamma(1) = (n-1)!$$

because $\Gamma(1) = \int_0^{+\infty} e^{-t} \, dt = 1$.

The function $\Gamma(z+1)$, defined for $x > -1$, which can also be written $z!$, takes the value $n!$ when z is a positive integer n. Notice that we also have $0! = \Gamma(1) = 1$.

Extension of Γ(z) by meams of the functional equation

Whenever $z \neq 0, -1, -2 \ldots$ the expression

$$\frac{\Gamma(z+n)}{(z+n-1)(z+n-2)\ldots(z+1)z}$$

is defined for all sufficiently large and integral $n > 0$. Its value is independent of n, because if we replace n by $n+p$, the equivalent expression becomes

$$\frac{\Gamma(z+n+p)}{(z+n+p-1)\ldots(z+1)z}$$
$$= \frac{\Gamma(z+n)(z+n+p-1)\ldots(z+n)}{(z+n+p-1)\ldots(z+n)(z+n-1)\ldots z}$$
$$= \frac{\Gamma(z+n)}{(z+n-1)\ldots z}.$$

Thus we *define* $\Gamma(z)$ for all z different from $0, -1, -2, \ldots$ by

(VIII, 1; 6)

$$\boxed{\Gamma(z) = \frac{\Gamma(z+n)}{(z+n-1)(z+n-2)\ldots(z+1)z}, \quad \text{all } n \text{ such that } x+n > 0.}$$

Naturally if $x > 0$ we have the same definition for $\Gamma(z)$ as before. This can easily be seen by putting $n = 0$. *Thus $\Gamma(z)$ is a holomorphic function of the complex variable z, provided $z \neq 0, -1, -2, \ldots$*

The poles of $\Gamma(z)$

Consider the behaviour of $\Gamma(z)$ in the neighbourhood of $z = -n$ where n is an integer $\geqslant 0$. Put $z = -n + u$. Then

$$\text{(VIII, 1; 7)} \quad \Gamma(z) = \frac{\Gamma(u+1)}{u(u-1)(u-2)\ldots(u-n)} \underset{(u \to 0)}{\sim} \frac{(-1)^n}{n!\,u}.$$

Thus at all the points $z = 0, -1, -2, \ldots, \Gamma(z)$ has *simple poles*, and the residue at the pole $z = -n$ is $(-1)^n/n!$ In particular we have the result

$$\text{(VIII, 1; 8)} \qquad \Gamma(z) \underset{(z \to 0)}{\sim} \frac{1}{z}.$$

We shall see later that $\Gamma'(1) = -\gamma$, where γ is Euler's constant. Then as $z \to 0$, $\Gamma(1 + z) = 1 - \gamma z + \cdots$, and

$$\text{(VIII, 1; 9)} \quad \Gamma(z) = \frac{\Gamma(z+1)}{z} = \frac{1}{z} - \gamma + \cdots \quad \text{for} \quad z \to 0.$$

Gaussian integral

By replacing t by t^2 in (VIII, 1; 1), we obtain

$$\text{(VIII, 1; 10)} \quad \Gamma(z) = 2\int_0^{+\infty} e^{-t^2} t^{2z-1}\,dt, \quad \text{for} \quad z = x + iy,\ x > 0.$$

In particular

$$\text{(VIII, 1; 11)} \qquad \int_0^{\infty} e^{-t^2}\,dt = \frac{1}{2}\Gamma\left(\frac{1}{2}\right) = \frac{1}{2}\sqrt{\pi},$$

as we shall see later [equation (VIII, 1; 15)].

2. THE BETA FUNCTION $B(p, q)$

Consider the product $\Gamma(p)\,\Gamma(q)$, where $\Re p > 0$, $\Re q > 0$. Now

$$\Gamma(p) = 2\int_0^{\infty} e^{-u^2}\,u^{2p-1}\,du,$$

$$\Gamma(q) = 2\int_0^{\infty} e^{-v^2} v^{2q-1}\,dv;$$

hence

$$\Gamma(p)\Gamma(q) = 4 \iint_{\substack{u \geqslant 0 \\ v \geqslant 0}} e^{-(u^2+v^2)} u^{2p-1} v^{2q-1}\, du\, dv.$$

Transforming to plane polar coordinates, $u = r\cos\theta$, $v = r\sin\theta$, we obtain

$$\Gamma(p)\,\Gamma(q) = 4 \iint_{\substack{r \geqslant 0 \\ 0 \leqslant \theta \leqslant \frac{\pi}{2}}} e^{-r^2} r^{2(p+q)-1} \cos^{2p-1}\theta\, \sin^{2q-1}\theta\, dr d\theta$$

$$= 2\int_0^\infty e^{-r^2} r^{2(p+q)-1}\, dr \cdot 2\int_0^{\pi/2} \cos^{2p-1}\theta\, \sin^{2q-1}\theta\, d\theta$$

$$= \Gamma(p+q)\cdot 2\int_0^{\pi/2} \cos^{2p-1}\theta\, \sin^{2p-1}\theta\, d\theta.$$

Hence we have finally

(VIII, 1; 12)

$$\boxed{B(p, q) = \frac{\Gamma(p)\Gamma(q)}{\Gamma(p+q)} = 2\int_0^{\pi/2} \cos^{2p-1}\theta \sin^{2q-1}\theta\, d\theta.}$$

A particular case of this formula is Wallis' integral

(VIII, 1; 13) $$W(r) = \int_0^{\pi/2} \cos^r\theta\, d\theta = \int_0^{\pi/2} \sin^r\theta\, d\theta = \frac{1}{2}\, B\left(\frac{r+1}{2}, \frac{1}{2}\right)$$

By putting $p = q = 1/2$ we find that

(VIII, 1; 14) $$(\Gamma(\tfrac{1}{2}))^2 = B\,(\tfrac{1}{2}, \tfrac{1}{2}) = 2\int_0^{\pi/2} d\theta = \pi.$$

Since $\Gamma(\frac{1}{2}) = \int_0^\infty e^{-t} \frac{dt}{\sqrt{t}} > 0$, we have

(VIII; 1, 15) $$\boxed{\Gamma(\tfrac{1}{2}) = \sqrt{\pi}.}$$

Thus we deduce

(VIII, 1; 16)

$$\Gamma(n + \tfrac{1}{2}) = (n - \tfrac{1}{2})\left(n - \frac{3}{2}\right) \cdots \frac{1}{2}\,\Gamma\,(\tfrac{1}{2})$$

$$= \frac{1\cdot 3\cdot 5 \cdots (2n-1)}{2^n}\sqrt{\pi}$$

$$= \frac{(2n)!}{2^{2n} n!}\sqrt{\pi},$$

where n is a positive integer.

Equation (VIII, 1; 12) can be transformed by putting $\cos^2\theta = t$ and gives

(VIII, 1; 17) $$\boxed{B(p, q) = \frac{\Gamma(p)\Gamma(q)}{\Gamma(p+q)} = \int_0^1 t^{p-1}(1-t)^{q-1}\,dt, \quad \text{for } \mathcal{R}p > 0,\ \mathcal{R}q > 0,}$$

and we find again that

(VIII, 1; 18) $$B\left(\frac{1}{2}, \frac{1}{2}\right) = \left(\Gamma\left(\frac{1}{2}\right)\right)^2 = \int_0^1 \frac{dt}{\sqrt{t(1-t)}} = \pi.$$

Finally, as a result of the preceding equations,

$$\int_\alpha^\beta (x-\alpha)^{p-1}(\beta - x)^{q-1}\,dx = (\beta-\alpha)^{p+q-1}\frac{\Gamma(p)\Gamma(q)}{\Gamma(p+q)},$$

$$\int_0^\infty \frac{r^\alpha}{(1+r^2)^\beta}\,dr = \frac{1}{2}\,\frac{\Gamma\left(\frac{\alpha+1}{2}\right)\Gamma\left(\beta - \frac{\alpha+1}{2}\right)}{\Gamma(\beta)},$$

$\left(\text{putting } \frac{r^2}{1+r^2} = t\right)$.

3. THE COMPLEMENTARY FORMULA

(VIII, 1; 19) $$B(z, 1-z) = \Gamma(z)\,\Gamma(1-z) = \int_0^1 t^{z-1}(1-t)^{-z}\,dt = \int_0^{+\infty} \frac{u^{z-1}}{1+u}\,du$$

putting $\frac{u}{1+u} = t, \qquad \frac{t}{1-t} = u,$

This integral (which is valid for $0 < x = \mathcal{R}z < 1$) can be evaluated by means of the calculus of residues, as given in the theory of functions of a complex variable.

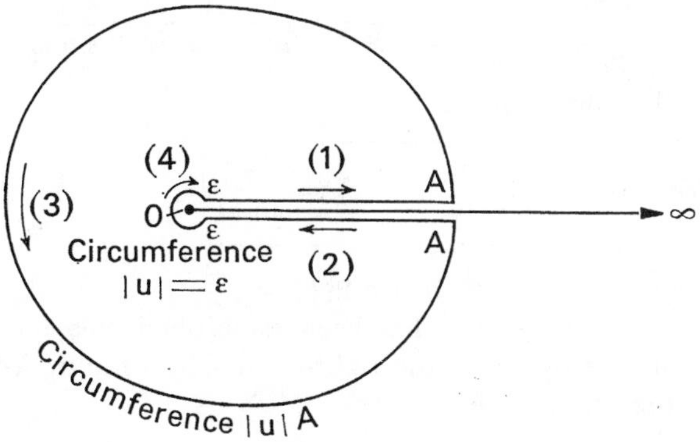

Figure (VIII, 1)

The function $u^{z-1}/(1+u)$ for fixed z, $0 < x = \mathcal{R}z < 1$, is many valued and has as its critical point $u = 0$. Consider the contour C (see Figure VIII, 1), where the parts labelled (1) and (2) are assumed to be infinitely close to the real axis. In the region bounded by the contour the function is composed of distinct uniform branches; we shall choose that branch for which, in (1), $u^{z-1} = e^{(z-1)\log u}$ and $\log u$ is real. Then

$$(a) \qquad \lim_{\varepsilon\to 0,\, A\to\infty} \int_{(1)} \frac{u^{z-1}}{1+u}\,du = \int_0^\infty \frac{u^{z-1}}{1+u}\,du,$$

$$(b) \qquad \lim_{\varepsilon\to 0,\, A\to\infty} \int_{(2)} = e^{2i\pi z}\int_\infty^0 = -e^{2i\pi z}\int_0^\infty \frac{u^{z-1}}{1+u}\,du,$$

$$(c) \qquad \lim_{A\to\infty} \int_{(3)} = 0, \quad \text{because} \quad \left|\frac{u^{z-1}}{1+u}\right| \leqslant \frac{e^{(x-1)\log A + 2\pi|y|}}{|1+u|} \lesssim C\frac{A^{x-1}}{A},$$

so that the integrals taken around a curve of length $2\pi A$ tend to zero as A^{x-1}, $x-1<0$.

$$d) \qquad \lim_{\varepsilon\to 0} \int_{(4)} = 0, \quad \text{because} \quad \left|\frac{u^{z-1}}{1+u}\right| < \frac{e^{(x-1)\log\varepsilon + 2\pi|y|}}{|1+u|} \lesssim C\varepsilon^{x-1},$$

so that the integral around a curve of length $2\pi\varepsilon$ tends to zero when ε^x, $x>0$. Finally we have

$$\lim_{\varepsilon\to 0,\, A\to\infty} \int_C \frac{u^{z-1}}{1+u}\,du = (1-e^{2i\pi z})\int_0^\infty \frac{u^{z-1}}{1+u}\,du.$$

But the integral around C can be evaluated by the method of residues. The only pole is $u = -1$, and the corresponding residue is the value of $z^{(z-1)\log u}$ for $u = -1$ which when $\log u = i\pi$ is $e^{i\pi(z-1)}$. Finally we have

$$\int_C \frac{u^{z-1}}{1+u}\,du = 2i\pi e^{i\pi(z-1)} = -2i\pi e^{i\pi z},$$

and

$$\int_0^\infty \frac{u^{z-1}}{1+u}\,du = \frac{-2i\pi e^{i\pi z}}{1-e^{2i\pi z}} = \pi\frac{2i}{e^{i\pi z}-e^{-i\pi z}} = \frac{\pi}{\sin \pi z}.$$

Hence the well-known formula

$$(\text{VIII, } 1;\ 20) \qquad \boxed{\Gamma(z)\Gamma(1-z) = \frac{\pi}{\sin \pi z}}.$$

This again gives $(\Gamma(\frac{1}{2}))^2 = \pi$, hence $\Gamma(\frac{1}{2}) = \sqrt{\pi}$.

The equation (VIII, 1; 20) has been established only for $0 < x < 1$. If we now replace z by $z+1$ the right-hand side is multiplied by (-1) and so also is the left-hand side, since

$$\Gamma(z+1)\Gamma(-z) = z\Gamma(z)\Gamma(-z) = -\Gamma(z)\Gamma(1-z).$$

Thus the equality of the two sides for $0 < x < 1$ implies their equality for all nonintegral z. For non-integral z the two sides are continuous. Finally, if z is an integer the two sides are infinite.

4. GENERALIZATION OF THE BETA FUNCTION

Since

$$\Gamma(p_i) = 2\int_0^\infty e^{-u_i^2} u_i^{2p_i-1} du_i,$$

we have

(VIII, 1; 21)

$$\Gamma(p_1)\Gamma(p_2)\ldots\Gamma(p_n) = 2^n \iint_{\substack{u_1\geqslant 0\\ \cdots\cdots \\ u_n\geqslant 0}} \cdots \int e^{-(u_1^2+\cdots+u_n^2)} u_1^{2p_1-1} u_2^{2p_2-1}\ldots u_n^{2p_n-1}\, du_1\, du_2 \cdots du_n.$$

We evaluate this multiple integral by integrating over the surface of a sphere of radius r, followed by an integration with respect to r. Putting $u_i = r\xi_i$, the integral over the sphere of radius r is

$$2^n e^{-r^2} r^{2(p_1+p_2+\cdots+p_n)-n} r^{n-1} \iint_{\substack{r=1\\ \xi_i\geqslant 0}} \cdots \int \xi_1^{2p_1-1}\ldots\xi_n^{2p_n-1}\, dS$$

and hence the right-hand side of (VIII, 1; 21) is equal to

$$2\int_0^\infty e^{-r^2} r^{2(p_1+p_2+\cdots+p_n)-1}\, dr \cdot 2^{n-1} \iint_{\substack{r=1\\ \xi_i\geqslant 0}} \cdots \int \xi_1^{2p_1-1}\ldots\xi_n^{2p_n-1}\, dS.$$

Hence

(VIII, 1; 22)
$$B(p_1, p_2, \ldots, p_n) = \frac{\Gamma(p_1)\Gamma(p_2)\ldots\Gamma(p_n)}{\Gamma(p_1+p_2+\cdots+p_n)} = 2^{n-1}\int_{\substack{r=1\\ \xi_i\geqslant 0}} \cdots \int \xi_1^{2p_1-1}\ldots\xi_n^{2p_n-1}\, dS.$$

In the case $n = 2$, we find that by putting $\xi_1 \geqslant 0$, $\xi_2 \geqslant 0$, $\xi_1 = \cos\theta$, $\xi_2 = \sin\theta$, $dS = d\theta$ on the unit circle $r = 1$, we recover the equation (VIII, 1; 12).

Equation (VIII, 1; 22) allows us to calculate the "moments" of the spherical quadrant, which are integrals of the same type as the surface integral on the right-hand side of equation (VIII, 1; 22). In particular, if all the p_i have the value $1/2$, then

(VIII, 1; 23)
$$B(\tfrac{1}{2}, \ldots, \tfrac{1}{2}) = \frac{(\Gamma(\frac{1}{2}))^n}{\Gamma\left(\dfrac{n}{2}\right)} = 2^{n-1}\frac{S_n}{2^n} = \frac{S_n}{2},$$

where S_n = area of the unit sphere, $\frac{S_n}{2^n}$ = area of the spherical quadrant. Hence

(VIII, 1; 24) $$S_n = \frac{2\left(\Gamma\left(\frac{1}{2}\right)\right)^n}{\Gamma\left(\frac{n}{2}\right)} = \frac{2\pi^{\frac{n}{2}}}{\Gamma\left(\frac{n}{2}\right)}.$$

In particular

(VIII, 1; 25) $$S_1 = 2 \quad \left[\text{which again gives } \Gamma\left(\frac{1}{2}\right) = \sqrt{\pi}\right], \quad S_2 = 2\pi,$$
$$S_3 = 4\pi, \quad S_4 = 2\pi^2, \quad S_5 = \frac{8}{3}\pi^2, \quad S_6 = \pi^3,$$

whilst from equation (VIII, 1; 22) by putting $p_i = (\alpha_i + 1)/2$ we have

(VIII, 1; 26)

$$\int_{\substack{r=1\\ \xi_i \geqslant 0}} \cdots \int \xi_1^{\alpha_1}\xi_2^{\alpha_2} \cdots \xi_n^{\alpha_n}\, dS = \frac{1}{2^{n-1}} \frac{\Gamma\left(\frac{\alpha_1+1}{2}\right)\Gamma\left(\frac{\alpha_2+1}{2}\right)\cdots\Gamma\left(\frac{\alpha_n+1}{2}\right)}{\Gamma\left(\frac{\alpha_1+\alpha_2+\cdots+\alpha_n+n}{2}\right)}.$$

In particular, if we require the centre of gravity of the octant of the sphere $r = 1$, $\xi_i \geqslant 0$ $(i = 1, 2, 3)$ in R^3, we have

(VIII, 1; 27) $$x_1 = \frac{\iint \xi_1\, dS}{\iint dS}, \qquad (i = 1, 2, 3,)$$

and $$\iint \xi_1\, dS = \frac{(\sqrt{\pi})^2}{4 \cdot 1} = \frac{\pi}{4}, \qquad \iint dS = \frac{(\sqrt{\pi^3})}{4 \cdot \frac{1}{2}\sqrt{\pi}} = \frac{\pi}{2},$$

hence $$x_i = \frac{1}{2}\, i(= 1, 2, 3), \qquad \text{whence} \qquad OG = \frac{\sqrt{3}}{2}.$$

5. GRAPHICAL REPRESENTATION OF THE FUNCTION $y = \Gamma(x)$ FOR REAL x

When $x > 0$, $\Gamma''(x)$ is always > 0 (VIII, 1; 3). Thus Γ is *convex*. Hence it is either a monotonic function, or it first decreases and then increases, and has a minimum value. Now $\Gamma(1) = \Gamma(2) = 1$. Thus we have a minimum at some $x = x_0$, $1 < x_0 < 2$, and the function decreases for $x < x_0$ and increases for $x > x_0$.

Now consider the case when $x < 0$. By taking the logarithmic derivative of both sides of equation (VIII, 1; 4) and then the ordinary derivative, we have

$$\frac{\Gamma'(x+1)}{\Gamma(x+1)} = \frac{\Gamma'(x)}{\Gamma(x)} + \frac{1}{x},$$

$$\frac{\Gamma''(x+1)\,\Gamma(x+1) - \Gamma'^2(x+1)}{\Gamma^2(x+1)} = \frac{\Gamma''(x)\,\Gamma(x) - \Gamma'^2(x)}{\Gamma^2(x)} - \frac{1}{x^2}.$$

Thus $\dfrac{\Gamma''\Gamma - \Gamma'^2}{\Gamma^2}$ decreases when we increase x by 1 and similarly it increases when we decrease x by 1. Now for $x > 0$,

$$\Gamma''\Gamma - \Gamma'^2 = \int_0^\infty e^{-t}t^{x-1}(\log t)^2\,dt \int_0^\infty e^{-t}t^{x-1}\,dt - \left(\int_0^\infty e^{-t}t^{x-1}\log t\,dt\right)^2$$

is always > 0, as can be seen by using Schwarz' inequality. Hence $\Gamma''\Gamma - \Gamma'^2$ is, *a fortiori*, greater than zero for $x < 0$. Again, *a fortiori*, $\Gamma''(x)\Gamma(x) > 0$, thus $\Gamma''(x)$ always has the same sign as $\Gamma(x)$. Consequently, where $\Gamma(x) > 0$ we have $\Gamma''(x) > 0$ and Γ is convex, and where $\Gamma(x) < 0$ we have $\Gamma''(x) < 0$ and Γ is concave*. Hence we obtain the curve shown in Figure (VIII, 2).

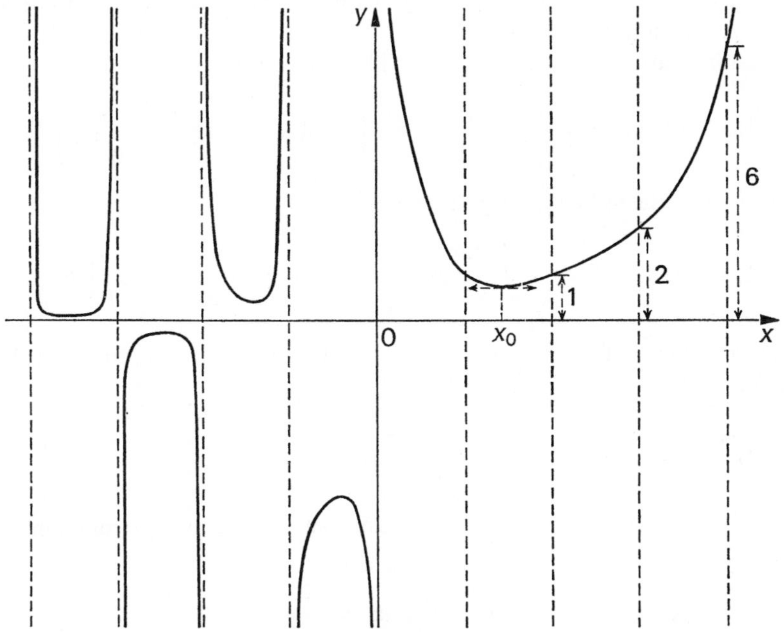

Figure (VIII, 2)

*The inequality $\Gamma''\Gamma - \Gamma'^2 > 0$ proves that $\log \Gamma$ is convex in each interval for which it is defined.

6. STIRLING'S FORMULA

(VIII, 1; 28) $\quad x! = \int_0^\infty e^{-t} t^x \, dt \sim x^x e^{-x} \sqrt{2\pi x}, \quad \text{as} \quad x \to +\infty.$

The logarithmic derivative of the function $e^{-t}t^x$ (for fixed x) is $-1 + x/t$ and this is > 0 for $0 \leqslant t \leqslant x$, zero for $t = x$ and < 0 for $t > x$. Hence this function has a maximum value at $t = x$ and the value of this maximum is $x^x e^{-x}$. Thus, it is natural to set $t = x + u$, which gives us

(VIII, 1; 29) $$x! = x^x e^{-x} \int_{-x}^{+\infty} e^{-u + x \log\left(1 + \frac{u}{x}\right)} du.$$

Provided that $|u|$ is always small compared with x, we have

$$-u + x \log\left(1 + \frac{u}{x}\right) = -u + x\left[\frac{u}{x} - \frac{u^2}{2x^2} + 0\left(\frac{|u|^3}{x^3}\right)\right] = -\frac{u^2}{2x} + 0\left(\frac{|u|^3}{x^2}\right).$$

For fixed x, $e^{-u+x\log\left(1+\frac{u}{x}\right)}$ is a function of u. It has a maximum value of 1, when $u = 0$, and as long as $|u|$ remains comparable with $\sqrt{x}$ it will never become infinitely small.

It seems natural to put $u = v\sqrt{x}$. Then we have

(VIII, 1; 30) $$x! = x^x e^{-x}\sqrt{x} \int_{-\sqrt{x}}^{+\infty} \exp\left[-v\sqrt{x} + x \log\left(1 + \frac{v}{\sqrt{x}}\right)\right] dv.$$

Thus it remains to be shown that the integral

(VIII, 1; 31) $$\mathrm{I}(x) = \int_{-\sqrt{x}}^{+\infty} \exp\left[-v\sqrt{x} + x \log\left(1 + \frac{\sqrt{x}}{v}\right)\right] dv$$

converges to $\sqrt{2\pi}$ for real x as $x \to +\infty$. We have, in fact, an integral of the type

(VIII, 1; 32) $$\mathrm{I}(x) = \int_{-\infty}^{+\infty} h(x, v) \, dv.$$

We will show that, for fixed v, $h(x, v)$ approaches a summable function $h(v)$ as $x \to \infty$, such that

(VIII, 1; 33) $$\int_{-\infty}^{+\infty} h(v) \, dv = \sqrt{2\pi}$$

and that $h(x, v)$ increases with x for $v < 0$ and decreases with x for $v > 0$. Thus for $v \leqslant 0$, $h(x, v)$ is bounded above by the summable limit $h(v)$, whilst for $v \geqslant 0$, $h(x, v)$ is bounded above (for $x \geqslant 1$) by the summable limit $h(1, v)$. Thus we shall have to apply Lebesgue's theorem (Chapter 1, Theorem 45), separately for $v \leqslant 0$ and $v \geqslant 0$.

Now, for fixed v and $x \to +\infty$ (hence $v > -\sqrt{x}$) we have

$$-v\sqrt{x} + x\log\left(1 + \frac{v}{\sqrt{x}}\right) = -v\sqrt{x} + x\left[\frac{v}{\sqrt{x}} - \frac{v^2}{2x} + 0\left(\frac{|v^3|}{x^{3/2}}\right)\right]$$

$$= -\frac{v^2}{2} + 0\left(\frac{|v^3|}{x}\right).$$

Thus

$$\lim_{x\to+\infty} h(x, v) = h(v) = e^{-\frac{v^2}{2}}$$

and

$$\int_{-\infty}^{+\infty} e^{-\frac{v^2}{2}}\, dv = \sqrt{2\pi} \quad \text{(Gauss integral).} \tag{VIII, 1; 34}$$

We shall now consider the case when the function increases. First, if $x_1 \leqslant x_2$, $h(x_1, v) \leqslant h(x_2, v)$ for $v \leqslant -\sqrt{x_1}$, because then $h(x_1, v) = 0$ and $h(x_2, v) \geqslant 0$. Thus it remains to be shown that $\log h(x, v)$ is an increasing function of $\sqrt{x}$ for $0 \geqslant v > -\sqrt{x}$, and a decreasing function of $\sqrt{x}$ for $v \geqslant 0$ or, alternatively, that

$$\frac{\partial}{\partial(\sqrt{x})}\log h(x, v) \geqslant 0 \quad \text{for} \quad -\sqrt{x} < v \leqslant 0,$$

$$\frac{\partial}{\partial(\sqrt{x})}\log h(x, v) \leqslant 0 \quad \text{for} \quad v \geqslant 0.$$

As $\frac{\partial}{\partial(\sqrt{x})}\log h(x, v)$ is zero for $v = 0$ [because $h(x, 0) \equiv 1$] it is sufficient to know that $\frac{\partial}{\partial(\sqrt{x})}\log h(x, v)$ is a decreasing function of v for $v > -\sqrt{x}$, or alternatively that

$$\frac{\partial}{\partial v}\frac{\partial}{\partial(\sqrt{x})}\log h(x, v) \leqslant 0 \quad \text{for} \quad v > -\sqrt{x}. \tag{VIII, 1; 35}$$

This derivative can be calculated immediately (and most conveniently by inverting the order of the differentiation):

$$\begin{cases} \log h(x, v) = -v\sqrt{x} + x\log\left(1 + \dfrac{v}{\sqrt{x}}\right), \\ \dfrac{\partial}{\partial v}\log h\,(x, v) = -\sqrt{x} + \dfrac{x}{\sqrt{x} + v} = \dfrac{-\sqrt{x}v}{\sqrt{x} + v}, \\ \dfrac{\partial}{\partial\sqrt{x}}\dfrac{\partial}{\partial v}\log h(x, v) = -\dfrac{v^2}{(v + \sqrt{x})^2} \leqslant 0. \end{cases} \tag{VIII, 1; 36}$$

Q.E.D.

7. APPLICATION TO THE EXPANSION OF $1/\Gamma$ AS AN INFINITE PRODUCT

Consider, for fixed x and $t \to +\infty$, the principal part of

(VIII, 1; 37) $$\frac{(x+t)!}{t!} \sim \frac{\Gamma(x+t)}{\Gamma(t)}.$$

Stirling's formula gives

(VIII, 1; 38) $$\frac{(x+t)!}{t!} \sim \frac{(x+t)^{x+t}\, e^{-x-t}\sqrt{2\pi(x+t)}}{t^t e^{-t}\sqrt{2\pi t}}$$

$$\sim t^x\left(1+\frac{x}{t}\right)^{x+t} e^{-x}.$$

Now, $$\left(1+\frac{x}{t}\right)^{x+t} = \left(1+\frac{x}{t}\right)^{x}\left(1+\frac{x}{t}\right)^{t}$$

approaches e^x as $t \to \infty$. Thus we find that

(VIII, 1; 39) $$\frac{\Gamma(x+t)}{\Gamma(t)} \sim t^x \quad \text{as} \quad t \to +\infty.$$

In particular, let $t = n$, an integer; then

(VIII, 1; 40) $$\frac{\Gamma(x+n)}{\Gamma(n)} = \frac{(x+n-1)(x+n-2)\cdots(x+1)x\Gamma(x)}{(n-1)(n-2)\cdots 1}$$

$$= \left(1+\frac{x}{n-1}\right)\left(1+\frac{x}{n-2}\right)\cdots(1+x)x\Gamma(x).$$

From (VIII, 1; 39) and (VIII, 1; 40) we can therefore deduce that

(VIII, 1; 41) $$\frac{1}{\Gamma(x)} = \lim_{n\to\infty}\left[x\left(1+\frac{x}{1}\right)\left(1+\frac{x}{2}\right)\cdots\left(1+\frac{x}{n-1}\right)n^{-x}\right].$$

But $$n^{-x} = e^{-x\log n} = \exp\left[-x\left(1+\tfrac{1}{2}+\cdots+\frac{1}{n-1}\right)+\gamma_n x\right]$$

where γ_n approaches the Euler constant γ as $n \to \infty$. Thus we have

(VIII, 1; 42)
$$\frac{1}{\Gamma(x)} = xe^{\gamma x}\lim_{n\to\infty}\left[\left(1+\frac{x}{1}\right)e^{-\frac{x}{1}}\left(1+\frac{x}{2}\right)e^{-\frac{x}{2}}\cdots\left(1+\frac{x}{n-1}\right)e^{-\frac{x}{n-1}}\right]$$

or

(VIII, 1; 43) $$\frac{1}{\Gamma(x)} = xe^{\gamma x}\prod_{n=1}^{n=\infty}\left(\left(1+\frac{x}{n}\right)e^{-\frac{x}{n}}\right).$$

Concerning infinite products

Just as for series, where we have the concepts of summability (absolute convergence) and of convergence (but only when the set of indices is the set of non-negative integers N) we say that infinite products can be "multipliable" and "convergent".

Only the multipliable infinite products have any practical interest, even when the set of indices is N. We often call them absolutely convergent or simply convergent.

A product $\prod_{i\in I} u_i$ *is multipliable if all the* u_i *are* $\neq 0$ *and if there exists a number* $P \neq 0$, *which is the value of the infinite product, satisfying the following property. Given* $\varepsilon > 0$, *there exists a finite set of indices* $J \subset I$ *such that for every finite set of indices* $K \supset J$,

$$\left|\frac{\Pi_K}{P} - 1\right| \leqslant \varepsilon \qquad \left(\Pi_K = \prod_{i\in K} u_i\right). \tag{VIII, 1; 44}$$

We notice from this definition that an infinite product which is "convergent to zero" (in the obvious way) must be considered as being *divergent*. The properties of multipliable infinite products are the same as those for summable series, the role of zero in the case of series being replaced by 1 in the case of products. The main results are :

1. For $\prod_{i\in I} u_i$ to be multipliable it is necessary that u_i *should tend to* 1. That is, given $\varepsilon > 0$, there exists a set $J \subset I$ such that

$$|u_i - 1| \leqslant \varepsilon \text{ whenever } i \notin J.$$

2. *Cauchy's criterion.* For $\prod_{i\in I} u_i$ to be multipliable it is necessary and sufficient that all the $u_i \neq 0$ and that given $\varepsilon > 0$ there exist a finite set $J \subset I$ such that for every finite subset K of I, where $K \cap J = \emptyset$,

$$|\Pi_K - 1| \leqslant \varepsilon.$$

3. *Repeated multiplication.* Let A be a set of indices, $(I_\alpha)_{\alpha\in A}$ a subdivision of I into subsets, finite or infinite, no two having a common element. Then if $\prod_{i\in I} u_i$ is multipliable, so also is $\prod_{i\in I_\alpha} u_i$ for all $\alpha \in A$, and if it has the value P_α then $\prod_{\alpha\in A} P_\alpha$ is also multipliable and we have

$$\prod_{i\in I} u_i = \prod_{\alpha\in A} P_\alpha = \prod_{\alpha\in A}\left(\prod_{i\in I_\alpha} u_i\right). \tag{VIII, 1; 45}$$

Further, the two sides are always equal, even if $\prod_{i\in I} u_i$ is not multipliable, provided that the u_i are $\geqslant 1$ (when $u_i \geqslant 1$, we say that the value of a divergent product is $+\infty$) or provided that all the u_i are > 0 and $\leqslant 1$. When $0 < u_i \leqslant 1$ the value of a divergent product is said to be zero.

These theorems can be proved directly, as for series. However, it is easier to get back first to series by taking logarithms. Admittedly, $\log u_i$ is not determined in a unique manner for complex u_i or even real $u_i < 0$. But if the u_i do not tend to 1 we know immediately that the product is divergent. If the u_i tend to 1 we have $|u_i - 1| < 1$ except for a finite number of terms *which we can neglect*.

Now the function $\log z$ is continuous and uniform for $|z - 1| < 1$, and we take that branch for which it is zero at $z = 1$. With these conditions, in order that $\prod_{i \in I} u_i$ should converge, it is necessary and sufficient that the series $\sum_{i \in I} \log u_i$ be summable. In fact :

(*a*) Given $\varepsilon > 0$, there exists $\eta > 0$ such that $|e^{\xi} - 1| \leqslant \varepsilon$ whenever $|\xi| \leqslant \eta$, by the continuity of the exponential function. If $\sum_{i \in I} \log u_i$ is summable, there exists a finite subset J of I such that $|\Sigma_K| \leqslant \eta$ for all finite $K \subset J$ for which $K \cap J = \emptyset$. Hence $|\Pi_K - 1| = |e^{\Sigma_K} - 1| \leqslant \varepsilon$, and therefore $\prod_{i \in I} u_i$ is multipliable.

(*b*) Given $\varepsilon > 0$, there exists $\eta > 0$ such that $|\xi - 1| \leqslant \eta$ implies $|\log \xi| \leqslant \varepsilon$ for that branch of the logarithm chosen above, by the continuity of the logarithmic function. If $\prod_{i \in I} u_i$ is multipliable, there exists a finite subset J of I such that $|\Pi_K - 1| \leqslant \eta$, for all finite $K \subset I$ for which $K \cap J = \emptyset$. Hence $|\Sigma_K| = |\log \Pi_K| \leqslant \varepsilon$; therefore $\sum_{i \in I} \log u_i$ is summable.

An important consequence of this is the following. In order that the infinite product $\prod_{i \in I} (1 + v_i)$ should be multipliable, it is necessary and sufficient that the series $\sum_{i \in I} v_i$ be summable, and that all the $1 + v_i$ are nonzero.

First, it is necessary that $v_i \to 0$, in order that $1 + v_i \to 1$. This condition being satisfied, we have seen that it is necessary and sufficient that the series $\sum_{i \in I} \log (1 + v_i)$ be summable. But for $v_i \to 0$, $\log (1 + v) \sim v$ and we know that when it is summable we can replace the general term $\log (1 + v_i)$ by the equivalent term v_i. Thus it is necessary and sufficient that the series $\sum v_i$ be summable. Q.E.D.

Examples

(VIII, 1; 46)
$$\prod_{n=1}^{\infty} \left(1 \pm \frac{1}{n^{\alpha}}\right)$$

is convergent if $\alpha > 1$ and divergent if $0 < \alpha \leqslant 1$.

In particular,

$$\prod_{n=1}^{\infty}\left(1+\frac{1}{n}\right)=+\infty \qquad \text{(divergent)},$$

$$\prod_{n=2}^{\infty}\left(1-\frac{1}{n}\right)=0 \qquad \text{(divergent)}.$$

These last two formulae are immediately obvious from the divergence of the series $\sum_{n=1}^{\infty}\frac{1}{n}$, or as follows:

$$\left(1+\frac{1}{1}\right)\left(1+\frac{1}{2}\right)\cdots\left(1+\frac{1}{n}\right)=\frac{2}{1}\cdot\frac{3}{2}\cdots\frac{n+1}{n}=n+1\to\infty,$$

$$\left(1-\frac{1}{2}\right)\left(1-\frac{1}{3}\right)\cdots\left(1-\frac{1}{n}\right)=\frac{1}{2}\cdot\frac{2}{3}\cdots\frac{n-1}{n}=\frac{1}{n}\to 0.$$

Convergence of the product representing $1/\Gamma$

The infinite product $e^{\gamma z}z\prod_{n=1}^{\infty}\left(\left(1+\frac{z}{n}\right)e^{-\frac{z}{n}}\right)$ is multipliable for all values of z for which no term is zero (thus $z\neq 0, -1, -2, \cdots, -n, \ldots$) and uniformly so on every bounded set in the complex plane not containing any of these points, for which $1/\Gamma(z)=0$. In fact, for bounded z we have

$$\begin{aligned}\log\left[\left(1+\frac{z}{n}\right)e^{-\frac{z}{n}}\right] &= \log\left(1+\frac{z}{n}\right)-\frac{z}{n}\\ &= \frac{z}{n}+0\left(\frac{|z|^2}{n^2}\right)-\frac{z}{n}=0\left(\frac{|z|^2}{n^2}\right);\end{aligned}$$

and $\sum_{n=1}^{\infty}\frac{1}{n^2}$ is a summable series.

Hence this infinite product represents a function of a complex variable z, which is everywhere holomorphic, and coincides with $1/\Gamma(z)$, which itself is meromorphic, for $z = x$ where x is real and greater than zero, and thus for all z. This proves, first of all, that $1/\Gamma(z)$ can be represented, for all complex z, by the infinite product above (to which we will give the value 0 when $z = 0, -1, -2, \ldots, -n \ldots$, since for these values one of the terms of the product is zero and all the others are finite). Secondly, it proves that $1/\Gamma(z)$ is everywhere holomorphic and thus that Γ never vanishes, for all z.

8. THE FUNCTION $\Psi(z) = \dfrac{\Gamma'(z)}{\Gamma(z)}$

Take the logarithmic derivative of both sides of the above product expansion

$$\text{(VIII, 1; 47)} \quad -\frac{\Gamma'(z)}{\Gamma(z)} = \frac{1}{z} + \left[\sum_{k=1}^{\infty}\left(\frac{1}{z+k} - \frac{1}{k}\right)\right] + \gamma.$$

Such an equation will be valid if the series obtained is uniformly convergent on compact sets in the z-plane not including the points $z = 0$, $-1, -2, \ldots$, and this is, in fact, the case because

$$\left|\frac{1}{z+k} - \frac{1}{k}\right| = \left|\frac{-z}{k(z+k)}\right| = \left|\frac{z}{k^2[1 + (z/k)]}\right|$$

is bounded above by $\dfrac{2 \sup |z|}{k^2}$ for $k > 2 \sup |z|$.

What we have just obtained is a decomposition of $-\Gamma'/\Gamma$ into partial fractions, for which the poles are $z = 0, -1, -2, \ldots$ with residue 1. Let

$$\text{(VIII, 1; 48)} \qquad \Psi(z) = \frac{\Gamma'(z)}{\Gamma(z)}.$$

By using equation (VIII, 1; 17) and making $p \to 0$ for fixed $q = z$ we obtain the equation

$$\text{(VIII, 1; 49)} \quad \Psi(z) = -\gamma + \int_0^1 \frac{1 - (1-t)^{z-1}}{t}\, dt, \quad \text{for} \quad \mathcal{R}z > 0.$$

This integral can be evaluated by elementary means for all rational values of z.

In particular

$$-\Gamma'(1) = 1 + \left[\left(\frac{1}{2} - 1\right) + \left(\frac{1}{3} - \frac{1}{2}\right) + \cdots\right] + \gamma$$

or

$$\Gamma'(1) = -\gamma.$$

Thus we have

$$\text{(VIII, 1; 50)} \qquad \boxed{\Gamma'(1) = \int_0^{\infty} e^{-t} \log t\, dt = -\gamma.}$$

Next, we have

$$-\frac{\Gamma'(2)}{\Gamma(2)} = \frac{1}{2} + \left(\frac{1}{3} - 1\right) + \left(\frac{1}{4} - \frac{1}{2}\right) + \cdots + \cdots + \gamma,$$

$$\frac{-\Gamma'(2)}{\Gamma(2)} = -1 + \gamma.$$

Thus

$$\text{(VIII, 1; 51)} \qquad \boxed{\Gamma'(2) = \int_0^\infty e^{-t}\, t \log t\, dt = 1 - \gamma.}$$

For any integer $p \geqslant 2$, we have

(VIII, 1; 52)

$$\Gamma'(p) = (p-1)!\left(1 + \frac{1}{2} + \frac{1}{3} + \cdots + \frac{1}{p-1} - \gamma\right) = \int_0^\infty e^{-t} t^{p-1} \log t\, dt.$$

Differentiating (VIII, 1; 47), we obtain

$$\text{(VIII, 1; 53)} \qquad \frac{\Gamma\Gamma'' - \Gamma'^2}{\Gamma^2} = + \frac{1}{z^2} + \sum_{k=1}^{\infty}\left(\frac{1}{z+k}\right)^2 = \sum_{k=0}^{\infty}\left(\frac{1}{z+k}\right)^2.$$

This series is uniformly convergent on compact sets in the z-plane not containing any poles of the function.

Note. For $z = x$, where x is real and is not a pole, we have $\Gamma\Gamma'' - \Gamma'^2 > 0$ as we have seen in Section 6. In particular,

$$\text{(VIII, 1; 54)} \qquad \Gamma''(1) = \gamma^2 + \sum_{k=0}^{\infty} \frac{1}{(k+1)^2} = \gamma^2 + \sum_{n=1}^{\infty} \frac{1}{n^2}$$

$$= \gamma^2 + \frac{\pi^2}{6} = \int_0^\infty e^{-t} (\log t)^2\, dt,$$

where the third equality is obtained from (VIII, 1; 63).

9. APPLICATIONS

Expansion of Γ'/Γ in ascending powers of z in the neighbourhood of $z = 0$

From (VIII, 1; 47) we deduce

$$\text{(VIII, 1; 55)} \qquad \frac{\Gamma'(z)}{\Gamma(z)} = -\frac{1}{z} - \gamma - \sum_{n=1}^{\infty}\left[\sum_{p=1}^{\infty} (-1)^p \frac{z^p}{n^{p+1}}\right].$$

If we can interchange the order of the summation we shall obtain

$$\text{(VIII, 1; 56)} \qquad \boxed{\frac{\Gamma'(z)}{\Gamma(z)} = -\frac{1}{z} - \gamma + \sum_{p=1}^{\infty} (-1)^{p+1} z^p \zeta(p+1),}$$

where

$$\zeta(\alpha) = \sum_{n=1}^{\infty} \frac{1}{n^\alpha}.$$

We are dealing with a double series (two indices) and we can interchange the order of the summation only if the double series is absolutely convergent. Now, the general term of the double series has as its modulus

$$u_{n,\,p} = \frac{|z|^p}{n^{p+1}} \leqslant \frac{|z|^p}{n^2}.$$

The series represented by the last term is convergent for $|z| < 1$, since it is the product of the series $\Sigma |z|^p$ (which is a convergent series for $|z| < 1$) and the series $\Sigma 1/n^2$. The radius of convergence of (VIII, 1; 56) is 1. We shall see later (VIII, 1; 63) what values $\zeta(\alpha)$ takes for an even integer α.

Expansion of sin z as an infinite product

We have

(VIII, 1; 57) $$\frac{\sin \pi z}{\pi} = \frac{1}{\Gamma(z)} \frac{1}{\Gamma(1 - z)} = -\frac{1}{z} \frac{1}{\Gamma(z)\Gamma(-z)}.$$

By taking the product of the various expansions, we get

(VIII, 1; 58) $$\sin \pi z = \pi z \prod_{n=1}^{\infty} \left[\left(1 + \frac{z}{n}\right) e^{-\frac{z}{n}} \left(1 - \frac{z}{n}\right) e^{\frac{z}{n}}\right]$$
$$= \pi z \prod_{n=1}^{\infty} \left(1 - \frac{z^2}{n^2}\right)$$

The expansion on the right-hand side is multipliable for all $z \neq 0, \pm 1, \pm 2, \ldots, \pm n, \ldots$, and uniformly so for values of z which are bounded and distinct from these, since $\sum_{n=1}^{\infty} \frac{1}{n^2} < \infty$. As in the preceding case we say that the product has the value zero for all integral z, and thus the product always represents $\sin \pi z$. This is a decomposition of $\sin \pi z$ into factors in the same way as polynomials are factored. Thus we deduce

(VIII, 1; 59) $$\sin z = z \prod_{n=1}^{\infty} \left(1 - \frac{z^2}{n^2 \pi^2}\right)$$

Decomposition of cot z into partial fractions

Take the logarithmic derivative of the above product (this can be justified in a similar manner to that on page 326 and we get

(VIII, 1; 60) $$\cot z = \frac{1}{z} + \sum_{n=1}^{\infty} \left(\frac{1}{z - n\pi} + \frac{1}{z + n\pi}\right).$$

Notice that, as $n \to \infty$,

$$\frac{1}{z - n\pi} + \frac{1}{z + n\pi} \sim -\frac{2z}{n^2\pi^2};$$

this shows that the series is summable. However the series

$$\sum_{\substack{n=-\infty \\ n\neq 0}}^{+\infty} \frac{1}{z - n\pi}$$

will not be summable.

Equation (VIII, 1; 60) gives a decomposition of $\cot z$ into partial fractions which has poles at values of z given by $z = n\pi$ ($n = -\infty$ to $+\infty$), the residues at the poles being 1.

Expansion of cot z in powers of z

By expanding in powers of z (the justification for this is similar to that on page 327) we obtain

(VIII, 1; 61) $$\cot z = \frac{1}{z} - \frac{2}{\pi^2}\zeta(2)z - \cdots - \frac{2}{\pi^{2p}}\zeta(2p)z^{2p-1} \ldots$$

where this expansion is valid for $|z| < \pi$.

Now, the expansion for $\cot z$ can be obtained directly by dividing the expansions of $\cos z$ and $\sin z$, thus

(VIII, 1; 62) $$\cot z = \frac{1}{z} - \frac{z}{3} - \frac{z^3}{45} - \frac{2z^5}{945} - \frac{z^7}{4725} - \cdots$$

Every coefficient in this expansion is a rational number, thus $\dfrac{\zeta(2p)}{\pi^{2p}}$ is rational, and can be calculated sucessively to give

(VIII, 1; 63) $$\zeta(2) = \frac{\pi^2}{6}, \quad \zeta(4) = \frac{\pi^4}{90}, \quad \zeta(6) = \frac{\pi^6}{945}, \quad \zeta(8) = \frac{\pi^8}{9450}, \ \cdots$$

In the numerical evaluation of integrals, using the Euler-Maclaurin formula, we use the coefficients in the expansion of $\dfrac{1}{e^z - 1}$ in powers of z. Now

(VIII, 1; 64) $$\frac{1}{e^z - 1} = -\frac{1}{2} + \frac{1}{2}\coth\frac{z}{2} = -\frac{1}{2} + \frac{i}{2}\cot\frac{iz}{2}$$

$$= \frac{1}{z} - \frac{1}{2} + \sum_{p=1}^{\infty} \frac{(-1)^{p+1}}{2^{2p-1}\pi^{2p}}\zeta(2p)z^{2p-1} \quad \text{for } |z| < 2\pi,$$

which can be written

(VIII, 1; 65) $$\frac{1}{e^z - 1} = \frac{1}{z} - \frac{1}{2} + \sum_{p=1}^{\infty} \frac{(-1)^{p+1}B_p}{(2p)!} z^{2p-1},$$

where

(VIII, 1; 66) $$B_p = 2\frac{(2p)\,!\,\zeta(2p)}{(2\pi)^{2p}}.$$

The numbers B_p are known as the *Bernoulli numbers* :

(VIII, 1; 67) $$B_1 = \tfrac{1}{6}, \quad B_2 = \tfrac{1}{30}, \quad B_3 = \tfrac{1}{42}, \quad B_4 = \tfrac{1}{30}, \cdots$$

EXERCISES FOR CHAPTER VIII

Exercise VIII-1

Show, by using the definition of the Γ function, that

$$\int_0^\infty e^{-x} \log x \, dx = -\gamma,$$

where γ is Euler's constant. Remembering that $\Psi(1) = -\gamma$, where $\Psi(x)$ is the logarithmic derivative of $\Gamma(x)$, show that all differentiations under the integral sign are possible.

Exercise VIII-2

Given the set of functions defined by

$$f_n(x) = \frac{n}{\sqrt{\pi}} e^{-n^2x^2},$$

discuss the behaviour of the curves $y = f_n(x)$. Show that if $\varphi(x)$ is a continuous function which is zero outside the interval $(-A, +A)$ of the real axis, then

$$\lim_{n \to +\infty} \int_{-\infty}^{+\infty} \varphi(x) f_n(x) \, dx = \varphi(0).$$

From the equality

$$\int_{-\infty}^{+\infty} f_n(x) e^{-px} \, dx = e^{\frac{p^2}{4n^2}} \quad (p \text{ any real number}),$$

deduce the sum of the series

$$\frac{1}{\sqrt{\pi}} \sum_{m=0}^{\infty} \frac{\Gamma\left(\frac{m+1}{2}\right)}{\Gamma(m+1)}.$$

Exercise VIII-3

Let

$$f(p) = \Gamma\left(\frac{1}{p}\right)\Gamma\left(\frac{2}{p}\right)\cdots\Gamma\left(\frac{p-1}{p}\right)\Gamma(1).$$

By using the formula for $\Gamma(x)\,\Gamma(1-x)$, show that

$$f(p)=\sqrt{\frac{(2\pi)^{p-1}}{p}}.$$

Exercise VIII-4
Let V_n be Vandermonde's determinant, which is defined as

$$V_n=\begin{vmatrix} 1 & 1 & \dots & 1 \\ 1 & 2 & \dots & n \\ 1 & 2^2 & \dots & n^2 \\ \dots & \dots & \dots & \dots \\ 1 & 2^{n-1} & \dots & n^{n-1} \end{vmatrix}.$$

Show that, when $n\to\infty$,

$$\log V_n \sim \frac{n^2}{2}\log n-\frac{3n^2}{4}+\frac{n}{2}\log 2\pi-\frac{1}{12}\log n.$$

Exercise VIII-5
Evaluate

$$\iiint\frac{dx\,dy\,dz}{1+x^\alpha+y^\beta+z^\gamma}$$

over the domain $x>0, y>0, z>0$, by changing the variables according to

$$u=x^{\alpha/2},\qquad v=y^{\beta/2},\qquad w=z^{\gamma/2}.$$

For what values of α, β, γ is this integral defined?

Exercise VIII-6
Evaluate the integrals

$$\int_0^\infty e^{-t^2}\cosh t\,dt \quad\text{and}\quad \int_0^\infty e^{-t^2}\cos t\,dt$$

by using the series expansion for $\cosh t$ and $\cos t$.

Exercise VIII-7
Evaluate and compare the values of the two integrals

$$\int_{-\frac{1}{2}}^{+\frac{1}{2}} e^{-x^2}\,dx \quad\text{and}\quad \int_0^\infty e^{-x^2}\frac{\sin x}{x}\,dx.$$

Exercise VIII-8
Show that

$$\Gamma(2p)=\frac{2^{2p-1}}{\sqrt{\pi}}\Gamma(p)\,\Gamma(p+\tfrac{1}{2}).$$

Exercise VIII-9
$\Gamma(x)$ is defined by

$$\Gamma(x) = \int_0^\infty e^{-t}t^{x-1}\,dt \quad \text{for} \quad x > 0.$$

Let

$$\varphi(x) = \log \Gamma(x).$$

(*a*) Show that $\varphi(x)$ is a convex function of x, for $x > 0$. That is, show that

$$\frac{\varphi(x_1) + \lambda\varphi(x_2)}{1 + \lambda} \geqslant \varphi\left(\frac{x_1 + \lambda x_2}{1 + \lambda}\right)$$

for all x_1, x_2, $\lambda > 0$. Or, alternatively, show that $\varphi''(x) \geqslant 0$ for all $x > 0$, if this derivative exists.

(*b*) $\varphi(x)$ is the only convex function $g(x)$ defined for $x > 0$ that satisfies

(i) $$g(x + 1) - g(x) = \log x,$$

(ii) $$g(1) = 0.$$

(*1*) For $0 < x \leqslant 1$ and n an integer > 1 the convexity of g implies that

$$g(n) - g(n-1) \leqslant \frac{g(n+x) - g(n)}{x} \leqslant g(n+1) - g(n).$$

(*2*) If we put $U_n(x) = x \log \dfrac{n}{n-1} - \log(x + n - 1) + \log(n-1)$ and

$$g_n(x) = -\log x + \sum_{k=2}^{n} U_k(x),$$

show that

$$g_n(x) - x \log \frac{n}{n-1} \leqslant g(x) \leqslant g_n(x), \qquad 0 < x \leqslant 1.$$

(*3*) Show that

$$g(x) = -\log x + \sum_{n=2}^{\infty} U_n(x)$$

exists for all $x > 0$, is convex, and satisfies (1) and (2). Hence show that $g(x) = \varphi(x)$.

(*c*) (i) Show that, as a consequence of (2), we have for all $x > 0$

$$\Gamma(x) = \lim_{n\to\infty} \frac{n^x n!}{x(x+1)\dots(x+n)}.$$

Can this equation be used for values of $x < 0$?

(ii) Show that the series whose n-th term is

$$V_n = \frac{1}{n} - \log\frac{n+1}{n}$$

is absolutely convergent. Let

$$\sum_{1}^{\infty} V_n = C.$$

(iii) Show that

$$\Gamma(x) = e^{-Cx}\frac{1}{x}\prod_{1}^{\infty}\frac{e^{\frac{x}{n}}}{1+\frac{x}{n}},$$

the infinite product being absolutely and uniformly convergent on any closed interval $a \leqslant x \leqslant b$ in R that contains no negative integers.

(*d*) Show that $\Gamma(x)$ is infinitely differentiable and that, for $x > 0$,

$$\frac{\Gamma'(x)}{\Gamma(x)} = -C - \frac{1}{x} + \sum_{1}^{\infty}\left(\frac{1}{n} - \frac{1}{x+n}\right)$$

and

$$\frac{d^k\,[\log \Gamma\,(x)]}{dx^k} = \sum_{0}^{\infty}\frac{(-1)^k(k-1)\,!}{(x+n)^k} \quad \text{for} \quad k \geqslant 2.$$

Show that, for all integers $q < 1$ and all intergers k satisfying $1 \leqslant k \leqslant q-1$, we have

$$\sum_{p=1}^{q}\frac{\Gamma'\left(\frac{p}{q}\right)}{\Gamma\left(\frac{p}{q}\right)}e^{2ip\frac{k\pi}{q}} = -q\sum_{n=1}^{\infty}\frac{1}{n}e^{2in\frac{k\pi}{q}} = q\log\left(1-e^{\frac{2ik\pi}{q}}\right).$$

Bessel functions

1. Definitions and elementary properties

1. DEFINITION OF THE BESSEL, NEUMANN, AND HANKEL FUNCTIONS

For simplicity * we shall always take x to be real and greater than 0. The equation

$$y'' + \frac{1}{x}y' + \left(1 - \frac{\nu^2}{x^2}\right)y = 0 \tag{IX, 1; 1}$$

where ν is a given complex number, which can always be assumed to be such that its real part is $\geqslant 0$, is known as *Bessel's equation*. Look for solutions of this equation in the form

$$y = x^{\lambda} \sum_{k=0}^{\infty} a_k x^k. \tag{IX, 1; 2}$$

The left-hand side of (IX, 1; 1) is therefore equal to

$$\sum_{k=0}^{\infty} \{(k+\lambda)^2 - \nu^2\}\, a_k x^{k+\lambda-2} + \sum_{k=0}^{\infty} a_k x^{k+\lambda}. \tag{IX, 1; 3}$$

Thus in order that (IX, 1; 2) may be a solution of (IX, 1; 1) we must have

$$\sum_{k=0}^{\infty} \{(k+\lambda)^2 - \nu^2\}\, a_k x^{k+\lambda} + \sum_{k=0}^{\infty} a_k x^{k+\lambda+2} \equiv 0. \tag{IX, 1; 4}$$

By equating to zero the coefficients of the terms in x^{λ}, $x^{\lambda+1}$, $x^{k+\lambda}$ for $k \geqslant 2$ we obtain the equations

$$(\lambda^2 - \nu^2)a_0 = 0, \tag{IX, 1; 5}$$

$$[(1+\lambda)^2 - \nu^2]a_1 = 0 \tag{IX, 1; 6}$$

$$[(k+\lambda)^2 - \nu^2]a_k + a_{k-2} = 0 \quad \text{for} \quad k \geqslant 2. \tag{IX, 1; 7}$$

---

*We make this hypothesis in order to fix our ideas and to simplify the proofs. In fact nearly all the results established here, in particular those in nos. 1, 2, and 3, are valid for any complex x.

As we can always assume that $a_0 \neq 0$ in (IX, 1; 2) by agreeing to call $a_0 x^\lambda$ the first nonzero term in the expansion (IX, 1; 2), equation (IX, 1; 5) gives

$$\lambda = \pm \nu. \tag{IX, 1; 8}$$

We shall look, first of all, for solutions that correspond to $\lambda = \nu$. Equations (IX, 1; 6) and (IX, 1; 7) can then be written

$$(2\nu + 1)\, a_1 = 0, \tag{IX, 1; 9}$$

$$k(k + 2\nu)\, a_k + a_{k-2} = 0, \qquad k \geqslant 2. \tag{IX, 1; 10}$$

From equations (IX, 1; 9) and (IX, 1; 10) we see that all the coefficients having odd subscripts are zero. The coefficients having even subscripts are easily obtained from (IX, 1; 10):

$$a_{2n} = \frac{(-1)^n a_0}{2^{2n} n!\,(\nu + 1)(\nu + 2) \cdots (\nu + n)}, \qquad n \geqslant 1. \tag{IX, 1; 11}$$

By choosing

$$a_0 = \frac{1}{2^\nu \Gamma(\nu + 1)} \tag{IX, 1; 12}$$

we obtain

$$a_{2n} = \frac{1}{2^\nu} \frac{(-1)^n}{2^{2n} n!\, \Gamma(\nu + n + 1)}, \tag{IX, 1; 13}$$

and thus for the solution (IX, 1; 2) we have the expression

$$J_\nu(x) = \left(\frac{x}{2}\right)^\nu \sum_{n=0}^{\infty} \frac{(-1)^n \left(\frac{x}{2}\right)^{2n}}{n!\,\Gamma(n + \nu + 1)}. \tag{IX, 1; 14}$$

It is clear that this series on the right-hand side is convergent for all x and uniformly convergent on every bounded interval. $J_\nu(x)$ is known as the *Bessel function of order* ν, as defined in equation (IX, 1; 14).

Now consider the case when $\lambda = -\nu$. Equations (IX, 1; 6) and (IX, 1; 7) can then be written

$$(-2\nu + 1)a_1 = 0, \tag{IX, 1; 15}$$

$$k(k - 2\nu)a_k + a_{k-2} = 0, \qquad k \geqslant 2. \tag{IX, 1; 16}$$

If ν is not an integer, let

$$J_{-\nu}(x) = \left(\frac{x}{2}\right)^{-\nu} \sum_{n=0}^{\infty} \frac{(-1)^n \left(\frac{x}{2}\right)^{2n}}{n!\,\Gamma(n - \nu + 1)}. \tag{IX, 1; 17}$$

This is obtained by substituting $-\nu$ for ν in equation (IX, 1; 14). It is easily seen that the coefficients a_k of $J_{-\nu}(x)$ satisfy the relations (IX, 1; 15) and (IX, 1; 16), and it follows that $J_{-\nu}(x)$ is a solution of equation (IX, 1; 1). If ν is an integer $p \geqslant 1$, as the poles of the Γ function are the integers $\leqslant 0$, the equation becomes

$$\text{(IX, 1; 18)} \qquad J_{-p}(x) = \left(\frac{x}{2}\right)^{-p} \sum_{n=p}^{\infty} \frac{(-1)^n \left(\frac{x}{2}\right)^{2n}}{n!\,\Gamma(n-p+1)}.$$

But by putting $n - p = k$ we get

$$\text{(IX, 1; 19)} \quad J_{-p}(x) = (-1)^p \left(\frac{x}{2}\right)^{p} \sum_{k=0}^{\infty} \frac{(-1)^k \left(\frac{x}{2}\right)^{2k}}{k!\,\Gamma(k+p+1)} = (-1)^p J_p(x).$$

When ν is not an integer $\geqslant 0$ the solutions given by (IX, 1; 14) and (IX, 1; 17) are linearly independent. To see this, it is sufficient to notice that, as $x \to 0$, $J_\nu(x)$ tends to zero as x^ν and $J_{-\nu}(x)$ tends to infinity like $x^{-\nu}$, because $\mathcal{R}\nu > 0$. Thus, in this case, the general solution of (IX, 1; 1) is

$$\text{(IX, 1; 20)} \qquad AJ_\nu(x) + BJ_{-\nu}(x),$$

where A and B are constants.

When ν is an integer $\geqslant 0$, since J_p and J_{-p} are related by equation (IX, 1; 19), we must look for a second solution of equation (IX, 1; 1) which is independent of J_p.

Neumann functions

For ν which is not an integer we define the *Neumann function* by

$$\text{(IX, 1; 21)} \qquad N_\nu(x) = \frac{(\cos \nu\pi)\, J_\nu(x) - J_{-\nu}(x)}{\sin \nu\pi}.$$

When $\nu = p$, an integer, we put

$$\text{(IX, 1; 22)} \qquad N_p(x) = \lim_{\nu \to p} N_\nu(x).$$

Since, for all real $x > 0$, $\nu \to J_\nu(x)$ is an analytic function of ν, the numerator and denominator of the right-hand side of (IX, 1; 21) are also analytic functions of ν. Hence it follows from l'Hospital's rule that

$$N_p(x) = \lim_{\nu \to p} \left\{ \frac{\frac{\partial}{\partial \nu}(\cos \nu\pi J_\nu(x) - J_{-\nu}(x))}{\pi \cos \nu\pi} \right\}$$

so that

$$\text{(IX, 1; 23)} \qquad \pi N_p(x) = \lim_{\nu \to p} \left\{ \frac{\partial}{\partial \nu} J_\nu(x) - (-1)^p \frac{\partial}{\partial \nu} J_{-\nu}(x) \right\}.$$

It follows from equation (IX, 1; 23) that, for every integer $m \geqslant 0$,

(IX, 1; 24) $$\frac{d^m}{dx^m}(\pi N_p(x)) = \left[\frac{\partial}{\partial \nu}\left(\frac{\partial^m}{\partial x^m} J_\nu(x)\right)\right]_{\nu=p} - (-1)^p\left[\frac{\partial}{\partial \nu}\left(\frac{\partial^m}{\partial x^m} J_{-\nu}(x)\right)\right]_{\nu=p}.$$

First evaluate $\partial J_\nu(x)/\partial \nu$. We have

(IX, 1; 14) $$J_\nu(x) = \left(\frac{x}{2}\right)^\nu \sum_{k=0}^{\infty} \frac{(-1)^k\left(\frac{x}{2}\right)^{2k}}{k!\,\Gamma(\nu+k+1)},$$

hence

(IX, 1; 25) $$\frac{\partial}{\partial \nu} J_\nu(x) = \left(\log\frac{x}{2}\right) J_\nu(x) + \left(\frac{x}{2}\right)^\nu \sum_{k=0}^{\infty} (-1)^k \left(\frac{x}{2}\right)^{2k} \frac{1}{k!}\frac{\partial}{\partial \nu}\left(\frac{1}{\Gamma(\nu+k+1)}\right).$$

But

(IX, 1; 26) $$\frac{\partial}{\partial \nu}\left(\frac{1}{\Gamma(\nu+k+1)}\right) = -\frac{\Gamma'(\nu+k+1)}{\Gamma^2(\nu+k+1)},$$

and from (VIII, 1; 47) we have

(IX, 1; 27)
$$-\frac{\Gamma'(\nu+k+1)}{\Gamma(\nu+k+1)} = \frac{1}{\nu+k+1} + \sum_{n=1}^{\infty}\left\{\frac{1}{\nu+k+1+n} - \frac{1}{n}\right\} + \gamma.$$

Now let ν tend to some integral value p. If $p = 0$, the right-hand side of equation (IX, 1 ; 27) approaches γ for $k = 0$, and for $k \neq 0$ approaches.

(IX, 1; 28) $$-\sum_{n=1}^{k} \frac{1}{n} + \gamma.$$

If $p > 0$, then the right-hand side of equation (IX, 1 ; 27) approaches

(IX, 1; 29) $$-\sum_{n=1}^{p+k} \frac{1}{n} + \gamma.$$

Therefore, as $\nu \to p$, $\dfrac{\partial}{\partial \nu} J_\nu(x)$ tends to

(IX, 1; 30) $$\left(\log\frac{x}{2} + \gamma\right) J_0(x) - \sum_{k=1}^{\infty}\left\{(-1)^k\left(\frac{x}{2}\right)^{2k} \frac{1}{(k!)^2}\sum_{n=1}^{k}\frac{1}{n}\right\}$$

if $p = 0$; if $p \neq 0$, it tends to

(IX, 1; 31)

$$\left(\log\frac{x}{2}+\gamma\right)J_p(x)+\left(\frac{x}{2}\right)^p\sum_{k=0}^{\infty}\left\{(-1)^k\left(\frac{x}{2}\right)^{2k}\frac{1}{k!}\frac{(-1)}{\Gamma(p+k+1)}\sum_{n=1}^{p+k}\frac{1}{n}\right\}.$$

Now evaluate $\partial J_{-\nu}(x)/\partial\nu$. We easily obtain from equation (IX, 1; 17)

(IX, 1; 32)
$$\frac{\partial}{\partial\nu}J_{-\nu}(x) = -\log\frac{x}{2}J_{-\nu}(x) + \left(\frac{x}{2}\right)^{-\nu}\sum_{k=0}^{\infty}(-1)^k\left(\frac{x}{2}\right)^{2k}\frac{1}{k!}\frac{\Gamma'(-\nu+k+1)}{\Gamma^2(-\nu+k+1)}.$$

As before, if $\nu \to 0$, $\partial J_{-\nu}(x)/\partial\nu$ tends to

(IX, 1; 33)
$$-\log\frac{x}{2}J_0(x)-\gamma J_0(x)+\sum_{k=1}^{\infty}\left((-1)^k\left(\frac{x}{2}\right)^{2k}\frac{1}{(k!)^2}\sum_{n=1}^{k}\frac{1}{n}\right).$$

Thus we see from equations (IX, 1; 23), (IX, 1; 30), and (IX, 1; 33) that

(IX, 1; 34)

$$N_0(x) = \frac{2}{\pi}\left\{\left(\log\frac{x}{2}+\gamma\right)J_0(x)-\left[\sum_{k=1}^{\infty}(-1)^k\left(\frac{x}{2}\right)^{2k}\frac{1}{(k!)^2}\sum_{n=1}^{k}\frac{1}{n}\right]\right\}.$$

Now assume that ν approaches some integer $p > 0$. Consider the series on the right-hand side of (IX, 1; 32) and examine the terms corresponding to $k \geqslant p$. When $\nu \to p$ we find from the above that

$$\left(\frac{x}{2}\right)^{-\nu}\sum_{k=p}^{\infty}(-1)^k\left(\frac{x}{2}\right)^{2k}\frac{1}{(k!)\,\Gamma(-\nu+k+1)}\frac{\Gamma'(-\nu+k+1)}{\Gamma(-\nu+k+1)}$$

approaches

(IX, 1; 35)

$$(-1)^p\left\{-\gamma J_p(x)+\left(\frac{x}{2}\right)^p\sum_{k=1}^{\infty}(-1)^k\left(\frac{x}{2}\right)^{2k}\frac{1}{k!\,\Gamma(k+p+1)}\sum_{n=1}^{k}\frac{1}{n}\right\}.$$

We must now examine the terms

(IX, 1; 36)
$$\left(\frac{x}{2}\right)^{-\nu}\sum_{k=0}^{p-1}(-1)^k\left(\frac{x}{2}\right)^{2k}\frac{1}{k!}\frac{\Gamma'(-\nu+k+1)}{\Gamma^2(-\nu+k+1)}.$$

From equations (VIII, 1; 7) and (VIII, 1; 47) we see that, for $0 \leqslant k \leqslant p-1$, as $\nu \to p$,

(IX, 1; 37) $$\frac{\Gamma'(-\nu+k+1)}{\Gamma^2(-\nu+k+1)} \text{ approaches } (-1)^{p-k}(p-k-1)!$$

and it follows that the expression (IX, 1; 36) tends to

(IX, 1; 38) $$(-1)^p\left(\frac{x}{2}\right)^{-p}\sum_{k=0}^{p-1}\left(\frac{x}{2}\right)^{2k}\frac{1}{k!}(p-k-1)!.$$

Thus we deduce from equations (IX, 1; 23), (IX, 1; 31), (IX, 1; 32) (IX, 1; 35), and (IX, 1; 38) that

(IX, 1; 39)
$$\pi N_p(x) = 2\left(\log\frac{x}{2}+\gamma\right)J_p(x) - \left(\frac{x}{2}\right)^{-p}\sum_{k=0}^{p-1}\left(\frac{x}{2}\right)^{2k}\frac{1}{k!}(p-k-1)!$$
$$-\left(\frac{x}{2}\right)^p\frac{1}{p!}\left(\sum_{n=1}^{p}\frac{1}{n}\right) - \left(\frac{x}{2}\right)^p\sum_{k=1}^{\infty}\frac{(-1)^k\left(\frac{x}{2}\right)^{2k}}{k!(p+k)!}\left(\sum_{n=1}^{k}\frac{2}{n}+\sum_{n=k+1}^{p+k}\frac{1}{n}\right).$$

It now remains to be shown that, for an integer $p \geqslant 0$, $N_p(x)$ is a solution of equation (IX, 1; 1) which is linearly independent of $J_p(x)$. The first requirement follows from equation (IX, 1; 24); however, it can also be established in another way as follows. Differentiate with respect to ν the identity

(IX, 1; 40) $$\frac{d^2}{dx^2}J_\nu(x) + \frac{1}{x}\frac{d}{dx}J_\nu(x) + \left(1-\frac{\nu^2}{x^2}\right)J_\nu(x) = 0,$$

and we get

(IX, 1; 41) $$\frac{d^2}{dx^2}\frac{\partial}{\partial\nu}J_\nu(x) + \frac{1}{x}\frac{d}{dx}\frac{\partial}{\partial\nu}J_\nu(x) + \left(1-\frac{\nu^2}{x^2}\right)\frac{\partial}{\partial\nu}J_\nu(x) = \frac{2\nu}{x^2}J_\nu(x).$$

Similarly we have

(IX, 1; 42) $$\frac{d^2}{dx^2}\frac{\partial}{\partial\nu}J_{-\nu}(x) + \frac{1}{x}\frac{d}{dx}\frac{\partial}{\partial\nu}J_{-\nu}(x) + \left(1-\frac{\nu^2}{x^2}\right)\frac{\partial}{\partial\nu}J_{-\nu}(x) = \frac{2\nu}{x^2}J_{-\nu}(x).$$

Hence by putting

(IX, 1; 43) $$F_\nu(x) = \frac{1}{\pi}\left(\frac{\partial}{\partial\nu}J_\nu(x) - (-1)^p\frac{\partial}{\partial\nu}J_{-\nu}(x)\right)$$

we obtain

(IX, 1; 44)
$$\frac{d^2}{dx^2}F_\nu(x) + \frac{1}{x}\frac{d}{dx}F_\nu(x) + \left(1-\frac{\nu^2}{x^2}\right)F_\nu(x) = \frac{2\nu}{\pi x^2}\left(J_\nu(x) - (-1)^p J_{-\nu}(x)\right).$$

Now let ν tend to some integer p; then, taking into account equation (IX, 1; 23) and (IX, 1; 19), we obtain

$$\text{(IX, 1; 45)} \qquad \frac{d^2}{dx^2}N_p(x) + \frac{1}{x}\frac{d}{dx}N_p(x) + \left(1 - \frac{p^2}{x^2}\right)N_p(x) = 0.$$

Q.E.D.

The linear independence of J_p and N_p can easily be established. In fact, for $p = 0$, when $x \to 0$, $J_0(x) \to 0$, while $N_0(x) \to -\infty$ as $\log x/2$. When $p > 0$, as $x \to 0$, $J_p(x) \to 0$ like x^p and $N_p(x) \to -\infty$ as $-(x)^{-p}$. Thus it follows that when ν is an integer $p \geqslant 0$ the general solution of Equation (IX, 1; 1) has the form

$$\text{(IX, 1; 46)} \qquad AJ_p(x) + BN_p(x),$$

where A and B are constants.

Note. Equation (IX, 1; 21) defines $N_\nu(x)$ *for nonintegral* ν, while equation (IX, 1; 22) defines $N_p(x)$ for *all integers* p, positive, negative, or zero. It can easily be seen that in this last case

$$\text{(IX, 1; 47)} \qquad N_{-p}(x) = (-1)^p N_p(x).$$

In fact from equation (IX, 1; 23) we have

$$\text{(IX, 1; 48)} \qquad \pi N_{-p}(x) = \lim_{\nu \to -p} \left(\frac{\partial}{\partial\nu}J_\nu(x) - (-1)^p \frac{\partial}{\partial\nu}J_{-\nu}(x)\right).$$

Hence, by putting $\nu = -\lambda$,

(IX, 1; 49)
$$\pi N_{-p}(x) = \lim_{\lambda \to p} (-1)^p \left(\frac{\partial}{\partial\lambda}J_\lambda(x) - (-1)^p \frac{\partial}{\partial\lambda}J_{-\lambda}(x)\right) = (-1)^p \pi N_p(x).$$

Hankel functions

For all complex numbers ν, the two *Hankel functtons* of order ν, $H_\nu^{(1)}$ and $H_\nu^{(2)}$, are defined by

$$\text{(IX, 1; 50)} \qquad H_\nu^{(1)} = J_\nu + iN_\nu,$$

$$\text{(IX, 1; 51)} \qquad H_\nu^{(2)} = J_\nu - iN_\nu.$$

These are again solutions of equation (IX, 1; 1). It will be left to the reader to verify the following two formulae:

$$\text{(IX, 1; 52)} \qquad H_{-\nu}^{(1)} = e^{i\nu\pi} H_\nu^{(1)},$$

$$\text{(IX, 1; 53)} \qquad H_{-\nu}^{(2)} = e^{-i\nu\pi} H_\nu^{(2)},$$

which are valid for all ν, and which when p is an integer can be written as

(IX, 1; 54) $$H^{(k)}_{-p} = (-1)^p H^{(k)}_p, \qquad k = 1,2.$$

Note. From equations (IX, 1; 50) and (IX, 1; 51) we deduce that

(IX, 1; 55) $$J_\nu = \frac{H^{(1)}_\nu + H^{(2)}_\nu}{2}$$

and

(IX, 1; 56) $$N_\nu = \frac{H^{(1)}_\nu - H^{(2)}_\nu}{2i},$$

and we see that if in equations (IX, 1; 50), (IX, 1; 51), (IX, 1; 55) and (IX, 1; 56) we replace

$$H^{(1)}_\nu \text{ by } e^{ix}, \qquad H^{(2)}_\nu \text{ by } e^{-ix},$$
$$J_\nu \text{ by } \cos x, \qquad N_\nu \text{ by } \sin x,$$

then we recover well-known formulae.

2. INTEGRAL REPRESENTATIONS OF BESSEL FUNCTIONS

Consider the complex function $f_\nu(t)$ of the real variable t which is defined by

(IX, 1; 57) $$f_\nu(t) = \begin{cases} (1 - t^2)^{\nu - 1/2} & \text{for} \quad |t| < 1, \\ 0 & \text{otherwise.} \end{cases}$$

When $\mathcal{R}\left(\nu + \frac{1}{2}\right) > 0$, $f_\nu(t)$ is summable, and has as its Fourier transform (Chap. V, § 2, no. 7)

(IX, 1; 58) $$\mathcal{F} f_\nu = \int_{-1}^{1} (1 - t^2)^{\nu - 1/2} e^{-itx}\, dt.$$

We will now show that, with the condition

(IX, 1; 59) $$\mathcal{R}\left(\nu + \frac{1}{2}\right) > 0,$$

the function

(IX, 1; 60) $$Z_\nu(x) = x^\nu (\mathcal{F} f_\nu)$$

is a solution of equation (IX, 1; 1). Now, for any integer $m > 0$, we see from (IX, 1; 59) that $t^m f_\nu(t)$ is summable and it follows that $\mathcal{F} f_\nu$ is infinitely differentiable. In particular

(IX, 1; 61) $$(\mathcal{F} f_\nu)' = \mathcal{F}(-it f_\nu(t)),$$
$$(\mathcal{F} f_\nu)'' = \mathcal{F}(-t^2 f_\nu(t)),$$

Now we can easily obtain

(IX, 1; 62) $$Z_\nu'' + \frac{1}{x} Z_\nu' + \left(1 - \frac{\nu^2}{x^2}\right) Z_\nu = x^\nu(\mathcal{F} f_\nu + (\mathcal{F} f_\nu)'') + (2\nu + 1)x^{\nu-1}(\mathcal{F} f_\nu)'$$

But, taking account of equation (IX, 1; 61), we have

(IX, 1; 63) $$\mathcal{F} f_\nu + (\mathcal{F} f_\nu)'' = \mathcal{F} f_{\nu+1},$$

where

(IX, 1; 64) $$f'_{\nu+1} = -(2\nu + 1)t f_\nu(t).$$

Thus $f'_{\nu+1}$ is summable and we have

(IX, 1; 65) $$x\mathcal{F} f_{\nu+1} = -i\mathcal{F}(f'_{\nu+1}).$$

The right-hand side of equation (IX, 1; 62) can now be written

(IX, 1; 66) $$-ix^{\nu-1}\{\mathcal{F}[-(2\nu + 1)t f_\nu(t) + (2\nu + 1)t f_\nu(t)]\} = 0$$

by taking account of the first relation in (IX, 1; 61) and equations (IX, 1; 65) and (IX, 1; 64). Q.E.D.

When $x \to 0$, $Z_\nu(x)$ behaves like x^ν and it follows that $Z_\nu(x)$ is proportional to $J_\nu(x)$. Let a_ν be the coefficient of proportionality, then

(IX, 1; 67) $$\mathcal{F} f_\nu = a_\nu \frac{1}{2^\nu} \sum_{k=0}^{\infty} \frac{(-1)^k \left(\frac{x}{2}\right)^{2k}}{k!\,\Gamma(\nu + k + 1)}.$$

Hence for $x = 0$

(IX, 1; 68) $$\int_{-1}^{1} (1 - t^2)^{\nu - \frac{1}{2}} dt = a_\nu \frac{1}{2^\nu \Gamma(\nu + 1)}.$$

But the left-hand side of this is $B(\frac{1}{2}, \nu + \frac{1}{2})$, hence by using equations (VIII, 1; 12) and (VIII, 1; 15) we have

(IX, 1; 69) $$a_\nu = 2^\nu \sqrt{\pi}\, \Gamma(\nu + \tfrac{1}{2}).$$

Finally, we obtain the important formula

(IX, 1; 70) $$J_\nu(x) = \frac{\left(\frac{x}{2}\right)^\nu}{\Gamma(\nu + \frac{1}{2})\sqrt{\pi}} \int_{-1}^{1} (1 - t^2)^{\nu - \frac{1}{2}} e^{-itx}\, dt,$$

which is valid for $\mathcal{R}(\nu + \frac{1}{2}) > 0$.

Note. Throughout, we assume that the condition (IX, 1; 59) is satisfied. Since

$$\text{(IX, 1; 71)} \qquad \int_{-1}^{1} (1 - t^2)^{\nu - 1/2} \sin tx\, dt = 0,$$

we also have

$$\text{(IX, 1; 72)} \qquad J_\nu(x) = \left(\frac{x}{2}\right)^\nu \frac{1}{\sqrt{\pi}\,\Gamma(\nu + \frac{1}{2})} \int_{-1}^{1} (1 - t^2)^{\nu - 1/2} e^{itx}\, dt$$

$$= \left(\frac{x}{2}\right)^\nu \frac{1}{\sqrt{\pi}\,\Gamma(\nu + \frac{1}{2})} \int_{-1}^{1} (1 - t^2)^{\nu - 1/2} \cos tx\, dt.$$

Another integral representation

In the relation (IX, 1; 70), put $t = \cos\theta$. Then

$$\text{(IX, 1; 73)} \qquad J_\nu(x) = \frac{\left(\frac{x}{2}\right)^\nu}{\sqrt{\pi}\,\Gamma\left(\nu + \frac{1}{2}\right)} \int_0^\pi \sin^{2\nu}\theta\, e^{-ix\cos\theta}\, d\theta,$$

which is valid for $\mathcal{R}\left(\nu + \frac{1}{2}\right) > 0$.

It will be left to the reader to obtain the formulae that follow from equation (IX, 1; 72) as a result of the same change of variable.

3. RECURRENCE RELATIONS

We shall first show that

$$\text{(IX, 1; 74)} \qquad -\frac{1}{x}\frac{d}{dx}(x^{-\nu} J_\nu(x)) = x^{-(\nu+1)} J_{\nu+1}(x),$$

which can also be written as

$$\text{(IX, 1; 75)} \qquad J'_\nu(x) = \frac{\nu}{x} J_\nu(x) - J_{\nu+1}(x).$$

It will be sufficient to establish equation (IX, 1; 74), for sufficiently large $\mathcal{R}\nu$: each side of equation (IX, 1; 74) is an analytic function and hence, by analytic continuation with respect to ν, the equation will be true everywhere, that is for all ν. Thus assuming the inequality (IX, 1; 59), we notice that a_ν, given by equation (IX, 1; 69), satisfies

$$\text{(IX, 1; 76)} \qquad \frac{1}{a_\nu} = \frac{2\nu + 1}{a_{\nu+1}}.$$

Thus we deduce from equations (IX, 1; 61) and (IX, 1; 64) that

$$\text{(IX, 1; 77)} \qquad \frac{i\mathcal{F}(f'_{\nu+1})}{a_{\nu+1}} = \frac{(\mathcal{F} f_\nu)'}{a_\nu}$$

or, again by using equation (IX, 1; 65)

$$\text{(IX, 1; 78)} \qquad -x\frac{\mathcal{F} f_{\nu+1}}{a_{\nu+1}} = \frac{d}{dx}\frac{\mathcal{F} f_\nu}{a_\nu}.$$

Finally, from equation (IX, 1; 67) we have

$$\text{(IX, 1; 79)} \qquad -x(x^{-(\nu+1)}J_{\nu+1}(x)) = \frac{d}{dx}(x^{-\nu}J_\nu(x)),$$

which is simply equation (IX, 1; 74). A particular case of equation (IX, 1; 75) is obtained from equation (IX, 1; 79) in the form

$$\text{(IX, 1; 80)} \qquad J'_0(x) = -J_1(x).$$

We shall now show that the equation

$$\text{(IX, 1; 81)} \qquad \frac{1}{x}\frac{d}{dx}(x^\nu J_\nu(x)) = x^{\nu-1}J_{\nu-1}(x)$$

can be written in the form

$$\text{(IX, 1; 82)} \qquad J'_\nu(x) = -\frac{\nu}{x}J_\nu(x) + J_{\nu-1}(x).$$

As in the above case it will be sufficient to establish equation (IX, 1; 81) for sufficiently large $\mathcal{R}\nu$. We shall assume here that $\mathcal{R}(\nu - 1 + \frac{1}{2}) > 0$. Thus we have

$$\text{(IX, 1; 83)} \qquad \frac{2\nu - 1}{a_\nu} = \frac{1}{a_{\nu-1}}.$$

On the other hand

$$\text{(IX, 1; 84)} \qquad (2\nu - 1)f_\nu - tf'_\nu = (2\nu - 1)f_{\nu-1},$$

so that

$$\text{(IX, 1; 85)} \qquad \frac{\mathcal{F}((2\nu - 1)f_\nu - tf'_\nu)}{a_\nu} = \frac{\mathcal{F} f_{\nu-1}}{a_{\nu-1}}.$$

But the left-hand side is

$$\text{(IX, 1; 86)} \qquad x\frac{d}{dx}\left(\frac{\mathcal{F} f_\nu}{a_\nu}\right) + 2\nu\frac{\mathcal{F} f_\nu}{a_\nu}.$$

Thus finally we have

$$\text{(IX, 1; 87)} \qquad x(x^{-\nu}J_\nu)' = -2\nu x^{-\nu}J_\nu + x^{-\nu+1}J_{\nu-1},$$

which is simply equation (IX, 1; 82).

Application of equations (IX, 1; 70) *and* (IX, 1; 81). The evaluation of $J_{\frac{1}{2}}$ and $J_{-\frac{1}{2}}$.

When $\nu = \frac{1}{2}$, equation (IX, 1; 70) can be written as

(IX, 1; 88)

$$J_{\frac{1}{2}}(x) = \sqrt{\frac{x}{2\pi}} \int_{-1}^{1} e^{-itx}\, dt = \sqrt{\frac{x}{2\pi}}\, \frac{1}{x}\, \frac{e^{ix} - e^{-ix}}{i} = \sqrt{\frac{2}{\pi x}} \sin x.$$

But using equation (IX, 1; 81) we get

(IX, 1; 89) $$x^{-\frac{1}{2}} J_{-\frac{1}{2}}(x) = \frac{1}{x}\left(\frac{d}{dx}\sqrt{\frac{2}{\pi}} \sin x\right) = \frac{1}{x}\sqrt{\frac{2}{\pi}} \cos x;$$

hence

(IX, 1; 90) $$J_{-\frac{1}{2}}(x) = \sqrt{\frac{2}{\pi x}} \cos x.$$

We also notice that

(IX, 1; 91) $$N_{-\frac{1}{2}}(x) = J_{\frac{1}{2}}(x) = \sqrt{\frac{2}{\pi x}} \sin x$$

and

(IX, 1; 92) $$N_{\frac{1}{2}}(x) = -J_{-\frac{1}{2}}(x) = -\sqrt{\frac{2}{\pi x}} \cos x.$$

Generalization of equations (IX, 1; 74) *and* (IX, 1; 81). Let D_1 be the operator

$$-\frac{1}{x}\frac{d}{dx}$$

and D_2 the operator

$$\frac{1}{x}\frac{d}{dx}$$

and let l be any integer $\geqslant 1$. By operating repeatedly on equation (IX, 1; 74) or equation (IX, 1; 81) we easily obtain

(IX, 1; 93) $$D_1^l(x^{-\nu} J_\nu) = x^{-(\nu+l)} J_{l+\nu}$$

or

(IX, 1; 94) $$D_2^l(x^{\nu} J_\nu) = x^{\nu-l} J_{\nu-l}.$$

4. OTHER PROPERTIES OF BESSEL FUNCTIONS

Bessel functions of integral order

We now expand the function $e^{ix \sin\theta}$, in a Fourier series, where x is a parameter assumed to be real and positive; thus

(IX, 1; 95) $$e^{ix\sin\theta} = \sum_{n=-\infty}^{+\infty} a_n(x)\, e^{ni\theta},$$

where

$$\text{(IX, 1; 96)} \qquad a_n(x) = \frac{1}{2\pi}\int_0^{2\pi} e^{ix\sin\theta}\, e^{-ni\theta}\, d\theta.$$

Now we can expand $e^{ix \sin\theta}$ in series :

$$\text{(IX, 1; 97)} \qquad e^{ix\sin\theta} = \sum_{k=0}^{\infty} \frac{i^k x^k \sin^k\theta}{k!}.$$

This series, for fixed x and θ varying between 0 and 2π, is uniformly convergent and therefore can be integrated term by term to give

$$\text{(IX, 1; 98)} \qquad a_n(x) = \sum_{k=0}^{\infty} \left(\frac{x}{2}\right)^k \frac{1}{k!}\,\frac{1}{2\pi}\int_0^{2\pi} (e^{i\theta} - e^{-i\theta})^k\, e^{-ni\theta}\, d\theta.$$

Suppose first that n is positive or zero. Expand $(e^{i\theta} - e^{-i\theta})^k$ by the Binomial Theorem. Then the integral

$$\text{(IX, 1; 99)} \qquad \frac{1}{2\pi}\int_0^{2\pi} (e^{i\theta} - e^{-i\theta})^k\, e^{-ni\theta}\, d\theta$$

is zero, except when $k = n + 2m$, for $m \geqslant 0$, in which case it has the value

$$\text{(IX, 1; 100)} \qquad (-1)^m C_{n+2m}^m \frac{1}{2\pi}\int_0^{2\pi} d\theta = (-1)^m \frac{(n+2m)\,!}{m\,!\,(n+m)\,!}.$$

Hence

$$\text{(IX, 1; 101)} \qquad a_n(x) = \left(\frac{x}{2}\right)^n \sum_{m=0}^{+\infty} (-1)^m \left(\frac{x}{2}\right)^{2m} \frac{1}{m\,!\,(n+m)\,!} = J_n(x).$$

When n is negative we can easily see that $a_n(x) = J_n(x)$. Thus, for integer n,

$$\text{(IX, 1; 102)} \qquad J_n(x) = \frac{1}{2\pi}\int_0^{2\pi} e^{ix\sin\theta}\, e^{-ni\theta}\, d\theta.$$

$J_n(x)$ is therefore the mean value of the function $e^{i(x\sin\theta - n\theta)}$ over one period and

$$\text{(IX, 1; 103)} \qquad e^{ix\sin\theta} = \sum_{n=-\infty}^{+\infty} J_n(x)\, e^{ni\theta}.$$

Equation (IX, 1; 103) can also be written as

(IX, 1; 104)

$$e^{ix\sin\theta} = J_0(x) + 2\sum_{n=1}^{\infty} J_{2n}(x)\cos 2n\theta + 2i\sum_{n=0}^{\infty} J_{2n+1}(x)\sin(2n+1)\theta,$$

whence

(IX, 1; 105) $$\cos(x \sin\theta) = J_0(x) + 2\sum_{n=1}^{\infty} J_{2n}(x)\cos 2n\theta$$

and

(IX, 1; 106) $$\sin(x\sin\theta) = 2\sum_{n=0}^{\infty} J_{2n+1}(x)\sin(2n+1)\theta.$$

Asymptotic expansions

We now propose to study the behaviour of $J_\nu(x)$ as $x \to \infty$ for real ν. Let

(IX, 1; 107) $$u_\nu(x) = \sqrt{x}J_\nu(x).$$

Then it is easy to see that $u_\nu(x)$ is a solution of the differential equation

(IX, 1; 108) $$u'' + \left(1 - \frac{\nu^2 - \frac{1}{4}}{x^2}\right)u = 0.$$

Intuitively then we would expect that, as $x \to \infty$, $u_\nu(x)$ would tend asymptotically to a trigonometrical solution of the equation

(IX, 1; 109) $$u'' + u = 0.$$

We can verify our assumption in a precise manner, because for real ν, as $x \to \infty$

(IX, 1; 110) $$J_\nu(x) = \sqrt{\frac{2}{\pi x}}\cos\left(x - (2\nu + 1)\frac{\pi}{4}\right) + 0\left(\frac{1}{x^{\frac{3}{2}}}\right)^{(*)}.$$

Thus, when $x \to \infty$, $J_\nu(x)$ behaves approximately like

$$\sqrt{\frac{2}{\pi x}}\cos\left[x - (2\nu + 1)\frac{\pi}{4}\right].$$

Notice that, for $\nu = \pm\frac{1}{2}$, by using equations (IX, 1; 88) and (IX, 1; 90), $J_\nu(x)$ is exactly and not approximately equal to

(IX, 1; 111) $$\sqrt{\frac{2}{\pi x}}\cos\left(x - (2\nu + 1)\frac{\pi}{4}\right) = \begin{cases} \sqrt{\frac{2}{\pi x}}\sin x & \text{for } \nu = \frac{1}{2}, \\ \sqrt{\frac{2}{\pi x}}\cos x & \text{for } \nu = -\frac{1}{2}, \end{cases}$$

and this is true for all x and not only sufficiently large x.

(*) $g(x) = 0(f(x))$ when $x \to \infty$ means that $\left|\frac{g(x)}{f(x)}\right|$ is bounded when $x \to \infty$.

Moreover, we see that for $\nu = \pm 1/2$ equation (IX, 1; 108) becomes

(IX, 1; 112) $$u'' + u = 0.$$

We can deduce from equation (IX, 1; 110), by using (IX, 1; 21), that

(IX, 1; 113) $$N_\nu(x) = \sqrt{\frac{2}{\pi x}} \sin\left(x - (2\nu + 1)\frac{\pi}{4}\right) + 0\left(\frac{1}{x^{\frac{3}{2}}}\right)$$

as $x \to \infty$. By using equations (IX, 1; 50) and (IX, 1; 51) we obtain

(IX, 1; 114) $$H_\nu^{(1)}(x) = \sqrt{\frac{2}{\pi x}}\, e^{i\left(x-(2\nu+1)\frac{\pi}{4}\right)} + 0\left(\frac{1}{x^{\frac{3}{2}}}\right)$$

when $x \to \infty$, and

(IX, 1; 115) $$H_\nu^{(2)}(x) = \sqrt{\frac{2}{\pi x}}\, e^{-i\left(x-(2\nu+1)\frac{\pi}{4}\right)} + 0\left(\frac{1}{x^{\frac{3}{2}}}\right)$$

as $x \to \infty$.

Positive zeros of Bessel functions

We now propose to examine the distribution of the zeros of $J_\nu(x)$ for real ν and sufficiently large x. We shall show that *as $x \to \infty$, the zeros of $J_\nu(x)$ are approximately the zeros of*

$$\cos\left(x - (2\nu + 1)\frac{\pi}{4}\right).$$

Notice first of all that, as $x \to \infty$, it follows from equation (IX, 1; 110) that the zeros of $J_\nu(x)$ can only be situated in the neighbourhood of the zeros of $\cos\left(x - (2\nu + 1)\frac{\pi}{4}\right)$.

We must now show that, conversely, in any neighbourhood of a zero of

$$\cos\left(x - (2\nu + 1)\frac{\pi}{4}\right)$$

there is only one zero of $J_\nu(x)$. The zeros of $\cos\left(x - (2\nu + 1)\frac{\pi}{4}\right)$ are

(IX, 1; 116) $$x = (2\nu + 1)\frac{\pi}{4} + (2n + 1)\frac{\pi}{2},$$ n an integer.

We are interested here only in those $n > 0$. Given $\varepsilon > 0$, there exists $\eta > 0$, independent of n, such that the condition

$$\text{(IX, 1; 117)} \qquad \left| x - (2\nu + 1)\frac{\pi}{4} - (2n + 1)\frac{\pi}{2} \right| < \eta$$

implies

$$\text{(IX, 1; 118)} \qquad \left| \cos\left(x - (2\nu + 1)\frac{\pi}{4} \right) \right| < \varepsilon.$$

Denote by I_n the interval

$$\left[(2\nu + 1)\frac{\pi}{4} + (2n + 1)\frac{\pi}{2} - \eta, \quad (2\nu + 1)\frac{\pi}{4} + (2n + 1)\frac{\pi}{2} + \eta \right].$$

When x traverses this interval, $\cos\left(x - (2\nu + 1)\frac{\pi}{4} \right)$ takes on values from $(-1)^n\varepsilon$ to $(-1)^{n+1}\varepsilon$. When $n \to \infty$, $J_\nu(x)$ has the same sign as $(-1)^n\varepsilon$ at the left-hand end of the interval I_n, while at the right-hand end it has the sign of $(-1)^{n+1}\varepsilon$. Thus $J_\nu(x)$ changes sign as x describes I_n,. Thus it follows that $J_\nu(x)$ must possess an odd number of zeros in this interval. To show that $J_\nu(x)$ possesses only one in this interval it is sufficient to show that $J'_\nu(x)$ is never zero in the interval.

Now, from equation (IX, 1; 75) we have

(IX, 1; 119)

$$J'_\nu(x) = -\sqrt{\frac{2}{\pi x}} \sin\left(x - (2\nu + 1)\frac{\pi}{4} \right) + 0\left(\frac{1}{x^{\frac{3}{2}}} \right) \quad \text{for} \quad x \to \infty.$$

Thus the zeros of $J'_\nu(x)$ can only be situated in the neighbourhood of $\sin\left(x - (2\nu + 1)\frac{\pi}{4} \right)$, and in any sufficiently small neighbourhood of $\cos\left(x - (2\nu + 1)\frac{\pi}{4} \right)$ there are no zeros of $\sin\left(x - (2\nu + 1)\frac{\pi}{4} \right)$.

Q.E.D.

Notes. Assuming that ν is always real, then :

1. If we know the positive zeros of $J_\nu(x)$ then we also know the negative zeros, since $J_\nu(-x) = e^{i\nu\pi} J_\nu(x)$.

2. We can show that for $\nu > -1$ all the zeros of $J_\nu(x)$ are real.

3. Notice that equations (IX, 1; 88) and (IX, 1; 90) give us *exactly every* zero of $J_{\frac{1}{2}}(x)$ and $J_{-\frac{1}{2}}(x)$.

The zeros of $J_{\frac{1}{2}}(x)$ are

(IX, 1; 120) $n\pi$, for any integer n, positive, negative, or zero.

The zeros of $J_{-\frac{1}{2}}(x)$ are

(IX, 1; 121) $(2n + 1)\pi/2$, for any integer n, positive, negative, or zero

Modified Bessel functions

Let

(IX, 1; 122) $$I_\nu(x) = e^{-i\nu\frac{\pi}{2}} J_\nu(ix) = \left(\frac{x}{2}\right)^\nu \sum_{k=0}^{\infty} \left(\frac{x}{2}\right)^{2k} \frac{1}{k!\,\Gamma(\nu + k + 1)}$$

The function

(IX, 1; 123) $$K_\nu(x) = \frac{\pi}{2 \sin \nu\pi} (I_{-\nu}(x) - I_\nu(x))$$

is called a *a Kelvin function of order* ν. Notice that

(IX, 1; 124) $$K_{-\nu} = K_\nu \quad \text{for} \quad \text{all } \nu$$

and

(IX, 1; 125) $$I_{-n} = I_n \quad \text{for} \quad \text{integer } n.$$

2. Formulae

We recapitulate in this section the formulae established above.

Bessel's differential equation

(IX, 2; 1) $$y'' + \frac{1}{x} y' + \left(1 - \frac{\nu^2}{x^2}\right) y = 0, \quad \nu \text{ complex.}$$

Bessel function of order ν

(IX, 2; 2) $$J_\nu(x) = \left(\frac{x}{2}\right)^\nu \sum_{k=0}^{\infty} (-1)^k \left(\frac{x}{2}\right)^{2k} \frac{1}{k!} \frac{1}{\Gamma(\nu + k + 1)},$$

and the particular case

(IX, 2; 3) $$J_0(x) = \sum_{k=0}^{\infty} (-1)^k \left(\frac{x}{2}\right)^{2k} \frac{1}{(k!)^2}.$$

Neumann function of order ν

(IX, 2; 4) $N_\nu(x) = \dfrac{(\cos\nu\pi)J_\nu(x) - J_{-\nu}(x)}{\sin\nu\pi}$ for $\nu \neq$ an integer.

If ν is an integer n then

(IX, 2; 5) $$N_n(x) = \lim_{\nu\to n} N_\nu(x),$$

and in particular

(IX, 2; 6)

$$N_0(x) = \frac{2}{\pi}\left[\left(\log\frac{x}{2}+\gamma\right)J_0(x) - \sum_{k=1}^{\infty}(-1)^k\left(\frac{x}{2}\right)^{2k}\frac{1}{(k!)^2}\left(\sum_{n=1}^{k}\frac{1}{n}\right)\right]$$

where γ is Euler's constant.

Hankel function of order ν

(IX, 2; 7) $$H_\nu^{(1)} = J_\nu + iN_\nu,$$

(IX, 2; 8) $$H_\nu^{(2)} = J_\nu - iN_\nu.$$

Hence

(IX, 2; 9) $$J_\nu = \frac{H_\nu^{(1)} + H_\nu^{(2)}}{2},$$

(IX, 2; 10) $$N_\nu = \frac{H_\nu^{(1)} - H_\nu^{(2)}}{2i}.$$

Bessel, Neumann, and Hankel functions of integral order

For n an integer we have

(IX, 2; 11) $$J_{-n} = (-1)^n J_n,$$

(IX, 2; 12) $$N_{-n} = (-1)^n N_n,$$

(IX, 2; 13) $$H_{-n}^{(1} = (-1)^n H^{(1)},\quad H_{-n}^{(2)} = (-1)^n H_n^{(2)}.$$

Independent solutions of Bessel's equation

For all ν : $H_\nu^{(1)}$ and $H_\nu^{(2)}$.
For ν not an integer : J_ν and $J_{-\nu}$.
For $\nu = n$ an integer : J_n and N_n.

Integral representations of the Bessel functions

For $\mathcal{R}\left(\nu + \frac{1}{2}\right) > 0$, we have

(IX, 2; 14) $$J_\nu(x) = \left(\frac{x}{2}\right)^\nu \frac{1}{\sqrt{\pi}\,\Gamma\left(\nu+\frac{1}{2}\right)}\int_{-1}^{1}(1-t^2)^{\nu-\frac{1}{2}}e^{\pm itx}\,dt.$$

(IX, 2; 15) $$J_\nu(x) = \left(\frac{x}{2}\right)^\nu \frac{2}{\sqrt{\pi}\Gamma\left(\nu + \frac{1}{2}\right)} \int_0^1 (1 - t^2)^{\nu - \frac{1}{2}} \cos tx \, dt,$$

(IX, 2; 16) $$J_\nu(x) = \left(\frac{x}{2}\right)^\nu \frac{1}{\sqrt{\pi}\Gamma\left(\nu + \frac{1}{2}\right)} \int_0^\pi e^{\pm ix\cos\theta} \sin^{2\nu}\theta \, d\theta,$$

(IX, 2; 17) $$J_\nu(x) = \left(\frac{x}{2}\right)^\nu \frac{2}{\sqrt{\pi}\Gamma\left(\nu + \frac{1}{2}\right)} \int_0^{\frac{\pi}{2}} \cos(x\cos\theta) \sin^{2\nu}\theta \, d\theta$$

and, in particular,

(IX, 2; 18) $$J_0(x) = \frac{2}{\pi} \int_0^{\frac{\pi}{2}} \cos(x\cos\theta) \, d\theta.$$

Recurrence formulae

Let

(IX, 2; 19) $$D_1 = -\frac{1}{x}\frac{d}{dx} \quad \text{and} \quad D_2 = \frac{1}{x}\frac{d}{dx}$$

Then, for any integer $n \geqslant 0$,

(IX, 2; 20) $$D_1^n(x^{-\nu}J_\nu) = x^{-(\nu+n)}J_{\nu+n},$$

(IX, 2; 21) $$D_2^n(x^\nu J_\nu) = x^{\nu-n}J_{\nu-n}.$$

Generating function

We have

(IX, 2; 22) $$e^{ix\sin\theta} = \sum_{n=-\infty}^{+\infty} J_n(x)\, e^{ni\theta}$$

and thus

(IX, 2; 23) $$J_n(x) = \frac{1}{2\pi}\int_0^{2\pi} e^{ix\sin\theta}\, e^{-ni\theta}\, d\theta.$$

Asymptotic expansion for real ν as $x \to \infty$

(IX, 2; 24) $$J_\nu(x) = \sqrt{\frac{2}{\pi x}} \cos\left[x - (2\nu + 1)\frac{\pi}{4}\right] + 0\left(\frac{1}{x^{\frac{3}{2}}}\right),$$

(IX, 2; 25) $$N_\nu(x) = \sqrt{\frac{2}{\pi x}} \sin\left[x - (2\nu + 1)\frac{\pi}{4}\right] + 0\left(\frac{1}{x^{\frac{3}{2}}}\right),$$

(IX, 1; 26) $$H_\nu^{(1)}(x) = \sqrt{\frac{2}{\pi x}}\, e^{i\left[x-(2\nu+1)\frac{\pi}{4}\right]} + 0\left(\frac{1}{x^{\frac{3}{2}}}\right),$$

(IX, 2; 27) $$H_\nu^{(2)}(x) = \sqrt{\frac{2}{\pi x}}\, e^{-i\left[x-(2\nu+1)\frac{\pi}{4}\right]} + 0\left(\frac{1}{x^{\frac{3}{2}}}\right).$$

Notice that for *all* x we have *exactly*

$$\text{(IX, 2; 28)} \qquad J_{\frac{1}{2}}(x) = \sqrt{\frac{2}{\pi x}} \sin x = N_{-\frac{1}{2}}(x),$$

$$\text{(IX, 2; 29)} \qquad J_{-\frac{1}{2}}(x) = \sqrt{\frac{2}{\pi x}} \cos x = -N_{\frac{1}{2}}(x).$$

Laplace transformation of the Bessel functions

$$\text{(IX, 2; 30)} \qquad Y(t)\left(\frac{t}{2}\right)^{\nu-\frac{1}{2}} \frac{\sqrt{\pi}}{\Gamma(\nu)} J_{\nu-\frac{1}{2}}(t) \sqsupset \left(\frac{1}{p^2+1}\right)^{\nu}$$

for $\mathcal{R}\nu > 0$, $\mathcal{R}p > 0$. In particular

$$\text{(IX, 2; 31)} \qquad Y(t) J_0(t) \sqsupset \frac{1}{\sqrt{p^2+1}}.$$

Modified Bessel functions

$$\text{(IX, 2; 32)} \quad I_\nu(x) = e^{-i\nu\frac{\pi}{2}} J_\nu(ix) = \left(\frac{x}{2}\right)^{\nu} \sum_{k=0}^{\infty} \left(\frac{x}{2}\right)^{2k} \frac{1}{k!\,\Gamma(\nu+k+1)}.$$

$$\text{(IX, 2; 33)} \quad K_\nu(x) = \frac{\pi}{2 \sin \nu\pi} (I_{-\nu}(x) - I_\nu(x)) \qquad \text{(Kelvin function)}.$$

We also have

$$\text{(IX, 2; 34)} \qquad K_{-\nu} = K_\nu \quad \text{for all } \nu$$

and

$$\text{(IX, 2; 35)} \qquad I_{-n} = I_n \quad \text{for an integer } n.$$

EXERCISES FOR CHAPTER IX

Exercise IX-1
Prove equations (IX, 1; 70) and (IX, 1; 72) by expanding e^{itx}, e^{-itx} and $\cos tx$.

Exercise IX-2
Prove equations (IX, 1; 74) and (IX, 1; 81) by expanding $J_\nu(x)$.

Exercise IX-3
(*a*) Show that, for any integer n,

$$J_n(a+b) = \sum_{p=-\infty}^{+\infty} J_p(a)\, J_{n-p}(b).$$

(*b*) Show that

$$J_0(\sqrt{a^2 + b^2 + 2ab\cos\alpha}) = J_0(a)\,J_0(b) + 2\sum_{p=1}^{\infty} J_p(a)J_{-p}(b)\cos p\alpha.$$

Exercise IX-4
Let R be any given positive number and ω_1 and ω_2 two numbers > 0 such that

$$\omega_1 \neq \omega_2 \qquad \text{and} \qquad J_0(\omega_1 R) = J_0(\omega_2 R) = 0.$$

Show that

$$\int_0^R J_0(\omega_1 r) J_0(\omega_2 r)\, r\, dr = 0.$$

Exercise IX-5
By using a similar method to that used for evaluating $\Delta\left(\frac{1}{r^{n-2}}\right)$ in $\mathbf{R}^n$, evaluate, *in the distribution sense*, in $\mathbf{R}^2$

$$(\Delta + k^2)N_0(kr),$$

where N_0 is Neumann's function of zero order and $r = \sqrt{x^2 + y^2}$.

Exercise IX-6

(*a*) Evaluate directly, by using the series expansion for J_0 and J_1, the convolution

$$S = Y(x)\,J_0(kx) * Y(x)\,J_0(kx)$$

where $Y(x)$ is Heaviside's function, assuming the relation

$$\sum_{m+n=l} \frac{(2m)!}{2^{2m}(m!)^2}\,\frac{(2n)!}{2^{2n}(n!)^2} = 1.$$

(*b*) Find a differential operator D, having constant coefficients, such that $DS = \delta$. Evaluate

$$D(Y(x)\,J_0(kx))$$

and hence evaluate

$$Y(x)\,J_0(kx) * \frac{Y(x)\,J_1(kx)}{x}.$$

Exercise IX-7
Has Bessel's equation

$$x^2y'' + xy' + (x^2 - \nu^2)y = 0$$

any solutions, in the distribution sense, of the form $\delta^{(m)}_{(0)}$? When $\nu = 0$, is N_0 a solution to this equation in the distribution sense?

Exercise IX-8

Prove the formula

$$\int_0^z \cos(z - t) J_0(t)\, dt = z J_0(z).$$

(Let U be the unknown integral, satisfying the differential equation

$$U'' + U = -J_1$$

so that

$$U = zJ_0 + A \sin z + B \cos z.$$

Then show that $A = B = 0$.)

Exercise IX-9

By using the series expansion for J_0, prove that

$$\int_0^{\frac{\pi}{2}} J_0(z \sin \theta) J_0(Z \cos \theta) \sin \theta \cos \theta \, d\theta = \frac{J_1(\sqrt{Z^2 + z^2})}{\sqrt{Z^2 + z^2}}.$$

Exercise IX-10

By using equation (IX, 2; 16) with $\nu = 0$, evaluate the integral

$$\int_0^\infty e^{-at} J_0(bt)\, dt.$$

Exercise IX-11

Show that

$$J_{\mu+\nu+1}(a) = \frac{a^{\mu+1}}{2^\mu \Gamma(\mu + 1)} \int_0^{\pi/2} J_\nu(a \sin \theta) \sin^{\nu+1}\theta \cos^{2\mu+1}\theta \, d\theta.$$

Exercise IX-12

(*a*) Expand $\cos(z \sin \theta)$ and $\sin(z \sin \theta)$ as Fourier series in θ, using Parseval's formula. Then deduce an upper bound for $|J_n(z)|$ which is independent of the real part of z.

(*b*) Show that

$$(m - 1) \int_0^x t^m J_{n+1}(t) J_{n-1}(t)\, dt = x^{m+1}(J_{n+1}(x) J_{n-1}(x) - J_n^2(x)) + (m + 1) \int_0^x t^m J_n^2(t) dt,$$

where n is a positive integer and $m + 2n + 1 > 0$.

(*c*) Show that $J_{-n}(z) J_{n-1}(z) + J_{-n+1}(z) J_n(z) = \dfrac{2 \sin n\pi}{\pi z}$ for all n.

(*d*) Given that

$$\varphi_n(z) = \frac{J_{n+1}(z)}{z J_n(z)},$$

show that φ_n satisfies the Riccati equation.

(*e*) Given that $R^2 = r^2 + r_1^2 - 2rr_1 \cos\theta$, where $r_1 > r > 0$, show that

$$J_0(R) = J_0(r)\,J_0(r_1) + 2\sum_{n=1}^{\infty} J_n(r)\,J_n(r_1)\cos n\theta,$$

$$N_0(R) = J_0(r)\,N_0(r_1) + 2\sum_{n=1}^{\infty} J_n(r)\,N_n(r_1)\cos n\theta.$$

(*f*) Show that

$$z^{2n} J_{2n+2}(z) = AJ_2(z) + BJ_0(z),$$

where A and B are polynomials in z of degree $2n$. Evaluate $J_4(\sqrt{6}) + 3J_0(\sqrt{6})$ and $3J_6(\sqrt{30}) + 5J_2(\sqrt{30})$.

Abel's theorem for integrals, 43.
  for series, 24.
  for uniform convergence, 50.
Abscissa of convergence (or summability) of the Laplace integral, 216.
Absolute convergence, 19.
Almost everywhere, 27.
Alternating series, 25.
Approximation theorem for continuous functions, 72.
Associativity of convolution, 121.
Associativity of the sum, 19.

Backward wave-cone, 254, 288.
Banach space, 70.
Bernoulli numbers, 330.
Bessel functions, 335.
  modified, 350.
Bessel's equation, 334.
Bessel-Parseval equation, 161.
Beta function, 313.
Bounded membranes, 289.

Cauchy principal value, 45.
Cauchy sequence (in a normed vector space), 69.
Cauchy's criterion for multipliable products, 323.
  summable series, 15.
  uniform convergence and uniform summability, 48.
Cauchy's problem for the equation of vibrating membranes, 308.
  equation of vibrating strings, 253, 269.
  heat conduction equation, 206, 298.
  wave equation, 283.
Change of variables (in a multiple integral), 32.
Characteristic function of a set, 26.
Complementary formula for the gamma function, 316.
Continuity of convolution, 120.
Convergence of distributions, 94.
Convergence of the Fourier series of a distribution, 154.
  function, 155.
Convolution algebra, 123.
Convolution equations, 123, 132.
Convolution of distributions, 112.

Derivative of a distribution, 80.
Descent, method of, 285.
Dirac distribution, 77.
Dirichlet's theorem, 292.
Distribution, 73.
  of order *m*, 92.
  with bounded support, 98.

Elementary solution (of a convolution equation), 124.
Equation of vibrating strings, 242.

Feynman's method of integration, 68.
Forward wave-cone, 255, 288.
Fourier coefficients of a function, 147.
Fourier expansion in an interval (a, b), 148.
Fourier series of a periodic function, 146.
  distribution, 152.
Fourier transform (or Fourier image) of a function, 179.
  tempered distribution, 188.
Fubini's theorem, 34.
Functional equation of the gamma function, 312.
Fundamental period, 145.

Gamma function, 311.
Gauss' integral, 313.
Green's function, 102.
Green's theorem, 88.

Hankel functions, 340.
Heat conduction equation, 206, 298.
Heaviside function, 82, 105.

Heaviside's operational calculus, 128.
Hermite polynomials, Hermite functions, 212, 310.
Hilbert bases (of a Hilbert space), 159.
Hilbert space, 158.

Improper semi-convergent Lebesgue integral, 43.
Indefinite integral, 42.
Integral equations, 130.
Integration over a sphere, 38.
Inversion of the Laplace transform, 224.
Inversion theorem for Fourier transforms, 192.

Laplace transformation (or Laplace image) of a function, 215.
  distribution, 217.
Lebesgue integral, 26.
Lebesgue's theorem, 53.
Locally summable function, 74.
Longitudinal vibrations of a solid bar, 246.
  column of liquid or gas, 249.

Mean square convergence, 161.
Measurable function, 27.
Measurable set, 30.
Measure of an open set, 26.
Minkowski's inequality, 159.
Multipliable infinite product, 322.
Multiplication of distributions, 91.

Neumann functions, 336.
Norm (on a vector space), 69.
Normally summable, 49.

Operational calculus, 128, 230.

Parseval-Plancherel equation, 195.
Period of a function, 145.
Periodic function, 145.
Poisson summation formula, 196.
Poisson's equation, 64, 122.
Positive type, 173.
Potential, 61, 122.
Primitive of order *n*, 36.
Principal value, Cauchy, 45.
Propagation, method of, 298.
Pseudo-function, 104.

Regularization of a distribution, 117.
Repeated summation, 19.
  multiplication, 323.
Riesz-Fischer theorem, 161.

Schwarz' inequality, 159.
Semi-convergent series, 23.
Set of measure zero, 27.
Stationary mode, 275.
Stirling's formula, 320.
Summable function, 26.
  series, 13.
  set, 30.
Support of a distribution, 79.
  function, 71.
System of convolution equations, 132.

Taylor's theorem, 36.
Tempered distribution, 188.
Tensor product of distributions, 111.
Transverse vibrations of a stretched string, 242.
Travelling waves, method of, 251, 281.
Trigonometric integrals, 44.
  series, 25.

Unification principle, 79.
Uniform convergence, 47.

Vibrating strings, 242.
  membranes, 280.
Volterra's integral equations, 130.

Wallis's integral, 314.

A CATALOG OF SELECTED

DOVER BOOKS

IN SCIENCE AND MATHEMATICS

Astronomy

BURNHAM'S CELESTIAL HANDBOOK, Robert Burnham, Jr. Thorough guide to the stars beyond our solar system. Exhaustive treatment. Alphabetical by constellation: Andromeda to Cetus in Vol. 1; Chamaeleon to Orion in Vol. 2; and Pavo to Vulpecula in Vol. 3. Hundreds of illustrations. Index in Vol. 3. 2,000pp. 6⅛ x 9¼.
Vol. I: 0-486-23567-X
Vol. II: 0-486-23568-8
Vol. III: 0-486-23673-0

EXPLORING THE MOON THROUGH BINOCULARS AND SMALL TELESCOPES, Ernest H. Cherrington, Jr. Informative, profusely illustrated guide to locating and identifying craters, rills, seas, mountains, other lunar features. Newly revised and updated with special section of new photos. Over 100 photos and diagrams. 240pp. 8¼ x 11. 0-486-24491-1

THE EXTRATERRESTRIAL LIFE DEBATE, 1750–1900, Michael J. Crowe. First detailed, scholarly study in English of the many ideas that developed from 1750 to 1900 regarding the existence of intelligent extraterrestrial life. Examines ideas of Kant, Herschel, Voltaire, Percival Lowell, many other scientists and thinkers. 16 illustrations. 704pp. 5⅜ x 8½. 0-486-40675-X

THEORIES OF THE WORLD FROM ANTIQUITY TO THE COPERNICAN REVOLUTION, Michael J. Crowe. Newly revised edition of an accessible, enlightening book recreates the change from an earth-centered to a sun-centered conception of the solar system. 242pp. 5⅜ x 8½. 0-486-41444-2

A HISTORY OF ASTRONOMY, A. Pannekoek. Well-balanced, carefully reasoned study covers such topics as Ptolemaic theory, work of Copernicus, Kepler, Newton, Eddington's work on stars, much more. Illustrated. References. 521pp. 5⅜ x 8½. 0-486-65994-1

A COMPLETE MANUAL OF AMATEUR ASTRONOMY: TOOLS AND TECHNIQUES FOR ASTRONOMICAL OBSERVATIONS, P. Clay Sherrod with Thomas L. Koed. Concise, highly readable book discusses: selecting, setting up and maintaining a telescope; amateur studies of the sun; lunar topography and occultations; observations of Mars, Jupiter, Saturn, the minor planets and the stars; an introduction to photoelectric photometry; more. 1981 ed. 124 figures. 25 halftones. 37 tables. 335pp. 6½ x 9¼. 0-486-40675-X

AMATEUR ASTRONOMER'S HANDBOOK, J. B. Sidgwick. Timeless, comprehensive coverage of telescopes, mirrors, lenses, mountings, telescope drives, micrometers, spectroscopes, more. 189 illustrations. 576pp. 5⅝ x 8¼. (Available in U.S. only.) 0-486-24034-7

STARS AND RELATIVITY, Ya. B. Zel'dovich and I. D. Novikov. Vol. 1 of *Relativistic Astrophysics* by famed Russian scientists. General relativity, properties of matter under astrophysical conditions, stars, and stellar systems. Deep physical insights, clear presentation. 1971 edition. References. 544pp. 5⅝ x 8¼. 0-486-69424-0

Chemistry

THE SCEPTICAL CHYMIST: THE CLASSIC 1661 TEXT, Robert Boyle. Boyle defines the term "element," asserting that all natural phenomena can be explained by the motion and organization of primary particles. 1911 ed. viii+232pp. 5⅜ x 8½. 0-486-42825-7

RADIOACTIVE SUBSTANCES, Marie Curie. Here is the celebrated scientist's doctoral thesis, the prelude to her receipt of the 1903 Nobel Prize. Curie discusses establishing atomic character of radioactivity found in compounds of uranium and thorium; extraction from pitchblende of polonium and radium; isolation of pure radium chloride; determination of atomic weight of radium; plus electric, photographic, luminous, heat, color effects of radioactivity. ii+94pp. 5⅜ x 8½. 0-486-42550-9

CHEMICAL MAGIC, Leonard A. Ford. Second Edition, Revised by E. Winston Grundmeier. Over 100 unusual stunts demonstrating cold fire, dust explosions, much more. Text explains scientific principles and stresses safety precautions. 128pp. 5⅜ x 8½. 0-486-67628-5

THE DEVELOPMENT OF MODERN CHEMISTRY, Aaron J. Ihde. Authoritative history of chemistry from ancient Greek theory to 20th-century innovation. Covers major chemists and their discoveries. 209 illustrations. 14 tables. Bibliographies. Indices. Appendices. 851pp. 5⅜ x 8½. 0-486-64235-6

CATALYSIS IN CHEMISTRY AND ENZYMOLOGY, William P. Jencks. Exceptionally clear coverage of mechanisms for catalysis, forces in aqueous solution, carbonyl- and acyl-group reactions, practical kinetics, more. 864pp. 5⅜ x 8½. 0-486-65460-5

ELEMENTS OF CHEMISTRY, Antoine Lavoisier. Monumental classic by founder of modern chemistry in remarkable reprint of rare 1790 Kerr translation. A must for every student of chemistry or the history of science. 539pp. 5⅜ x 8½. 0-486-64624-6

THE HISTORICAL BACKGROUND OF CHEMISTRY, Henry M. Leicester. Evolution of ideas, not individual biography. Concentrates on formulation of a coherent set of chemical laws. 260pp. 5⅜ x 8½. 0-486-61053-5

A SHORT HISTORY OF CHEMISTRY, J. R. Partington. Classic exposition explores origins of chemistry, alchemy, early medical chemistry, nature of atmosphere, theory of valency, laws and structure of atomic theory, much more. 428pp. 5⅜ x 8½. (Available in U.S. only.) 0-486-65977-1

GENERAL CHEMISTRY, Linus Pauling. Revised 3rd edition of classic first-year text by Nobel laureate. Atomic and molecular structure, quantum mechanics, statistical mechanics, thermodynamics correlated with descriptive chemistry. Problems. 992pp. 5⅜ x 8½. 0-486-65622-5

FROM ALCHEMY TO CHEMISTRY, John Read. Broad, humanistic treatment focuses on great figures of chemistry and ideas that revolutionized the science. 50 illustrations. 240pp. 5⅜ x 8½. 0-486-28690-8

Engineering

DE RE METALLICA, Georgius Agricola. The famous Hoover translation of greatest treatise on technological chemistry, engineering, geology, mining of early modern times (1556). All 289 original woodcuts. 638pp. 6¾ x 11. 0-486-60006-8

FUNDAMENTALS OF ASTRODYNAMICS, Roger Bate et al. Modern approach developed by U.S. Air Force Academy. Designed as a first course. Problems, exercises. Numerous illustrations. 455pp. 5⅜ x 8½. 0-486-60061-0

DYNAMICS OF FLUIDS IN POROUS MEDIA, Jacob Bear. For advanced students of ground water hydrology, soil mechanics and physics, drainage and irrigation engineering and more. 335 illustrations. Exercises, with answers. 784pp. 6⅛ x 9¼. 0-486-65675-6

THEORY OF VISCOELASTICITY (Second Edition), Richard M. Christensen. Complete consistent description of the linear theory of the viscoelastic behavior of materials. Problem-solving techniques discussed. 1982 edition. 29 figures. xiv+364pp. 6⅛ x 9¼. 0-486-42880-X

MECHANICS, J. P. Den Hartog. A classic introductory text or refresher. Hundreds of applications and design problems illuminate fundamentals of trusses, loaded beams and cables, etc. 334 answered problems. 462pp. 5⅜ x 8½. 0-486-60754-2

MECHANICAL VIBRATIONS, J. P. Den Hartog. Classic textbook offers lucid explanations and illustrative models, applying theories of vibrations to a variety of practical industrial engineering problems. Numerous figures. 233 problems, solutions. Appendix. Index. Preface. 436pp. 5⅜ x 8½. 0-486-64785-4

STRENGTH OF MATERIALS, J. P. Den Hartog. Full, clear treatment of basic material (tension, torsion, bending, etc.) plus advanced material on engineering methods, applications. 350 answered problems. 323pp. 5⅜ x 8½. 0-486-60755-0

A HISTORY OF MECHANICS, René Dugas. Monumental study of mechanical principles from antiquity to quantum mechanics. Contributions of ancient Greeks, Galileo, Leonardo, Kepler, Lagrange, many others. 671pp. 5⅜ x 8½. 0-486-65632-2

STABILITY THEORY AND ITS APPLICATIONS TO STRUCTURAL MECHANICS, Clive L. Dym. Self-contained text focuses on Koiter postbuckling analyses, with mathematical notions of stability of motion. Basing minimum energy principles for static stability upon dynamic concepts of stability of motion, it develops asymptotic buckling and postbuckling analyses from potential energy considerations, with applications to columns, plates, and arches. 1974 ed. 208pp. 5⅜ x 8½. 0-486-42541-X

METAL FATIGUE, N. E. Frost, K. J. Marsh, and L. P. Pook. Definitive, clearly written, and well-illustrated volume addresses all aspects of the subject, from the historical development of understanding metal fatigue to vital concepts of the cyclic stress that causes a crack to grow. Includes 7 appendixes. 544pp. 5⅜ x 8½. 0-486-40927-9

Mathematics

FUNCTIONAL ANALYSIS (Second Corrected Edition), George Bachman and Lawrence Narici. Excellent treatment of subject geared toward students with background in linear algebra, advanced calculus, physics and engineering. Text covers introduction to inner-product spaces, normed, metric spaces, and topological spaces; complete orthonormal sets, the Hahn-Banach Theorem and its consequences, and many other related subjects. 1966 ed. 544pp. 6⅛ x 9¼. 0-486-40251-7

ASYMPTOTIC EXPANSIONS OF INTEGRALS, Norman Bleistein & Richard A. Handelsman. Best introduction to important field with applications in a variety of scientific disciplines. New preface. Problems. Diagrams. Tables. Bibliography. Index. 448pp. 5⅜ x 8½. 0-486-65082-0

VECTOR AND TENSOR ANALYSIS WITH APPLICATIONS, A. I. Borisenko and I. E. Tarapov. Concise introduction. Worked-out problems, solutions, exercises. 257pp. 5⅜ x 8¼. 0-486-63833-2

AN INTRODUCTION TO ORDINARY DIFFERENTIAL EQUATIONS, Earl A. Coddington. A thorough and systematic first course in elementary differential equations for undergraduates in mathematics and science, with many exercises and problems (with answers). Index. 304pp. 5⅜ x 8½. 0-486-65942-9

FOURIER SERIES AND ORTHOGONAL FUNCTIONS, Harry F. Davis. An incisive text combining theory and practical example to introduce Fourier series, orthogonal functions and applications of the Fourier method to boundary-value problems. 570 exercises. Answers and notes. 416pp. 5⅜ x 8½. 0-486-65973-9

COMPUTABILITY AND UNSOLVABILITY, Martin Davis. Classic graduate-level introduction to theory of computability, usually referred to as theory of recurrent functions. New preface and appendix. 288pp. 5⅜ x 8½. 0-486-61471-9

ASYMPTOTIC METHODS IN ANALYSIS, N. G. de Bruijn. An inexpensive, comprehensive guide to asymptotic methods–the pioneering work that teaches by explaining worked examples in detail. Index. 224pp. 5⅜ x 8½ 0-486-64221-6

APPLIED COMPLEX VARIABLES, John W. Dettman. Step-by-step coverage of fundamentals of analytic function theory–plus lucid exposition of five important applications: Potential Theory; Ordinary Differential Equations; Fourier Transforms; Laplace Transforms; Asymptotic Expansions. 66 figures. Exercises at chapter ends. 512pp. 5⅜ x 8½. 0-486-64670-X

INTRODUCTION TO LINEAR ALGEBRA AND DIFFERENTIAL EQUATIONS, John W. Dettman. Excellent text covers complex numbers, determinants, orthonormal bases, Laplace transforms, much more. Exercises with solutions. Undergraduate level. 416pp. 5⅜ x 8½. 0-486-65191-6

RIEMANN'S ZETA FUNCTION, H. M. Edwards. Superb, high-level study of landmark 1859 publication entitled "On the Number of Primes Less Than a Given Magnitude" traces developments in mathematical theory that it inspired. xiv+315pp. 5⅜ x 8½. 0-486-41740-9

Math–Geometry and Topology

ELEMENTARY CONCEPTS OF TOPOLOGY, Paul Alexandroff. Elegant, intuitive approach to topology from set-theoretic topology to Betti groups; how concepts of topology are useful in math and physics. 25 figures. 57pp. 5⅜ x 8½. 0-486-60747-X

COMBINATORIAL TOPOLOGY, P. S. Alexandrov. Clearly written, well-organized, three-part text begins by dealing with certain classic problems without using the formal techniques of homology theory and advances to the central concept, the Betti groups. Numerous detailed examples. 654pp. 5⅜ x 8½. 0-486-40179-0

EXPERIMENTS IN TOPOLOGY, Stephen Barr. Classic, lively explanation of one of the byways of mathematics. Klein bottles, Moebius strips, projective planes, map coloring, problem of the Koenigsberg bridges, much more, described with clarity and wit. 43 figures. 210pp. 5⅜ x 8½. 0-486-25933-1

THE GEOMETRY OF RENÉ DESCARTES, René Descartes. The great work founded analytical geometry. Original French text, Descartes's own diagrams, together with definitive Smith-Latham translation. 244pp. 5⅜ x 8½. 0-486-60068-8

EUCLIDEAN GEOMETRY AND TRANSFORMATIONS, Clayton W. Dodge. This introduction to Euclidean geometry emphasizes transformations, particularly isometries and similarities. Suitable for undergraduate courses, it includes numerous examples, many with detailed answers. 1972 ed. viii+296pp. 6⅛ x 9¼. 0-486-43476-1

PRACTICAL CONIC SECTIONS: THE GEOMETRIC PROPERTIES OF ELLIPSES, PARABOLAS AND HYPERBOLAS, J. W. Downs. This text shows how to create ellipses, parabolas, and hyperbolas. It also presents historical background on their ancient origins and describes the reflective properties and roles of curves in design applications. 1993 ed. 98 figures. xii+100pp. 6½ x 9¼.0-486-42876-1

THE THIRTEEN BOOKS OF EUCLID'S ELEMENTS, translated with introduction and commentary by Sir Thomas L. Heath. Definitive edition. Textual and linguistic notes, mathematical analysis. 2,500 years of critical commentary. Unabridged. 1,414pp. 5⅜ x 8½. Three-vol. set.
Vol. I: 0-486-60088-2 Vol. II: 0-486-60089-0 Vol. III: 0-486-60090-4

SPACE AND GEOMETRY: IN THE LIGHT OF PHYSIOLOGICAL, PSYCHOLOGICAL AND PHYSICAL INQUIRY, Ernst Mach. Three essays by an eminent philosopher and scientist explore the nature, origin, and development of our concepts of space, with a distinctness and precision suitable for undergraduate students and other readers. 1906 ed. vi+148pp. 5⅜ x 8½. 0-486-43909-7

GEOMETRY OF COMPLEX NUMBERS, Hans Schwerdtfeger. Illuminating, widely praised book on analytic geometry of circles, the Moebius transformation, and two-dimensional non-Euclidean geometries. 200pp. 5⅝ x 8¼. 0-486-63830-8

DIFFERENTIAL GEOMETRY, Heinrich W. Guggenheimer. Local differential geometry as an application of advanced calculus and linear algebra. Curvature, transformation groups, surfaces, more. Exercises. 62 figures. 378pp. 5⅜ x 8½. 0-486-63433-7

Physics

OPTICAL RESONANCE AND TWO-LEVEL ATOMS, L. Allen and J. H. Eberly. Clear, comprehensive introduction to basic principles behind all quantum optical resonance phenomena. 53 illustrations. Preface. Index. 256pp. 5⅜ x 8½. 0-486-65533-4

QUANTUM THEORY, David Bohm. This advanced undergraduate-level text presents the quantum theory in terms of qualitative and imaginative concepts, followed by specific applications worked out in mathematical detail. Preface. Index. 655pp. 5⅜ x 8½. 0-486-65969-0

ATOMIC PHYSICS (8th EDITION), Max Born. Nobel laureate's lucid treatment of kinetic theory of gases, elementary particles, nuclear atom, wave-corpuscles, atomic structure and spectral lines, much more. Over 40 appendices, bibliography. 495pp. 5⅜ x 8½. 0-486-65984-4

A SOPHISTICATE'S PRIMER OF RELATIVITY, P. W. Bridgman. Geared toward readers already acquainted with special relativity, this book transcends the view of theory as a working tool to answer natural questions: What is a frame of reference? What is a "law of nature"? What is the role of the "observer"? Extensive treatment, written in terms accessible to those without a scientific background. 1983 ed. xlviii+172pp. 5⅜ x 8½. 0-486-42549-5

AN INTRODUCTION TO HAMILTONIAN OPTICS, H. A. Buchdahl. Detailed account of the Hamiltonian treatment of aberration theory in geometrical optics. Many classes of optical systems defined in terms of the symmetries they possess. Problems with detailed solutions. 1970 edition. xv + 360pp. 5⅜ x 8½. 0-486-67597-1

PRIMER OF QUANTUM MECHANICS, Marvin Chester. Introductory text examines the classical quantum bead on a track: its state and representations; operator eigenvalues; harmonic oscillator and bound bead in a symmetric force field; and bead in a spherical shell. Other topics include spin, matrices, and the structure of quantum mechanics; the simplest atom; indistinguishable particles; and stationary-state perturbation theory. 1992 ed. xiv+314pp. 6⅛ x 9¼. 0-486-42878-8

LECTURES ON QUANTUM MECHANICS, Paul A. M. Dirac. Four concise, brilliant lectures on mathematical methods in quantum mechanics from Nobel Prize-winning quantum pioneer build on idea of visualizing quantum theory through the use of classical mechanics. 96pp. 5⅜ x 8½. 0-486-41713-1

THIRTY YEARS THAT SHOOK PHYSICS: THE STORY OF QUANTUM THEORY, George Gamow. Lucid, accessible introduction to influential theory of energy and matter. Careful explanations of Dirac's anti-particles, Bohr's model of the atom, much more. 12 plates. Numerous drawings. 240pp. 5⅜ x 8½. 0-486-24895-X

ELECTRONIC STRUCTURE AND THE PROPERTIES OF SOLIDS: THE PHYSICS OF THE CHEMICAL BOND, Walter A. Harrison. Innovative text offers basic understanding of the electronic structure of covalent and ionic solids, simple metals, transition metals and their compounds. Problems. 1980 edition. 582pp. 6⅛ x 9¼. 0-486-66021-4